9

Springer Series in Chemical Physics

Edited by Fritz Peter Schäfer

Springer Series in Chemical Physics

Editors: V. I. Goldanskii R. Gomer F. P. Schäfer J. P. Toennies

Secondary Ion Mass Spectrometry

SIMS II

Proceedings of the Second International Conference on
Secondary Ion Mass Spectrometry (SIMS II)
Stanford University, Stanford, California, USA
August 27–31, 1979

Editors
A. Benninghoven · C. A. Evans, Jr. · R. A. Powell
R. Shimizu · H. A. Storms

With 234 Figures

Springer-Verlag Berlin Heidelberg New York 1979

Editors

Prof. A. Benninghoven, Physikalisches Institut der Universität Münster, F.R.G.

Dr. C. A. Evans, Jr., Charles Evans & Associates, San Mateo, CA 94402, USA

Dr. R. A. Powell, Varian Associates, Palo Alto, CA 94303, USA

Prof. R. Shimizu, Osaka University, Osaka, Japan

Dr. H. A. Storms, General Electric Vallecitos Nuclear Center, Pleasanton, CA 94566, USA

International Organizing Committee

A. Benninghoven (Chairman), V. T. Cherepin, C. A. Evans, Jr., K. F. Y. Heinrich, J. Okano, G. Slodzian, H. W. Werner

Local Organizing Committee

R. A. Powell (Chairman), C. A. Evans, Jr., H. A. Storms, J. Stein

Supported by Grants from

Office of Naval Research (Grant N00014-79-G-0063)

Microbeam Analysis Society

Atomika

Balzers

Charles Evans & Associates

General Electric

Hitachi

Inficon Leybold-Heraeus

Instruments, SA

Perkin-Elmer

Riber

ULVAC

Varian

ISBN-13: 978-3-642-61873-4 e-ISBN-13: 978-3-642-61871-0
DOI: 10.1007/978-3-642-61871-0

Library of Congress Cataloging in Publication Data
International Conference on Secondary Ion Mass Spectrometry, 2d, Stanford University, 1979. Secondary ion mass spectrometry, SIMS-II.
(Springer series in chemical physics; v. 9)
Includes index.
1. Mass spectrometry—Congresses.
I. Benninghoven, A. II. Title. III. Series.
QD96.M3157 1979 543'.08 79-23997

Softcover reprint of the hardcover 1st edition 1979

2153-543210

Preface

The Second International Conference on Secondary Ion Mass Spectrometry (SIMS-II) was held at Stanford University, Stanford, California, from August 27-31, 1979. Over 80 invited and contributed papers, collected in the present volume, were presented to about 200 participants. It is the hope of the editors that these proceedings will meet the needs of the participants as well as those not able to attend, in order to make the best use of the information presented during the conference.

SIMS-II was the second conference of its type--the first having been held in Muenster, West Germany, in September 1977 with about the same number of papers and participants. This series of biannual conferences on secondary ion mass spectrometry has been created as a forum to bring together people who are working in the different fields of secondary ion emission. These conferences should encourage the exchange of results, ideas, and projects, with the final goal being a better understanding of the secondary ion emission process and its application--especially in the field of analysis.

In this regard, SIMS-II was a very successful conference: papers discussed fundamental aspects of SIMS as well as a wide range of analytical applications, supplying the participants with an excellent, up-to-date survey of the field. New ideas, recent results, experiments in progress, etc., were discussed following the presentations and also, on a more informal basis, in small groups. Such informal exchanges are of great value in this type of highly specialized conference.

The conference revealed very clearly that SIMS is still a very rapidly expanding field in both fundamental research as well as for its application. Today we are still far away from a complete understanding of the very complex processes which result in the emission of a charged atomic or molecular cluster. These processes include the sputtering process as well as the different electronic interactions which are responsible for the composition and final charge state of the emitted particles. There exist several approaches to a theoretical treatment of these processes. The challenging task for future theoretical work will be to go more and more in the direction of a combined consideration of these two groups of processes, thus approaching a real and complete understanding of the entire process of secondary ion formation.

Understanding ion formation is also of great importance for the optimum application of secondary ion emission, especially in its most important field of analysis. This analytical application, particularly for quantitative analysis, has made great progress during the last few years. On the other hand, new fields of application such as analysis of organic compounds, localized isotopic analysis, the use of negative secondary ions--especially from Cs covered surfaces--etc., are still in the formative stages.

Much of the success of SIMS-II resulted from the hospitality of Stanford University, its excellent facilities and attractive surroundings. Particular thanks are due the staff of the Stanford University conference scheduling office. We also want to acknowledge Mrs. Elfie Liebl and Mrs. Elaine Storms and family for their efforts during the conference; the session chairpersons for reviewing the submitted papers; and Ms. S. Sargentini for typing the final manuscript. The financial support of our sponsors is gratefully acknowledged. Finally, we want to thank all those who participated in SIMS-II, especially those who attended from outside the United States, for making the conference as productive and informative as it was.

Palo Alto, California
September 1979

A. *Benninghoven*
C. A. *Evans, Jr.*
R. A. *Powell*
R. *Shimizu*
H. A. *Storms*

Contents

I. Fundamentals

Chairpersons: D.E. Harrison and C.A. Evans, Jr.

II. Quantitation

III. Semiconductors

VI. Instrumentation

Chairpersons: D.S. Simons and F.G. Rüdenauer

VII. Geology

Chairpersons: J. Okano and C. Meyer

X. Combined Techniques

Chairpersons: C. Johnson and W.H. Christie

XI. Postdeadline Papers

I. Fundamentals

The Dynamics of Ion-Solid Interactions: A Basis for Understanding SIMS

Nicholas Winograd
Department of Chemistry
The Pennsylvania State University
University Park, Pennsylvania 16802

A fundamental understanding of the SIMS process requires first a knowledge of how the nuclear positions of the atoms or molecules in the sample change with time in response to the impact of the primary ion. Secondly, a quantitative description of the ionization processes is required for those species which leave the sample surface as ions and are subsequently detected by the mass spectrometer. Both of these tasks are exceedingly complex to treat theoretically, given the large number of perturbed particles and the tremendous range of kinetic energies and subsequent ejection mechanisms that need to be considered. For example, in Fig.1a is shown a model microcrystallite of Cu(001) about to be struck by a 600 eV Ar^+ ion at normal incidence. From the results of the classical dynamics calculations we have been exploring, the state of the crystallite after the last Cu atom has left the surface is shown in Fig. 1b. Note that virtually every atom has moved from its original position after a time of only 194 fsec (1.94 x 10^{-13} sec). In addition, the kinetic energy of the atoms in the solid varies from nearly 0 to as high as 250 eV.

In this paper, the focus of our objective is to probe the experimental consequences of this classical dynamics model. With SIMS there are many different types of experimental observables. These include yields, energy and angular distributions of the ejected particles, effect of energy and angle of incidence of the primary ion, cluster formation processes, as well

a 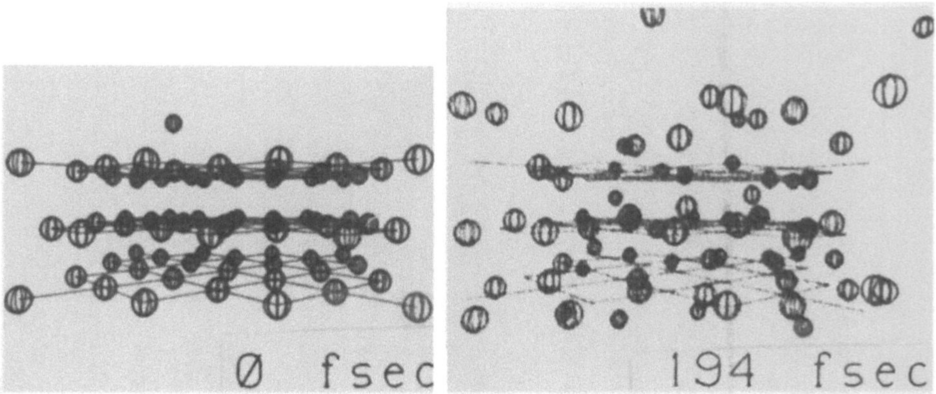

Fig.1 Graphical depiction of a section of the model microcrystallite before and after bombardment. The sizes of the atoms are drawn arbitrarily. The 4th layer and part of the crystal sides have been truncated for graphical clarity

as crystal structure effects. Although the model as yet ignores the critically important ionization process, it provides a unified theory from which all these observables can be determined from a single calculation. The comparisons to experiment, both from SIMS measurements and from experiments which detect the neutral particles, are at least in semi-quantitative agreement. This agreement gives us considerable evidence that the model provides a sufficient answer to the first requirement posed above, to gain a fundamental understanding of the SIMS process.

1. Procedures

A number of theoretical approaches which describe the crystal dynamics after the primary ion impact have been proposed and exploited in considerable detail [1-7]. These include mainly statistical models which involve solving transport equations for the momentum deposition [1,2], and models which utilize classical dynamics to calculate the actual trajectories that the atoms follow after being struck by the primary ion [3-7]. It is possible to perform the calculations using the binary collision approximation where it is assumed that only two particles interact at a time [3,4]. It is also possible to perform a complete calculation using as accurate an interaction potential as possible to calculate the forces between all the atoms in the microcrystallite [5-7]. The classical trajectory methods would appear to be more accurate and more generally applicable since there are fewer assumptions in the formalism. However, from the trajectory results the development of an anayltical expression to provide a simple means of calculating the experimental observables is not generally possible.

The results described in this paper have been performed using the complete interaction potential after the method developed by Harrison during the 1960's. The early calculations have been considerably extended by increasing the size of the microcrystallite by a factor of two to three so that all of the important events that lead to particle ejection are included [8]. In addition, provisions have been included to consider metal surfaces with atomic [9] and molecular [10] adsorbates. Statistically meaningful results are now available to provide quantitative insight into atom yields as a function of exposed crystal face [8], primary ion energy [11], angle of incidence [12] and type of metal. In addition, the ejected particle angular distributions [13,14] and energy distributions [15] can be computed accurately. And finally, it is possible to determine the mechanism of cluster formation [16,17] which includes a great deal of insight into how much surface structure information can be deduced from the composition of the ejected clusters.

For nearly all of these calculations, it is possible to make quantitative comparisons to various SIMS measurements. Several specific conditions are required. First, at this point in time measurements must be performed on single crystals. The calculations clearly show that relative atomic placement determine the ejection mechanisms and yields, and these placements must be experimentally well-defined. Secondly, particularly for the study of adsorbates, the static SIMS approach must be employed to insure that the surface modeled in the computer is the same as the one studied experimentally [18]. And finally, to minimize the uncertainties in the ionization process, an appropriate ion yield ratio must be found which can be directly compared to the ratio calculated for the corresponding neutral species.

2. Yields

The calculation of the number of ejected particles per incident ion is dependent on the knowledge of an accurate interaction potential. The yields are determined by averaging the yields of several hundred individual trajectories initiated over the symmetrically irreducible zone of the crystal surface. The averaging is important since the number of atoms sputtered per incident ion (ASI) [11] is strongly dependent on the impact coordinates. It is possible to find potential functions that give exact agreement with experimental yields, but the physical significance of the results are open to question. Our approach has been to select a given set of parameters that have been fit to the elastic constants of the solid and to other scattering data [7,8] and to probe the consequences of performing calculations with this potential.

The absolute yields from Cu are somewhat different than the measured absolute yields, although our value of 4.0 atoms/ion on (100) vs. the experimental value of 2.1 is not too unrealistic [8]. The relative yields between the (100), (110) and (111) faces are in excellent agreement with experiment. The yield of (111)/(100) is calculated to be 1.6 and has been measured to be 1.8 [8]. The assumption that the choice of potential parameters is not too critical is also illustrated by the fact that this ratio is 1.6 for Ag and 1.5 for Au [19]. In fact, when examined in this way, the importance of surface morphology becomes clear. The position of the atoms relative to the point of ion impact seems to have a lot to do with the mechanism of momentum deposition and suggests that ion bombardment methods will be a useful tool in surface structural analysis.

3. Energy Distributions

The energy distribution of the ejected particles can be easily determined by examining their final velocities. Without adjusting parameters, the model provides quantitative agreement with the experimental results for Cu [15]. In general, as shown in Fig.2, these curves rise quickly from zero at 0.0 eV to a maximum at an energy between 1-5 eV and then gradually fall off as $\sim E^{-1.6}$. The curves have long tails which extend to a significant fraction of the primary ion energy.

Of particular interest is the appearance of structure including a peak at ~2.7 eV and a shoulder at ~5 eV. The shoulder(s) on the high energy side of the peak have also been seen experimentally [20]. From a careful analysis of the dynamic motion of the atoms during each impact, it is possible to isolate certain ejection mechanisms that contribute to this structure. Using the labeling scheme shown in Fig.2b, the energy distribution of atom 1-5 is seen to largely contribute to the tail in the energy distribution while atoms 6-9 form the basis of the peak at 2.7 eV. Since atoms 1-5 are generally of higher kinetic energy, they leave the surface early in the collision sequence while considerable surface structure is still present.

4. Cluster Formation Mechanisms

During the ion bombardment of a solid, many types of clusters are ejected. For adsorbates on clean metals, these vary from pure metal clusters M_n where n can be as large as 12, to metal atoms attached to adsorbed species, e.g., CuO or NiCO, to large organic molecules that were originally adsorbed on the surface of the solid which eject retaining their molecular formula. From the classical dynamics treatment, it is possible to examine the cluster formation mechanism in detail, and to provide semi-quantitative information about cluster yields. In general, these calculations tell us that there are

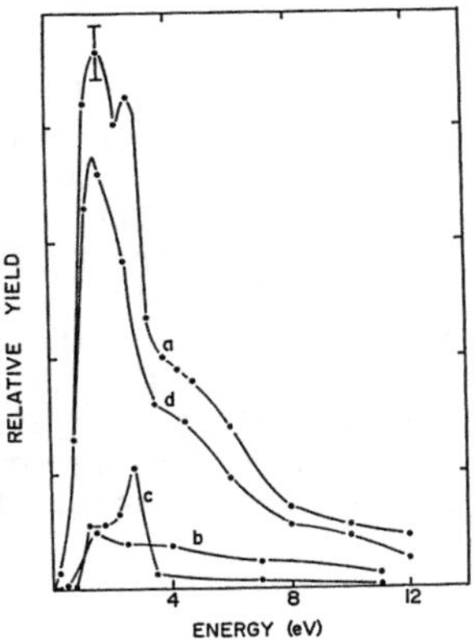

Fig.2a

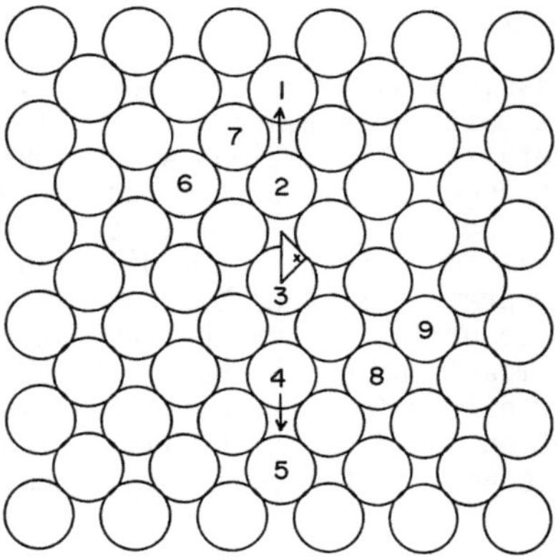

Fig.2b

Fig. 2 a) Energy distribution of Cu atoms from the clean (100) face: (a) is drawn for all Cu atoms; (b) is for atoms 1-5 of Fig. 2b; (c) is for atoms 6-9 in Fig. 2b while (d) is drawn for the remaining atoms not included in (b) and (c). b) Clean Cu(100) surface. The numbers are used to identify particular atoms (see text). The impact zone is denoted by the triangle

three basic mechanisms of cluster formation. First, for clean metals or metals covered with atomic absorbates, the ejected atoms can interact with each other in the near surface region above the crystal to form a cluster via a recombination type of process. This description would apply to clusters of metal atoms and of metal-oxygen clusters of the type M_nO_m observed in many types of SIMS experiments. For this case the atoms in the cluster do not need to arise from contiguous sites on the surface, although we do find that in the absence of long-range ionic forces, that most of them originate from a circular region of radius ~5 Å. A second type of cluster emission involves molecular adsorbates like CO adsorbed onto Ni. Here, the CO bond strength is ~11 eV, but the interaction with the surface is only about 1.3 eV. This energy difference is sufficient to allow CO to eject molecularly although we do find that ~20% of them can be dissociated by the ion beam or by energetic metal atoms [10]. Clearly, for the case of these molecular systems, it is easy to infer the original atomic configurations of the molecule and to determine the surface chemical state. If CO were dissociated into oxygen and carbon atoms, for example, our calculations suggest that the amount of CO observed should drop dramatically. This type of process undoubtedly applies to the adsorption of organic molecules on surfaces, since the strong carbon framework can soak up excess energy from violent collisions. The surface interactions are weak enough, however, to allow ejection intact. The final mechanism for cluster ejection is essentially a hybrid mechanism between the first two. For the case of CO on Ni again, we find that the observed NiCO and Ni_2CO clusters form by a recombination of ejecting Ni atoms with ejecting CO molecules. A similar mechanism ought to apply to the formation of cationized organic species [21]. The organic molecule ejects intact, but interacts with an ejecting metal ion to form a new cluster species.

5. Prospects

The classical dynamics approach provides an excellent base to quantitatively predict most experimental observables in sputtering. In the coming years, more detailed computations with experiments aimed to match, will provide an even stronger basis for the model. On the other hand, the computation of accurate absolute yields must await the availability of better interaction potential functions. The model also provides a good starting point in which to interpret SIMS spectra. Currently, the ionization phenomena is the lacking piece of the theory, but by taking ratios of ion yields, semi-quantitative comparisons can be made. In addition, the fact that atomic trajectories can be followed on a microscopic level is an advantageous feature in coupling ionization theories which will necessarily include the kinetic energy of the particle and its temporal local atomic environment.

Acknowledgement

I am particularly grateful to Don Harrison for willingly providing Barbara Garrison and myself with a working computer program and for all his direct efforts in teaching us his unique insight into the ion bombardment process. The financial support of the National Science Foundation (CHE78-08728) and the Air Force Office of Scientific Research (AF76-2974) are gratefully appreciated. Portions of the computations were supported by the National Resource for Computation in Chemistry under a grant from NSF and the U.S. Department of Energy (Contract No. W-7405-ENG-48).

References

1. M.W. Thompson, Phil. Mag. 18, 377 (1968).
2. P. Sigmund, Phys. Rev. 184, 383 (1969).
3. M. Robinson and I. Torrens, Phys. Rev. B 9, 5008 (1974).
4. R. Shimizu, Proc. Int. Vac. Congr., 7th, 2, 1417 (1977).
5. J.B. Gibson, A.N. Goland, M. Milgram and G.H. Vineyard, Phys. Rev. 120, 1229 (1960).
6. D.P. Jackson, Can. J. Phys. 53, 1513 (1975).
7. D. Harrison, Jr., W. Moore, Jr. and H. Holcombe, Rad. Eff. 17, 167 (1973).
8. D. Harrison, Jr., P. Kelly, B. Garrison and N. Winograd, Surface Sci. 76, 311 (1978).
9. B. Garrison, N. Winograd and D. Harrison, Jr., Phys. Rev. B 18, 6000 (1978).
10. N. Winograd, B. Garrison and D. Harrison, Jr., Phys. Rev. B, submitted.
11. D. Harrison, Jr., B. Garrison and N. Winograd, SIMS-II Conference Proceedings, in press.
12. K.E. Foley and B.J. Garrison, J. Chem. Phys., submitted.
13. N. Winograd, B. Garrison and D. Harrison, Jr., Phys. Rev. Lett. 41, 1121 (1978).
14. S. Holland, B. Garrison and N. Winograd, Phys. Rev. Lett. 43, 0000 (1979).
15. B. Garrison, N. Winograd and D. Harrison, Jr., Surface Sci. 85, 0000 (1979)
16. B. Garrison, N. Winograd and D. Harrison, Jr., J. Chem. Phys. 69, 1440 (1978).
17. N. Winograd, D. Harrison, Jr. and B. Garrison, Surface Sci. 78, 467 (1978).
18. A. Benninghoven, Surface Sci., 53, 596 (1975).
19. M.T. Robinson and A.L. Southern, J. Appl. Phys. 35, 1819 (1964).
20. P. Hucks, G. Stocklin, E. Vietgyke and K. Vegelbruch, J. Nucl. Mater. 76, 136 (1978).
21. H. Grade, R. Cooks, and N. Winograd, J. Am. Chem. Soc. 99, 7725 (1977).

Simultaneous Measurements of Photon and Secondary Ion Emissions from Ion-Bombarded Metal Surfaces

T. Okutani and R. Shimizu
Department of Applied Physics, Osaka University
Suita, Osaka 565, Japan

1. Introduction

A powerful approach to the understanding of secondary ion emission is the simultaneous (or in-situ) observations of ion-induced photon and secondary ion emissions [1,2,3,4,5,6]. These investigations have emphasized that some similarities exist between these emission phenomena.

As proposed in earlier work [1], if ion-induced photons and secondary ions were generated in a plasma in local thermal equilibrium (LTE) condition, one can independently determine the plasma temperature, T, and electronic density, Ne, from both the photon emission and secondary ion mass spectra. This approach would shed additional light into verification of LTE-quantitation by secondary ion mass spectrometry (SIMS) as proposed by ANDERSEN and HINTHORNE [7]. This method has been widely used though the physical basis has been in dispute.

MacDONALD and GARRETT [8] have recently applied this approach to quantitative analysis of NBS-steel samples with considerable success. They evaluated the plasma temperature, T, and electronic density, Ne, of the plasma assumed to be formed in and above an ion bombarded surface from ion-induced photon emission spectrum and then used these T and Ne in LTE-quantitation for the quantitative analysis by SIMS.

We have attempted to perform simultaneous measurements of ion-induced photon and secondary ion emissions from ion bombarded metal surfaces with a home-made type SIMS-SCANIIR apparatus [6]. This apparatus allows us to observe the variation of the photon and secondary ion emissions by changing experimental conditions, e.g., oxygen partial pressure for positive secondary ion emission or deposition rate of Cs-atoms on a sample surface under Ar^+ ion bombardment for negative secondary ion emission [9].

In this study we aimed at, firstly, examining whether any close relationship between ion-induced photon and secondary ion emissions exists and, finally, applying this technique to LTE-quantitation of an alloy sample, 92wt%Al-3wt%Mg-5wt%Zn, for verification of the LTE-model though only the latter investigation is briefly described in this paper.[1]

---

[1]A part of this paper was presented at the Second US-Japan Joint Seminar on SIMS held at Takarazuka, Japan, October 23-27, 1978.

2. Apparatus

Figure 1 shows a schematic diagram of a homemade type SIMS-SCANIIR apparatus. The primary ion beam system is that of the commercial Hitachi-IMA consisting of hollow cathodetype duoplasmatron ion source and two electrostatic lens systems with beam deflection plates. The ion beam is focussed to 10 µm in diameter on the sample surface. The vacuum chamber is equipped with a differential pumping system employing a diffusion pump with liquid nitrogen trap for the primary ion beam system. The specimen chamber is evacuated by an ion pump to pressure in the 10^{-9} Torr range. The secondary ion detection system consists of an extractor, Einzel lens, 90° sector electrostatic energy analyzer, and a quadrupole mass analyzer (QMA) of commercial type UTI-100C. Hence this apparatus allows us to make secondary ion measurements under the same condition as the Hitachi-IMA.

Photon emission is detected by an optical spectrometer with a chopping system followed by photomultiplier and lock-in amplifier. The optical system is of an off-planed Ebert type Echelett grating (1200 lines/mm) as described in detail elsewhere [1].

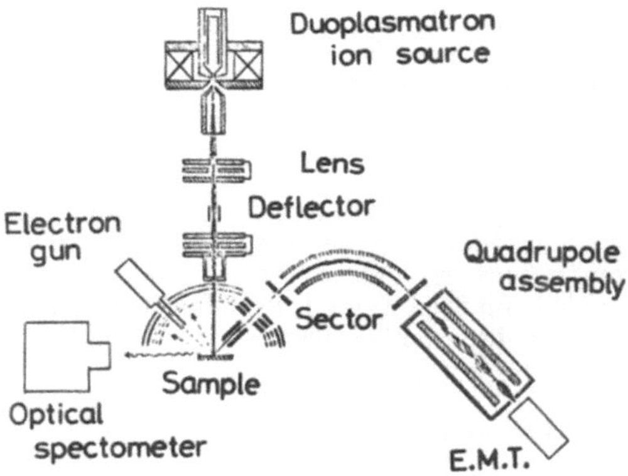

Fig.1 Schematic of the apparatus for simultaneous measurements of ion-induced photon and secondary ion emissions

3. LTE Quantitation by Simultaneous Measurements of Ion-Induced Photon and Secondary Ion Emissions

For verification of the LTE model we attempted to determine T and Ne independently from both the photon emission spectrum and LTE-quantitation by SIMS. For this we adopted 92wt%Al-3wt%Mg-5wt%Zn alloy as a sample since this ternary alloy of known concentration allows us to perform LTE-quantitation by SIMS and to observe strong optical emission lines from excited neutral atoms of Al and Mg and from Mg^+ ions. These emission lines from neutral atoms allow for determination of T while those from Mg atoms and Mg^+ ions enable us to determine Ne as proposed in earlier work [1] and actually done by MacDONALD and GARRETT [8] for the quantitative analysis.

The main procedure of this quantitation is, first, to determine T and Ne from ion-induced photon emission spectrum and, then, using these T and Ne the quantitative analysis based on the LTE model is made for secondary ion intensities obtained by SIMS. The measured secondary ion and ion-induced photon emission spectra are shown in Fig.2-a and -b respectively. As the spectrum of Mg^+ is superimposed on the tail of Al^+ owing to the rather poor mass resolution, we determined Mg^+ intensity by subtracting the Al^+ tail.

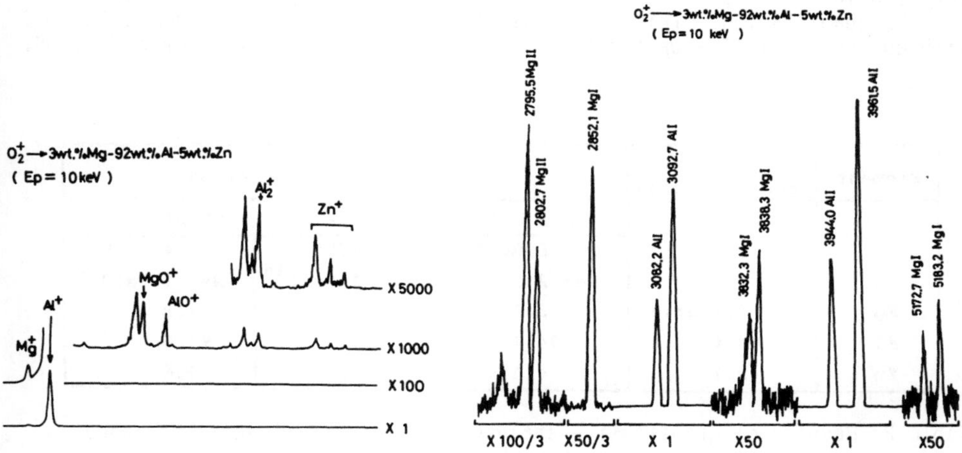

(a) (b)

Fig.2 Spectra of secondary ions (a) and ion-induced photons (b) from 92wt%Al-3wt%Mg-5wt%Zn alloy under 10 keV O_2^+ ion bombardment

Using relative intensities of two optical lines which originate from the same species, we can determine the optical temperature, T', according to spectral analysis of optical emission [10], which corresponds to plasma temperature T in the LTE quantitation.

We show these temperatures in Table 1, which were determined with neutral Al and Mg spectra, respectively.

Temperature	Electron density (cm^{-3})
4500 $\pm$ 400 °K(AlI) 4800 $\pm$1600 °K(MgI)	4×10^{13}

Table 1 The optical temperature and electronic density determined from ion-induced photon emission spectrum shown in Fig.2-b

Although the temperature obtained from Mg has a large error, both the optical temperatures can be considered to be in agreement. We also determined the electron density, Ne, from the relative intensities of neutral and singly charged Mg and show it in Table 1.

Although this Ne is subject to large error, using these values of T'and Ne shown in Table 1 we attempted to determine the ionization efficiency of each element and obtained the concentrations of constituent elements from secondary ion intensities without using the internal standards commonly used in the LTE-quantitation. This result is shown in the SCANIIR column of Table 2 with conventional LTE-quantitations done with Hitachi-IMA and homemade SIMS.

Element	True concentration	Calculated concentration		
		S I M S	I M A	SCANIIR
		T = 6900°K Ne= 2.5×10^{17}	T = 6750°K Ne= 1.6×10^{16}	T = 4500°K Ne= 1.2×10^{12}
Mg	3.5 at.%	3.5	3.5	2.8
Al	94.4	94.4	94.4	94.6
Zn	2.1	4.5	2.7	2.6

Table 2 Results of quantitative analyses by SIMS and Hitachi-IMA based on LTE model and by the present SIMS-SCANIIR approach

Considering that in the quantitation shown in SCANIIR column in Table 2 any known concentrations as those of internal standard elements in the LTE quantitation are not used, one can see that the concentration results are surprisingly good. Concerning the (T, Ne) values, however, we find large differences between the (T, Ne) values from optical lines and those from secondary ions. This low optical value of T is not a special case. Relative intensities of four lines from neutral Al from the same sample by 10 keV Ar^+ ion bombardment given an even lower value of T' (3500 K). Such a tendency can also be found in the results of MacDONALD and GARRETT [8] who obtained T' less than 4500 K for NBS-samples whilst the LTE quantitation with Hitachi-IMA, for instance, gave 6000 K [11] though one should take into account the difference between experimental conditions of both the measurements. Thus, both the values of T and Ne obtained by optical emission spectra appear to be rather low compared with those obtained by SIMS. This difference can be understood to some extent as follows: The minimum seeking method in LTE quantitation clearly indicates that the best fit values of T and Ne for the present sample are distributed over a wide range in T-Ne space covering Ne from 10^5 to 10^{19} and T from 2500 to 7000 K, forming a narrow band area. Thus, the values of (T, Ne) shown in Table 2 are considered quite reasonable because LTE quantitation may also provide similar values of T and Ne as those from SCANIIR if one extends the minimum seeking over a wider area in T-Ne space. This is a main reason why we got good agreement in quantitative analyses shown in Table 2 by using a different set of (T, Ne).

In this study the SIMS-SCANIIR approach for quantitative analysis by LTE model was made with considerable success for an aluminum based alloy. However, it is worth noting that the present experiment has also revealed an indication of a basic difference between secondary ion and photon emission spectra observed under O_2^+ and Ar^+ ion bombardments in oxygen atmosphere: With respect to Ar^+ ion bombardment in oxygen atmosphere, secondary ion spectrum changes as oxygen partial pressure becomes higher suggesting that T from LTE quantitation elevates, and secondary ion mass spectra become almost identical, as often reported, with those obtained by O_2^+ ion bombardment for sufficiently high oxygen partial pressure. On the contrary, in the photon emission spectrum the intensities of emission lines increase but their relative intensities among each emission line hardly change as oxygen partial pressure increases. This suggests that, according to LTE model, T obtained from photon emission spectra hardly changes for the change of oxygen partial pressure whilst T from LTE quantitation by SIMS does.

Consequently, the present result suggests that ion-induced photon and secondary ions are generated in quite similar mechanisms and the present SIMS-SCANIIR approach allows us to make quantitative analysis with considerable success as reported by MacDONALD and GARRETT [8]. It should, however, be emphasized that more systematic investigations, e.g., quantitative analysis by the present approach at different oxygen partial pressures, etc., are highly required before the above can accommodate to an argument for verification of the LTE model. For this an improvement of the present SCANIIR system is most necessary.

Acknowledgements

The authors wish to extend their thanks to Dr. T. Ishitani, Hitachi Central Research Laboratory, for stimulating discussions and for technical assistance in performing IMA-analysis. Helpful comments for the investigation given by Dr. H.A. Storms, Vallecitos Nuclear Center, G.E. Co., are gratefully acknowledged.

References

1. M. Kato, R. Shimizu and T. Ishitani, Technol. Repts., Osaka Univ., 24, 451 (1974).
2. I.S.T. Tsong and A.C. McLaren, Spectrochem. Acta 30B, 343 (1975).
3. G. Blaise, Surface Sci., 60, 65 (1976).
4. P.J. Martin and R.J. MacDonald, Surface Sci., 62, 551 (1977).
5. R.J. MacDonald and P.J. Martin, Surface Sci., 66, 423 (1977).
6. R. Shimizu, T. Okutani, T. Ishitani and H. Tamura, Surface Sci., 69, 349 (1977).
7. C.A. Andersen and J.R. Hinthorne, Anal. Chem., 45, 1421 (1973).
8. R.J. MacDonald and R.F. Garrett, Surface Sci., 78, 371 (1978).
9. T. Okutani, K. Shohno and R. Shimizu, to be presented at this Conference.
10. P.W.J.M. Boumans, Theory of Spectrochemical Exitation (Plenum Press, New York, 1966).
11. R. Shimizu, T. Ishitani, T. Kondo and H. Tamura, Anal. Chem. 47, 1020 (1975).

Atom Ejection Mechanisms and Models

Don E. Harrison, Jr.
Department of Physics and Chemistry
Naval Postgraduate School
Monterey, California 93940

Barbara J. Garrison
and
Nicholas Winograd
Department of Chemistry
Pennsylvania State University
University Park, Pennsylvania 16802

In a series of publications the authors have been exploring atom ejection from clean and chemically reacted copper and nickel surfaces under argon ion bombardment at 600 eV ion energy by classical trajectory simulations. The calculations generate a variety of experimental observables including relative sputtering yields, energy and angular distributions for the ejected atoms, multimer yields and surface damage information. The method provides a necessary first step in the development of rigorous ejection models, prior to the inclusion of ionization effects. A general conclusion from all of the calculations is that surface morphology - the location of specific atoms with respect to each other - effectively controls the ejection mechanisms for a specific system. A comprehensive comparison between the predictions of the simulations and experimental SIMS results obtained from clean and reacted surfaces has been published [1]. These comparisons are most successful when ratios of variables between different crystal faces or adsorbates in different coverages are used. Computed yields are sensitive to the crystal structure, coverage, binding energy and site symmetry of adsorbates. A recent investigation was concerned with the Cu(111)/Ar^+<111> system as a function of the ion energy [2]. This report summarizes and extends that work.

Details of the classical dynamical procedure have been described elsewhere [3]. In general, we solve Hamilton's equations of motion in time for a model microcrystallite consisting of four layers with ~90 atoms/layer. The Ar^+ ion is aimed toward an impact point in an irreducible symmetry zone of the surface, and the resultant collision cascade continues until the remaining atoms have insufficient energy to escape from the surface. The process is then repeated for a number of additional impact points. The original results were obtained with sets of 80 impact points trajectories. The model is discussed in greater detail in ref. 3.

1. Summary of Reported Results

Previous results [1-3] indicate that the computed yields compare quite favorably with the available experimental data. Most of the atom yield has been found to come from the first atomic layer of the target. The second layer contributes <10 percent and the third layer ~1 percent. The experimental ejected atom yield, atoms/ion, is a surprisingly poor measure of the yield from an individual ion trajectory. The distribution of atoms ejected/single ion (ASI) contains valuable information. Ejected atom energy distributions and various parameters associated with multimer formation were presented. Multimer yield ratios have been calculated which can be directly compared with experimental results.

All of the results support a theory of atom ejection which is a combination of the transparency of the lattice coupled with dynamic processes in the first few layers of the target surface. There is no support for a model which depends upon the flux of atoms or energy from below the surface.

2. Atoms Ejected/Single Ion

An examination of the details of individual ion collision events indicates that some ions eject surprisingly large numbers of atoms, even at low ion energies, and some eject none. A study of ASI as a function of ion energy gives valuable insight into the atom ejection mechanism.

The earlier ASI studies were limited by the small number of trajectories calculated at each energy. While 80 trajectories are sufficient for yield and multimer studies they give very 'noisy' ASI distributions. To remedy this situation 480 additional trajectories were calculated at 8.0 keV. The original impact points were distributed as uniformly as possible over the irreducible symmetry zone. The additional points were chosen to allow a sensitivity analysis of the original set. This was accomplished by shifting the entire grid of points by small displacements in six different directions. Thus, the relative orientation and separation between impact points was maintained and a set of points near each of the original points was examined. Standard deviations were calculated for many of the data previously reported from the seven data sets, (see Table 1).

Table 1 Data at 8.0 keV for the $Cu(111)/Ar^{+}$<111> System

Atom Yield	10.9	$\pm$ 0.3	atoms/ion
Atom fraction in multimers	0.21	$\pm$ 0.02	
Fractional yield, layer 2	0.087	$\pm$ 0.005	
Fractional yield, layer 3	0.0012	$\pm$ 0.0002	
Dimer/monomer ratio	0.085	$\pm$ 0.007	
Trimer/monomer ratio	0.011	$\pm$ 0.004	
Quadrimer/monomer ratio	0.006	$\pm$ 0.002	
Trimer/dimer ratio	0.13	$\pm$ 0.01	
Quadrimer/dimer ratio	0.075	$\pm$ 0.002	

The largest multimer encountered in this series was a nomomer, Cu_9, which was bound with 0.6 eV/atom. All of the smaller multimers also occurred, but the yields were too small to be meaningful.

The ASI distribution function at 8.0 keV is shown in Fig.1. Note that there are impact points which have no yield. The function appears to consist of two components, a narrow one centered near an ASI of 11, and a broad distribution from about 2 to 25. With this curve as a guide the 80 point ASI functions can also be interpreted. The first component appears in the energy range from 3.0 to 10.0 keV, where the yield is largest. The maximum ASI also occurs in this region; both 5.0 and 10.0 keV produced values which exceed the ASI of 25 shown in the Figure. The low energy and high energy distribution functions appear to contain only the broad component. At 0.6 keV the maximum is 19. The two functions are remarkably similar. Above 10.0 keV the maximum ASI is decreasing with energy, which is not the result expected from the statistical theories.

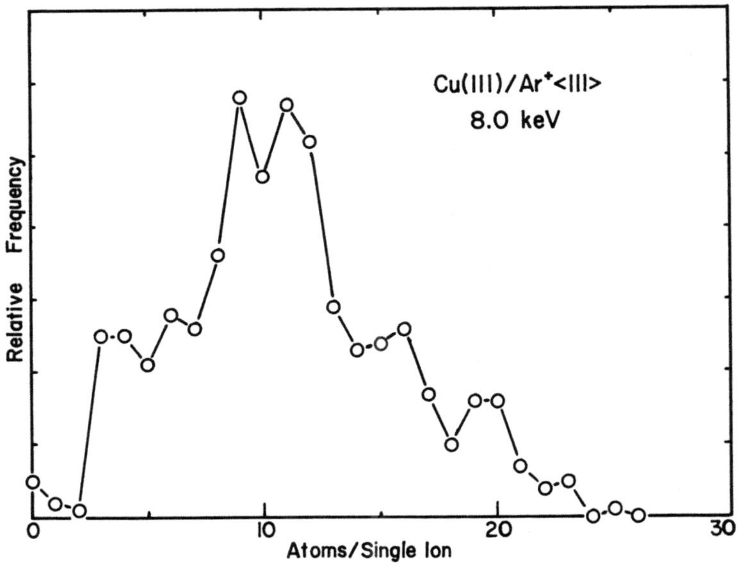

Fig.1 ASI Distribution Function for Cu(111)/Ar^+ 111 at 8.0 keV. The most probable ASI, 9, occurred 58 times in 560 trajectories

Acknowledgement

This research was supported by the National Science Foundation and by the Foundation Research Program of the Naval Postgraduate School.

References

1. N. Winograd, B.J. Garrison, T. Fleisch, W.N. Deglass and D.E. Harrison, Jr., J. Vac. Sci. and Technol., 16, 629 (1979).
2. Don E. Harrison, Jr., B.J. Garrison and N. Winograd, Phys. Rev. B., Submitted.
3. D.E. Harrison, Jr., P.W. Kelly, B.J. Garrison and N. Winograd, Surface Sci., 76, 311 (1978).

New Models of Sputtering and Ion Knock-On Mixing

S.A. Schwarz and C.R. Helms
Stanford Electronics Laboratories
Stanford, California 94305

We have recently developed new models of sputtering [1], ion emission [2], and ion knock-on mixing [3] which provide quick and accurate estimates of neutral and ion sputtering yields and of the broadening observed in a sputter-profiling experiment. In this abstract, we briefly summarize the major features of the sputtering and ion emission models.

We separate the sputtering process into bulk and surface effects. The relative sputtering yield, i.e., variations with ion energy, mass and incident angle, is determined by a bulk effect model which is illustrated in Fig.1.

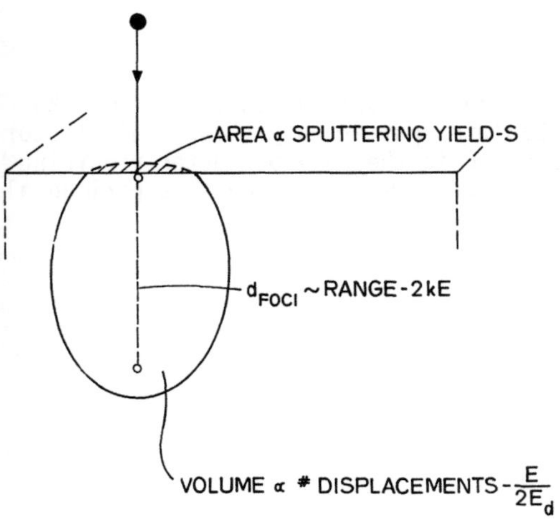

Fig.1 The incident ion creates a disturbed statistical region in the bulk. Its intersection with the surface defines an affected area which is proportional to the yield

This model leads to a simple algebraic equation [1] which accurately predicts the sputtering yield as a function of the ion energy, mass, and incident angle over a broad range of these parameters. In Fig.2, the mass dependence of the model is checked against the experimental data of EERNISSE [4].

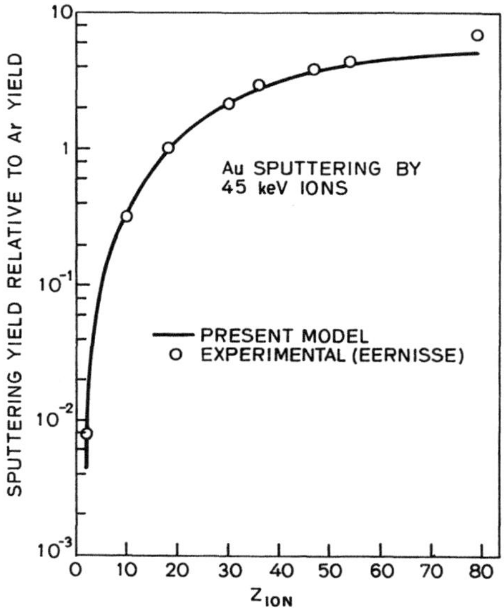

Fig.2 Au is bombarded by 45keV ions (data of EERNISSE [4]). Solid curve is the prediction of the present model

A scale factor B, defined as the fraction of displaced surface atoms that actually escape, is needed to obtain absolute sputtering yields. This factor is a property of the target surface only and is the only parameter requiring adjustment in the model. B may be determined from comparison to experimental data in the literature. See Fig.3.

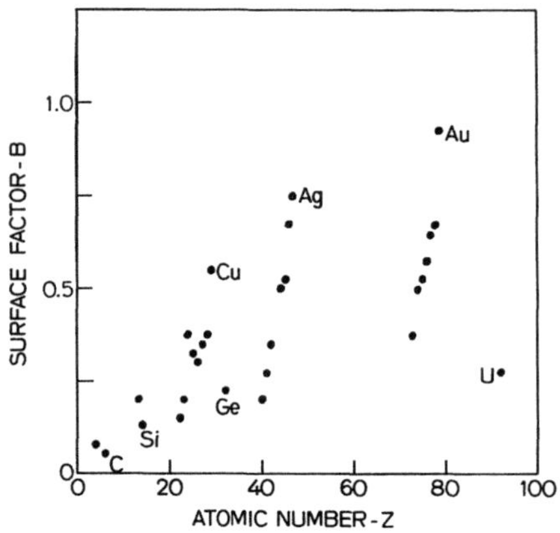

Fig.3 Values of B, the fraction of displaced surface atoms that escape, derived from the experimental data of ROSENBERG and WEHNER [5]

It appears that B is proportional to $n\sqrt{M}$ where n is the number of valence electrons and M is the atomic mass. This is consistent with an ion neutralization model (e.g. [6]) in which displaced ion cores at the surface are neutralized with a probability inversely proportional to their emitted velocity.

We propose that unneutralized ions must overcome an additional image force. The majority of emitted ions have energies of only a few eV and cannot surmount the image force barrier. This model may now be used to predict the ion yield I and sputtering yield S as follows: We let N_T be the total number of displaced surface atoms as predicted by our algebraic equation; N_N is the number of neutrals and N_I the number of ions such that $N_T = N_N + N_I$. We now have $S = BN_T \cong N_N$ and we assume that $I = \delta N_I = \delta(1 - B)N_T$ with $\delta \ll 1$. We thus obtain the following equation for the yields:

$$I/S = \delta(1 - B)/B \qquad (1)$$

The factor (1 - B)/B compares very favorably with the values of I/S measured by PRIVAL [6]. This model provides a simple explanation of the correlation between high sputtering yields and low ion yields.

The observed increase of both positive and negative ion yields upon oxidation [7] may be related to a reduction in the image force. Ions which surmount the image potential come from the broad high energy tail of the emitted atom distribution. For this reason, the average energy of emitted ions is greater than the average energy of emitted neutrals. The factor δ varies inversely with the image potential. We note also that the I/S dependence of the ion yield observed by DELINE [8] is similar to the (I - B)/B dependence observed here.

Finally, we comment on the physical content of the model described herein. The ellipsoid illustrated in Fig.1 is a mathematical construct which allows us to estimate the amount of energy deposited near the surface. The remarkable agreement of the model with experiment over a broad range of ion energies, masses, and incident angles indicates that we are using a valid statistical approach. Crystalline and other second order effects are, of course, not included in the model. The validity of the sputtering model is discussed further in [1].

Acknowledgements

We would like to thank Professor Bill Spicer for useful discussions and financial support. This work was supported in part by NBS Contract No. 5-35944 and by ARPA Contract No. DAAB-2684.

References

1. S.A. Schwarz, C.R. Helms, J. Appl. Phys., Aug. 1979, in press.
2. S.A. Schwarz, C.R. Helms, to be published.
3. S.A. Schwarz, C.R. Helms, J. Vac. Sci. Technol., 16, 781 (1979).
4. E.P. EerNisse, Appl. Phys. Lett. 29, 14 (1976).
5. D. Rosenberg, G.K. Wehner, J. Appl. Phys. 33, 1842 (1962).
6. H.G. Prival, Surf. Sci., 76, 443 (1978).
7. P. Williams, C.A. Evans, Jr., Surf. Sci., 78, 324 (1978).
8. V.R. Deline, C.A. Evans, Jr., and P. Williams, Appl. Phys. Lett., 33, 578 (1978).

Clustering Distances in Secondary Ion Mass Spectrometry

F. Honda[1], Y. Fukuda and J. W. Rabalais
Department of Chemistry
University of Houston
Houston, TX 77004 USA

1. Introduction

The successful application of SIMS to chemical structure analysis is at present restricted by our rudimentary knowledge of two basic processes, i.e., the degree to which the detected secondary ion clusters reflect the virgin surface structure and the mechanism of cluster formation during sputtering. Some proofs have been offered for direct emanation of dimers and multimers from the surface [1]. Other works have shown that non-adjacent atoms on a surface can combine to form dimers and multimers upon sputtering [2]. Very low ion current and/or low impinging kinetic energy have been applied to avoid inducing molecular rearrangement during sputtering [3]. Our recent SIMS work shows that some molecular rearrangement occurs along with the characteristic fragmentation of the solid structure [4-7]. On the basis of these results, we propose that the bound secondary clusters are formed through potential interactions during irreversible adiabatic expansion of an activated region near the surface surrounding the impact site. The purpose of this work is to investigate the distances over which single atoms in a lattice can combine to form dimer clusters during sputtering.

2. Experimental methods and results

The spectrometer chamber is pumped by a turbomolecular pump and a Ni-Ti sublimation pump providing a base pressure of 2 X 10^{-10} Torr. A primary ion current of 1 X 10^{-8} A over the energy range 0.2 - 3.0 keV was generated from a Varian ion gun. A binary system consisting of the mixed salt KCl-CsCl was chosen for the experiment; this system forms a solid solution in all proportions. For an ideal homogeneous solution, the average distance between metal atoms can be calculated from the concentration. The experiments were performed by monitoring the Cs_2^+/Cs^+ secondary ion ratio, R. With decreasing [Cs] the average distance between Cs atoms, r, will increase (and the average number of Cs atoms in the region over which the momentum is dissipated will decrease) and the clustering probability to form the Cs_2^+ dimer should decrease. When r is large enough that Cs atoms cannot be neighbors, all dimers observed are assumed to come from a recombination process. The SIMS intensity ratios Cs_2^+/Cs^+, R, as a function of [Cs] in the solid solution for different E_k's are shown in Fig.1. The average distances between Cs atoms, r, at the indicated concentration (assuming ideal mixing of the components) are shown at the top of the figure. It shows that R increases from 0.2 to 3.0 KeV as [Cs] increases. At constant voltage, these points form a reasonably straight line

[1]On leave from The Res. Inst. Iron, Steel & Other Metals, Tohoku University, Sendai, JAPAN.

for low [Cs]; in this region R could be expressed as $R = a \cdot \log[Cs] + b$, where a and b are constants. Extrapolating these straight lines results in intercepts with the R=0 abscissa in the region of r∼200 Å for He^+ and r∼400 Å for Ar^+ bombardment. The R values for [Cs/K] > 0.1 deviate significantly from straight lines; these deviations are most likely due to the fact that the intensity of the monomer is >10^6cps. Nonlinearities can occur at such high count rates because the modular gain of the multiplier decreases and the amplification electronics beyond the channeltron begin to saturate as a result of pulse pile up. The Cs^+ and Cs_2^+ intensities from contamination on the sample holder were about 0.5 and 0.1 that observed from the most dilute sample ($[Cs]/[K] = 6.3 \times 10^{-4}$), respectively. Therefore the contribution to the Cs_2^+ signal from stray dimers due to migration along the surface was considered to be negligible.

3. Discussion

If sputtering is dominantly a momentum transfer process, the distances over which atoms combine during sputtering to form dimers should provide a measure of the size of the region over which the momentum is dissipated. Such a process can be studied by monitoring the intensity of secondary dimer clusters as a function of i) the distances r between the atoms in the lattice that make up the dimer and ii) the primary ion kinetic energy E_k and mass M. The experimental results suggest that the observed dimers are formed by combination of Cs atoms from non-adjacent sites during the sputtering process. If the detected dimers come dominantly from sites where two Cs atoms are accidentally adjacent, R would be independent of [Cs] because the probability of such sites existing is proportional to $[Cs]^2$. The results show that R is strongly dependent on [Cs] suggesting that the Cs_2^+ intensity obtained from accidentally adjacent Cs atoms or sputtering induced Cs migration is negligible. The convergence of the R versus [Cs]/[K] lines for different E_k

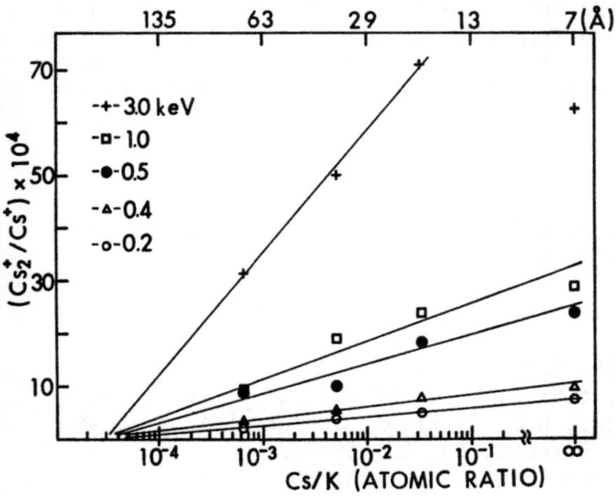

Fig.1 SIMS intensity ratio Cs_2^+/Cs^+ as a function of the Cs/K atomic ratio in the mixed salt solution KCl-CsCl. The primary ions are He^+ with E_k's ranging from 0.2 - 3.0 keV. The point for pure CsCl at 3 keV is neglected in the least squares fitting of the straight lines

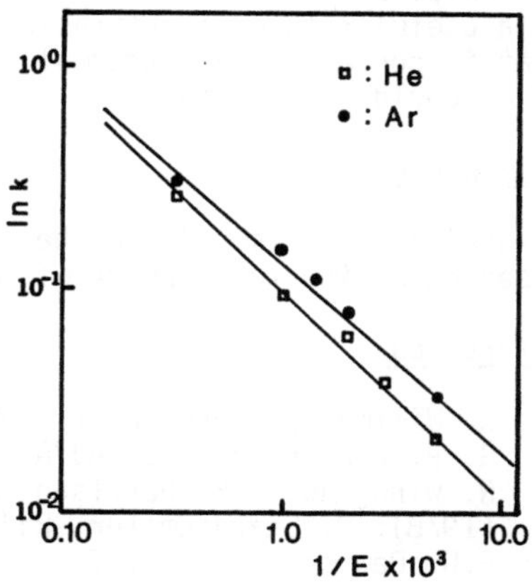

Fig.2 Plots of ln k versus $1/E_k$ assuming that k is proportional to the slope of the lines in Fig.1

suggests that there is a maximum distance between Cs atoms, r_{max}, above which recombination does not occur; r_{max} is independent of E_k within the range 200 - 3000 eV. This clustering distance defines a clustering region around the primary collision site. The increase in R with E_k at constant [Cs] is therefore attributed to the increased momentum dissipation within the clustering region. The R values obtained from Ar^+ and He^+ bombardment show that R for Ar^+ bombardment, R_{Ar}, is several times larger than R for He^+ bombardment R_{He}, at the same E_k and that r_{max} for Ar^+ is about twice that for He^+.

If the Cs_2^+ formation probability is proportional to the effective rate of dimer formation,

$$Cs^+ + Cs \xrightarrow{\overline{k}} Cs_2^+,$$

can be expressed as $\overline{k} = A \cdot \exp(-E_a/kT)$, where E_a is the activation energy, T is the effective localized transit temperature of the reaction region, k is the Boltzman constant, and $\overline{k}$ and A are the rate constant and pre-exponential factor. If we assume that the effective local temperature is proportional to the thermal energy transferred to the surface, then $kT = KE_k$ and the above equation becomes $\ln \overline{k} = \ln A - (E_a/K)(1/E_k)$. If the slopes of the lines in Fig.1 are proportional to $\overline{k}$, then plots of ln(slope) versus $1/E_k$ should yield straight lines with slopes $(-E_a/K)$. Such plots are shown in Fig.2 with the results $E_a/K^{Ar} = 2.06 \times 10^2$ eV and $E_a/K^{He} = 3.48 \times 10^2$ eV. Estimates of the percent of primary ion energy that is converted into thermal energy and the localized effective temperature can be obtained if we know the activation energy for dimer formation. Assuming E_a 20 kcal/mole, in keeping with activation energies for normal chemical reactions, we obtain $K^{Ar} = 4.21 \times 10^{-3}$ and $K^{He} = 2.49 \times 10^{-3}$. This implies that only a few tenths of one percent of the impinging ion energy is converted into thermal energy. Using the expression $kT = KE_k$, we can calculate effective temperatures; for example, $T(He, 1000\ eV) = 2.9 \times 10^4$ °K. This result is rather high compared to the calculated electron temperature from LTE considerations [8], however it is a useful method of estimating the fraction of primary ion kinetic energy that is converted into thermal energy at a surface if more accurate values of E_a could be obtained.

Acknowledgement

Acknowledgement is made to the R. A. Welch Foundation and the U.S. Army Research Office for support of this research.

References

1. K. Wittmaack, Phys. Lett., 69A, 322 (1979).
2. G. P. Konnen, A. Tip and A.E. deVries, Rad. Effects., 26, 26 (1975): N. Winograd, D.E. Harrison, Jr., B.J. Garrison, Surface Sci., 78, 467 (1978): G. Staudenmaier, Rad. Effects, 13, 87 (1972).
3. P.H. Dawson and W.-C. Tam, Surface Sci., 81, 164, 464 (1979).
4. F. Honda, G.M. Lancaster, Y. Fukuda, and J.W. Rabalais, J. Chem. Phys., 69, 4931 (1978).
5. G.M. Lancaster, F. Honda, Y. Fukuda, J.W. Rabalais, J. Amer. Chem. Soc., 101, 1951 (1979).
6. J.A. Taylor and J.W. Rabalais, Surface Sci., 74, 229 (1978).
7. F. Honda, Y. Fukuda, and J.W. Rabalais, J. Chem. Phys., 70, 4834 (1979).
8. C.A. Andersen and J.R. Hinthorne, Anal. Chem., 45, 1421 (1973): Science, 175, 853 (1972).

Basic Aspects in the Sputtering of Atoms, Ions, and Excited States

Roger Kelly

Institute for Materials Research
McMaster University
Hamilton, Ontario, L8S 4M1, Canada

1. The Sputtering Process as an Idealization

It is possible to classify sputtering variants according to the time scale [1]. If an incident particle hits at t = 0, we have prompt collisional processes at roughly 10^{-15}-10^{-14} s. There is no unique model for these as one is dealing with specific, rather than statistical, trajectories. For example, one can assume the incident particle to strike a target atom and the latter to migrate with an integral distribution function $F(x,\theta)$, θ, being the angle of recoil [2]. The yield for target atoms located at $x = x'$ is then

$$S_{prompt} = N^{2/3} \int d\sigma\{F(-\infty,\theta) - F(-x',\theta)\}, \tag{1}$$

where N is the number density and d the differential scattering cross section. In practice, atoms sputtered according to Eq. (1) or variants of Eq. (1) should appear with higher energies than normal [3].

Letting the time move on to 10^{-14}-10^{-12} s we have slow collisional processes, due basically to an indirect interaction: ion → target atom → surface atom. A satisfactory estimate of the yield can be obtained in terms of $C_D(x)$, the differential damage distribution function [4].

$$S_{slow} = E_1 C_D(0)\lambda/4U, \tag{2}$$

where E_1 is the incident energy, λ is the mean atomic spacing, and U is the surface binding energy. (The corresponding expression deduced by SIGMUND [5] is $S_{slow} = 0.0420 E_1 C_D(0)/NU$, with N in units of $Å^{-3}$.) In practice, atoms sputtered according to Eq. (2) or variants of Eq. (2) should show a distribution of the form $E(E + U)^{-3}dE$.

For 10^{-12}-10^{-10} s one can expect prompt thermal processes, in so far as these exist. The assumption is that E_1 finally degenerates into heat (1000-4000 K) and this causes vaporization during the brief interval before the heat is dissipated by thermal conduction. The corresponding yield should be given by [6]

$$S_{thermal} = \iint p(2\pi mkT)^{-\frac{1}{2}} \cdot 2\pi y dy \cdot dt. \tag{3}$$

In practice atoms sputtered according to Eq. (3) or variants of Eq. (3) should show a distribution of the form $E\exp(-E/kT)$, with T large [7]. The evidence is very limited [7].

For t $\gtrsim 10^{-10}$ s an ensemble of processes takes over based on point defects. For example, interstitials may diffuse to the surface and expel surface atoms due to the internal energy of the interstitials [8,9]. With halides and oxides, surface anions can be neutralized by diffusing holes (V_K centers), by direct interaction with the beam, or [10] by interatomic Auger processes. Concurrent with this, metal accumulates and there is either a metal signal (if the metal is volatile) or else the target becomes metallized (if the metal vaporizes too slowly) [7]. The yield can, as with Eq. (2), be estimated in terms of $C_D(x)$. Introducing E*, the energy consumed per relevant defect, and $\exp(-x/[D\tau]^{\frac{1}{2}})$, the probability that the defect gets to the surface within its lifetime τ, we have [7,8]

$$S_{defect} \approx E_1 C_D(0)(D\tau)^{\frac{1}{2}}/E^*. \tag{4}$$

The close similarity between Eqs. (2) and (4) is worth noting.

2. The Sputtering of Ions

There is at present no generally accepted model for the sputtering of ions. The basic facts that must be contended with include the following: (a) the energy distributions are similar to $E(E + U)^{-3}$, as for slow collisional sputtering, though often with a lower power than "3" [11]. (b) The yields of ions (both positive and negative) from oxidized surfaces are 10^{-1}-10^{0} and thus very high [12]. (c) The yields of ions from clean metals are 10^{-3}-10^{-2} and thus significantly lower [12]. (d) Yields are strongly affected by alloying: for example, the yield of Au^- is proportional to 10^2 from Au but 10^6 from $SmAu_3$ [13]. (3) Positive ions are typically small, with mean radii of 0.7-1.3 Å, and therefore could conceivably exist in or on a solid surface. This would not be true of negative ions, however, in view of their size.

We would like to propose that, whatever the details of the mechanism leading to <u>ion emission</u>, the following is true: ions are formed by slow collisional sputtering, there is no threshold (though possibly an energy-dependent ionization probability), and there is a tendency for ion yields to be influenced by the surface chemistry as if they pre-existed on the surface (whether or not, in fact, they actually did). For example, surface oxygen enhances both positive and negative ions, suggesting that a practical surface consists of oxygen-metal dipoles in both possible senses [14]. Likewise, the behavior of Au-Sm alloys is easily rationalized taking into account that the electronegativity of Sm is 1.2 and of Au is 2.3 [13].

3. The Sputtering of Excited States

The situation with excited states is similar to that with ions in that there is no generally accepted model for their formation. The basic facts that must be contended with include the following: (1) the energy distributions appear to have a threshold of 10-100 eV, a result inferable from intensity-vs.-distance measurements [15]. (b) The yields of excited states from oxidized surfaces are 10^{-3}-10^{-2} and thus distinctly lower than ion yields [16]. (c) The yield of excited states from clean Al is about 10^{-4} and thus again lower than the corresponding ion yield, 10^{-2} [15]. (d) Yields are strongly affected by alloying: for example, the Be signal from $Be_{0.12}Cu_{0.88}$ is comparable to that from pure Be [17]. (e) Excited states are typically large, with mean radii of 10-40 Å and therefore quite incapable of existing in or on a solid surface. This follows from the expression for the mean radius of a hydrogenic atom [18].

$$\langle r \rangle = \frac{n^2 a_o}{Z}\{1 + \tfrac{1}{2}[1 - \frac{\ell(\ell + 1)}{n^2}]\},$$

where a_o is the Bohr radius, 0.529 Å.

We would like to propose that, whatever the details of the mechanism leading to excited-state emission, the following is true: excited states may or may not be formed by slow collisional sputtering, a significant threshold exists, and excited states cannot exist in or on a solid surface. A curve-crossing mechanism involving particles which encounter each other beyond the surface is a sufficient though not necessary model [19].

4. The Chemistry of a Bombarded Surface

Each of the idealized types of sputtering has a built-in capability for preferentiality, thence for changing the surface composition. Prompt collisional processes favor loss of the lighter component, slow collisional processes favor loss of the less tightly bound component, prompt thermal processes expel volatile components, and point-defect based processes will normally expel anions [1,7,20].

Previous discussion of alloys has in general involved a survey of the relevant literature in an attempt to find trends. Until recently two trends could be confirmed: loss of the lighter and at the same time the less tightly bound component, as with Ag-Au, Al-Au, Al-Cu, and Au-Cu [20]. A greater variety of systems has not been studied, including Ag-Au-Cu, Au-Ni, Co-Gd, and Ni-Pd, all of which lose the heavier component preferentially, the heavier component being, however, that with the lower surface binding energy [4]. This suggests that only the binding-energy correlation is valid, thence that slow collisional sputtering is involved.

Ion-bombarded oxides were found to follow at least two trends: loss of oxygen corresponding to it being less tightly bound and loss of oxygen corresponding to high volatility [21]. We have nothing new to add concerning these trends but would indicate that a remarkable body of evidence exists suggesting still a further trend: loss of oxygen corresponding to point-defect based processes, especially when bombardments are carried out with electrons [7]. The yields are very low, 10^{-6}-10^{-4}, often vary as I^2, where I is the incident current, and can be understood in terms of a two-step process such as

$$h^+ + O^{2-}(\text{surface}) = O^-(\text{surface}) \tag{5a}$$

$$h^+ + O^- \ (\text{surface}) = O^0(\text{surface}) = O^0(\text{gas}). \tag{5b}$$

Here h^+ represents a hole. The details in particular cases depend on the relative values of the rate of point-defect sputtering, $IS_{defect}\lambda^2$, and the rate of vaporization of the accumulating metal, $p(2\pi mkT)^{-\frac{1}{2}}\lambda^2$. If the latter is high enough, the sputtering is continuous (BaO, K_2O, MgO, SrO, ZnO). In the contrary case, the target becomes metallized (Al_2O_3, B_2O_3, GeO_2, MoO_3, Nb_2O_5, SiO_2, Ta_2O_5, TiO_2, V_2O_5, WO_3).

Halides have been studied mainly in terms of incident electrons. The yields are high (10^0-10^2), often vary as I, and can be understood in terms of a one step process like that of Eq. (5b). Again the details in individual cases depend on the relative rates of sputtering and vaporization.

5. The Topography of a Bombarded Surface

The remarkable aspect of the topography of a bombarded surface is the way in which blunt surface asperities, due normally to the preceding surface treatment, evolve into well-defined pyramids or cones [22]. The basic relation governing such evolution is [23]

$$\frac{\partial}{\partial t}\left(\frac{\partial y}{\partial x}\right) = -\frac{I}{N}\cdot\frac{\partial S}{\partial \theta}\cdot\frac{\partial \theta}{\partial x}, \qquad (6)$$

where I is the incident ion flux. If Eq. (6) is applied to a convex-up asperity, it is easily shown that the surface should rotate towards the angle, $\hat{\theta}$, at which S maximizes. Once the slope is near $\hat{\theta}$ one can expect fluctuations in the sputtering process to lead to a slope increase, thence a tendency for the incident ions to be reflected to the base of the apserity, now having a pronounced pyramidal or conical shape [22]. This gives a transient stability but at the same time creates a groove which directs sputtered and scattered atoms onto the sides of the pyramid or cone. The latter finally disappears for high enough doses, leaving a characteristic pit.

Theory has sometimes predicted pyramids or cones to have an indefinite stability, but this is probably a result of the neglect of the indirect processes involving the grooves beneath the pyramids or cones.

References

1. R. Kelly, in Proc. Int. Conf. on Ion-Beam Modification of Mat. (Budapest 1978) (in press). Also to be published in Rad. Effects.
2. S. Dzioba, R. Kelly, J. Nucl. Mat. 76, 175 (1978).
3. I. Reid, B.W. Farmery, M.W. Thompson, Nucl. Instr. Meth. 132, 317 (1976)
4. R. Kelly (to be published).
5. P. Sigmund, Phys. Rev. 184, 383 (1969).
6. R. Kelly, Rad, Effects, 32, 91 (1977).
7. R. Kelly, Surface Sci. (in press).
8. H. Overeijnder, M. Szymonski, A. Haring, A.E. de Vries, Rad. Effects 38, 21 (1978).
9. M.W. Thompson, I. Reid, B.W. Farmery, Phil. Mag. A38, 727 (1978).
10. M.L. Knotek, P.J. Feibelman, Phys. Rev. Lett. 40, 964 (1978).
11. A.R. Krauss, D.M. Gruen, Nucl. Instr. Meth. 149, 547 (1978).
12. K. Wittmaack, in Inelastic Ion-Surface Collisions (Academic Press, New York, 1977), p. 153.
13. J.J. Cuomo, R.J. Gambino, J.M.E. Harper, J.D. Kuptsis, J. Vac. Sci. Technol. 15, 281 (1978).
14. P. Williams, C.A. Evans, Jr., Surface Sci. 78, 324 (1978).
15. S. Dzioba, O. Auciello, R. Kelly, Rad. Effects (in press).
16. I.S.T. Tsong, N.A. Yusuf, Appl. Phys. Lett. 33, 999 (1978).
17. G. Pilon, Unpublished work at McMaster University (1979).
18. L. Pauling, E.B. Wilson, Introduction to Quantum Mechanics (McGraw-Hill, New York, 1935), p. 144.

19. N.H. Tolk, et al., Phys. Rev. A13, 969 (1976).
20. R. Kelly, Nucl. Instr. Meth. 149, 553 (1978).
21. H.M. Naguib, R. Kelly, Rad. Effects 25, 1 (1975).
22. O. Auciello, R. Kelly, R. Iricibar, Rad. Effects (in press).
23. M.J. Nobes, J.S. Colligon, G. Carter, J. Mat. Sci. 4, 730 (1969).

Cluster Formation in SIMS: CO on PdAg

George J. Slusser
IBM General Technology Division
Essex Junction, Vermont 05452

The technique of Secondary Ion Mass Spectrometry (SIMS) reportedly has great promise in the determination of local order structure of the top surface layer. A key possibility of this technique is that the composition and abundance of the cluster ions reflect to some degree the geometrical arrangement of reacted atoms on the surface. However, the information available in the cluster ions lies unused for lack of a coherent theory capable of explaining the way in which the cluster ions are formed. Probably the most important missing part of our understanding lies in the determination of whether or not the cluster ions leave the surface as intact species.

In an attempt to elucidate the true theory, a study of CO adsorption on the metals Pd and Ag and a 79% PdAg alloy was performed. Several studies using other techniques are available on these systems and a consistent picture of CO adsorption on these materials has been formulated. No adsorption of CO on Ag foils is found at temperatures above 200K [1] while CO adsorption on Pd foils is found to be molecular and very fast [2-4]. The exposure of PdAg to CO results in adsorption of CO in a linear bond only to the Pd species on the surface of the alloy [4-6]. If the cluster ions observed in the SIMS technique arise entirely from emission of intact species from the surface, then we should find no evidence for any Ag-CO clusters without the presence of Pd in the cluster.

The components of the XPS/SIMS instrumentation have been described in detail elsewhere [7]. Ar^+ is used as the primary beam with a flux <10 namps/cm^2 at an energy of 3 keV. Polycrystalline foils of Ag, Pd and a 79% PdAg alloy were cleaned by alternating cycles of argon ion bombardment and heating. Cleanliness and composition of the samples were monitored by both XPS and SIMS.

The Ag foil exposed to CO pressures up to 10^{-6} Torr showed no evidence for any CO adsorption. No Ag-CO type clusters were observed for up to two hours exposure and only negligible increases in Ag^+ were noted. In contrast exposure of the Pd foil to CO produced an immediate rise in Pd^+ and Pd_2^+ ion intensities. Concomitant rises in $PdCO^+$ and Pd_2CO^+ were noted with no PdO^+ or PdC^+ observed in agreement with a previous study [2].

Exposure of the 79% PdAg alloy foil resulted in a slow rise in metal ion intensities. The rise in intensity of these ions was extremely slow when compared with pure Pd and was found to be dependent on the CO pressure. Saturation of ion intensities at 10^{-6} Torr occurred after approximately 50 minutes exposure time. This is similar to the time required for saturation of the work function change of the alloy to occur when exposed to this same pressure of CO [8].

The CO exposure resulted in the rise of various metal-CO species. Included among these species were large amounts of $PdCO^+$ and $AgCO^+$. When ion intensities are corrected for isotope distribution, the $AgCO^+$ ion yield is approximately twice the $PdCO^+$ ion yield. This appearance of $AgCO^+$ in the SIMS spectrum is surprising in that IR studies [4,5] have shown that CO preferentially adsorbs onto Pd atoms in the alloy in a linear-bonded scheme. We have also seen that pressures up to 10^{-6} Torr CO do not cause the appearance of any Ag-CO type species on a pure Ag foil. The fact that the $AgCO^+$ ion intensity is greater than the $PdCO^+$ ion intensity is even more perplexing.

If it is to be believed that there is no adsorption of CO onto Ag in the alloy, then the presence of $AgCO^+$ in the SIMS spectrum dictates that local order information is not as easily available from the SIMS technique as was originally proposed. The clusters cannot be originating from nearest neighbor atoms or no $AgCO^+$ would be formed. The result, then, is that $AgCO^+$ is being formed from species not necessarily bonded to one another originally on the surface.

Recent calculations using a classical trajectory model of the ion bombardment process have found results similar to those reported here [9]. A CO-covered surface was found to generate metal-CO species above the surface of the metal. The metal and CO species were found to originate from locations on the surface within close proximity to one another, but not necessarily in nearest neighbor positions. This type of model would then allow CO to be bonded only to Pd on the alloy, but would still allow formation of $AgCO^+$ above the surface.

To pursue the reasons for the large amounts of $AgCO^+$ relative to the $PdCO^+$ ion intensities, the ratios of the $AgCO^+$ to the Ag^+ ion intensities were calculated and are shown in Fig. 1. This ratio immediately jumps to a value of ~ 0.04 and remains constant throughout the remaining CO exposure time. There appears to be no dependence of the ratio on the amount of CO available on the surface. Once CO is present, the formation of $AgCO^+$ is dependent mainly on the availability of Ag ions above the surface.

Also shown in Fig. 1 is the change in the $PdCO^+/Pd^+$ ratio with CO exposure time at 10^{-6} Torr. In contrast to the $AgCO^+/Ag^+$ ratio, this ratio follows the rise in ion intensities in that saturation occurs after approximately 50 minutes. It is intriguing that the $PdCO^+/Pd^+$ ratio has a direct dependence on the amount of CO present on the alloy surface while the $AgCO^+/Ag^+$ ratio appears to have no observable dependence at all on this value. It is entirely possible that this phenomenon is related to the fact that CO is adsorbed on the Pd and not the Ag atom.

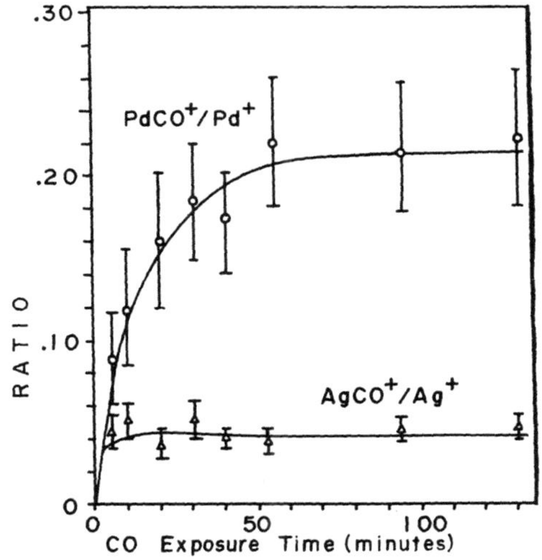

Fig.1 Dependence of $PdCO^+/Pd^+$ and $AgCO^+/Ag^+$ ratios on CO exposure time
CO pressure = 10^{-6} Torr

Acknowledgement

The support of Dr. Nicholas Winograd and Purdue University is greatly appreciated.

References

1. G.E. McElhiney, H.Papp and J.Pritchard, Surf.Sci. 54, 617 (1976).
2. M.Barber, J.C.Vickerman and J.Wolstenholme, Surf.Sci. 68, 130 (1977).
3. H.Conrad, G.Ertl, J.Koch and E.E.Latta, Surf.Sci. 43, 462 (1974).
4. Y.Soma-Noto and W.M.H.Sachtler, J.Catal. 32, 315 (1974).
5. M.Primet, M.V.Mathieu and W.M.H.Sachtler, J.Catal. 44, 324 (1976).
6. J.J.Stephan, P.L.Franke and V.Ponec, J. Catal. 44, 359 (1976).
7. T.Fleisch, A.T.Shepard, T.Y.Ridley, W.E.Vaughn, N. Winograd, W.E.Baitinger, G.L.Ott and W.N.Delgass, J.Vac.Sci.Technol. 15, 1756 (1978).
8. L.Whalley, D.H.Thomas and R.L.Moss, J.Catal. 22, 302 (1971).
9. N.Winograd, B.J.Garrison, T.Fleisch, W.N.Delgass and D.E. Harrison, Jr., J.Vac.Sci.Technol. 16, 629 (1979).

Effect of Partial Oxygen Pressure on Metal Single Crystals Bombarded by Noble Gas Ions

M. Bernheim and G. Slodzian
Université de Paris-Sud - Bât. 510
91 405 Orsay, France

It is well known that oxygen chemisorption on metallic surface being simultaneously bombarded with noble gas ions, modifies the sputtering yields and the ionization probabilities for positive ions. It is also well known that even when the surface is saturated by the oxygen flooding, the orientation of the crystal lattice with regards to the incident beams still has an influence on emission and sputtering processes.

In order to study these effects we have used the backscattering of the primary ions, the very same which sputter the crystal. The experimental set-up--the angles of incidence and backscattering are respectively 45° and 135°--is such that the elastic backscattering concerns either atoms in a first layer position (transparent orientation) or atoms in a first and second layer position (opaque orientation) [1,2].It should also be noted that sputtering yields (and thus ionic yields) and the rate of production of defects are quite different in opaque and transparent orientations. Various monocrystalline targets were investigated:

Aluminum (100) - When secondary emission is saturated by the oxygen flood, secondary ion emission as well as backscattering show the upper layers are nearly amorphous. However, once the oxygen flooding has been stopped, one can erode the target and observe that the amorphous film has different thicknesses according to which crystal orientation was set previously. This shows the influence of the rate of production defects on the film growth (remember that the sample was continuously sputtered while being flooded with oxygen).

Copper (100) - When secondary emission is saturated by the oxygen flood, the surface of the target maintains a crystalline structure. Backscattering suggests that the superficial layer is composed by CuO. Ionic erosion indicates that there is probably very little incorporation of oxygen.

Copper-Aluminum (2 atomic percent Al) (100) - As previously, the upper layers maintain an ordered superficial structure. Backscattering experiments show that, at low coverage, aluminum atoms in a second layer position are hidden by oxygen atoms. Then, at higher coverage, there is a reorganization of the superficial layer which results in a slight increase of the concentration of aluminum inducing a very strong enhancement of Cu ions as the coverage of oxygen is increased.

Nickel (100) - Two ordered structures can be observed, the first one is made up of chemisorbed oxygen (low coverage), the second one is probably formed by NiO. At a given primary ion density and in an opaque orientation, there is a given partial pressure of oxygen for which one can observe--either by secondary emission or by backscattering--a slowly progressing transition from the chemisorbed structure to the oxide structure, characteristic time of which for the primary density used here is about 10 mn (the reverse transition exhibits hysteresis behavior). The changes of Ni^+ and NiO^+ currents are due to the transition from a chemisorbed layer to an oxide layer. On these layers the ionization probabilities as well as the ratio between polyatomic clusters and single particles may be very different.

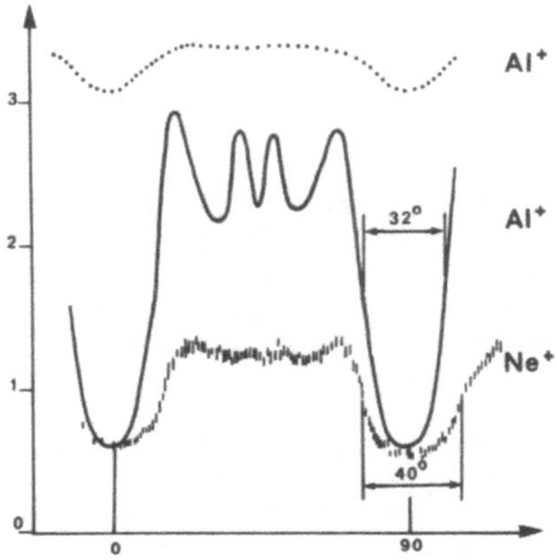

Fig.1 Partial azimuthal recording of Ne^+ ions backscattered from a 100 aluminum crystal and of Al^+ secondary ions sputtered by Ne^+ bombardment. The partial oxygen pressure nearly suppresses the crystalline transparency contrast on secondary ion emission (dashed curve) as on backscattered Ne^+ ions. At each oxygen partial pressure, the energy distribution of Ne^+ was recorded. This curve allowed to subtract the background due to multiple collisions and to obtain the intensity of the elastic peak due to binary collisions only. In all the experiments reported here, the primary Ne^+ density was kept constant

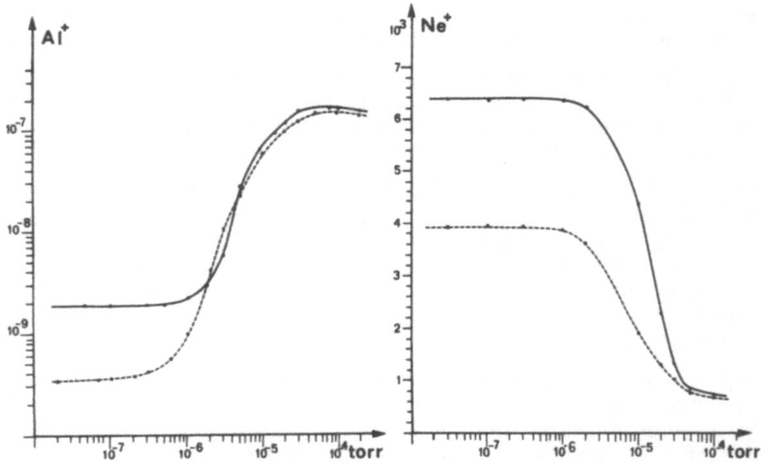

Fig.2 Al^+ secondary ions and Ne^+ backscattered ions from an (100) aluminum crystal set either in an opaque position (continuous lines) or in a transparent position (dashed curve). An oxygen partial pressure nearly suppresses the contrast

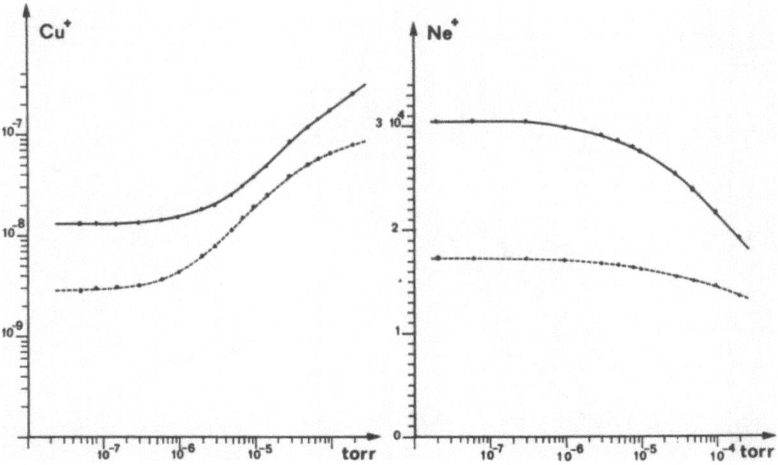

<u>Fig.3</u> Variation of Cu^+ secondary ions and Ne^+ backscattered ions on a copper (100) crystal with the oxygen partial pressure

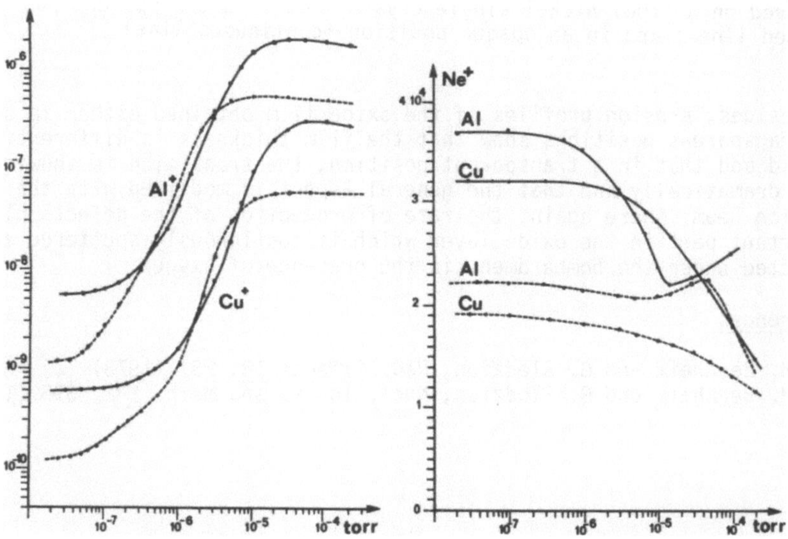

<u>Fig.4</u> 2 atomic per cent <u>copper</u> aluminum alloy; Al^+ and Cu^+ secondary ions and backscattered Ne^+ intensities in binary collision on aluminum or copper atoms were measured in transparent (dashed lines) or opaque position at different partial oxygen pressure

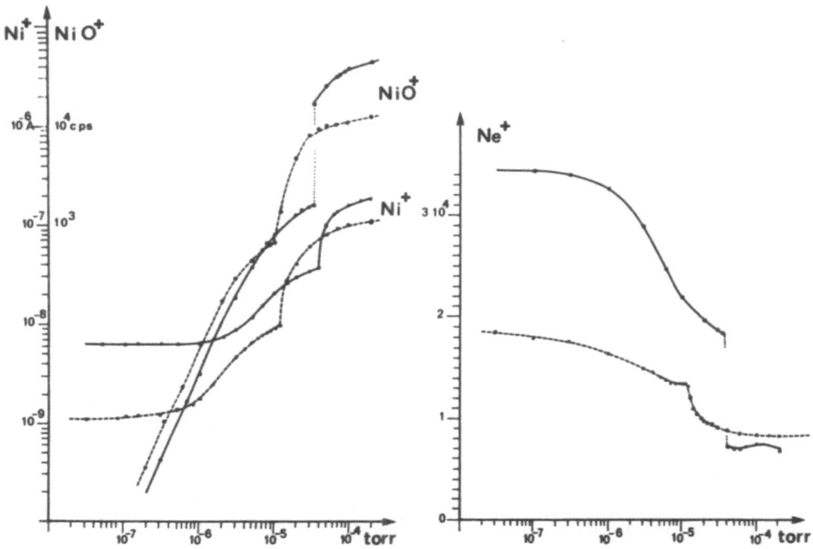

Fig.5 Variation of Ni^+, NiO^+ secondary ions and Ne^+ backscattered ions measured on a (100) nickel single crystal set in a transparent position (dashed lines) and in an opaque position (continuous line)

Besides, erosion profiles of the oxide film obtained either in opaque or transparent positions show that the film thickness is different. One should add that in a transparent position, the transition is showing up less dramatically and that the general aspect is modified with the mass of the ion beam. Here again, the rate of production of the defects plays an important part in the oxide layer which is continuously sputtered and reconstructed under the bombardment in the presence of oxygen.

References

1. M. Bernheim and G. Slodzian, Rad. Effects 18, 231 (1973)
2. M. Bernheim and G. Slodzian, Nucl. Instr. and Meth. 132, 695 (1976)

A Comparison of Absolute Yields of Excited Neutrals and Positive Ions from Ion-Bombarded Surfaces

P. Williams*†
Materials Research Laboratory
University of Illinois
Urbana, IL 61801

I.S.T. Tsong† and S. Tsuji†
Materials Research Laboratory
The Pennsylvania State University
University Park, PA 16802

Abstract

We have examined the yields of excited neutral atoms and of positive ions sputtered from argon ion-bombarded surfaces under conditions of saturation oxygen coverage. By determining the transmissions of the photon spectrometer and of the mass spectrometer used for these studies it was possible to estimate the absolute yields of the excited and ionized species. For silicon, under the conditions of the experiments, excited neutral yields were $\sim$ 0.1% and the positive ion yield was $\sim$ 1%. Optical emission from excited silicon ions was also monitored and it could be estimated that only $\sim$ 1% of these ions were in an excited state. These results will be discussed together with their implications for analysis and for theories of ion and excited state formation in the sputtering process.

*Work supported by the National Science Foundation MRL Grant DMR-77-23999
†and by the Office of Naval Research (L. Cooper).

Correlation Between the Spectral Ionization Probability of Sputtered Atoms and the Electron Density of States

T.R. Lundquist
Gatan, Inc.
3117 Babcock Blvd., Pittsburgh, PA 15237

Because of its sensitivity, the potential applications of secondary ion mass spectrometry (SIMS) are much greater than the problems to which it is currently applied. Its potential has not been achieved because the mechanisms by which atoms are sputtered as ions are not understood, even though a relationship between the chemical bonding and the ionization probability has been well documented. There has been some work, however, which suggests a relationship between the ionization probability and the electron density of states at the surface [1-5]. Ion yield measurements, in themselves, do not give sufficient information to test various correlations. Measurements of the energy distributions of sputtered ions give much more information on the ionization/neutralization process. I have made these measurements and also have obtained the energy distributions of neutral sputtered atoms [6,7]. This allowed the determination of the spectral ionization probability, i.e., the energy dependence of the ionization probability. By examining clean Ti, Ni and Cu and these same surfaces with some oxygen coverage, it has been possible to assess the significance of the electron density of states.

The spectral ionization probability P(E) is shown on log-log scales in Fig.1, 2 and 3. The similar fine structure in the curves of each figure is not significant; it is due to the same neutral energy distribution being used in each case. The spectral ionization probability was obtained by dividing the ion energy distribution by the neutral energy distribution. The flatness in some of the curves at very low energy is apparently an experimental artifact which could result from two factors: the over focusing of the primary

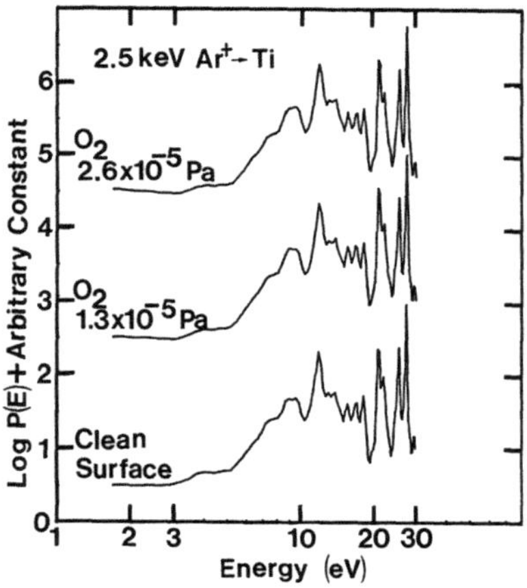

Fig.1 Energy dependence of the ionization probability P(E) for sputtered Ti for different oxygen coverages. The dependence on E for Ti is not affected by the oxygen exposures used here even though the ion yield increased by a factor of 12. The similarity in fine structure is due to the same neutral Ti energy distribution being used in each case

ion beam, which increased the second-order ionizing collisions among the sputtered atoms and the secondary ion optics. Except for the initially extended flat portion of the Ti curves, P(E) in each plot exhibits the tendency towards linearity--P(E)αE^N--inferred by others [8-10]. A model put forward by CINI [3] predicted this form of dependence on energy for the ionization probability at low energies. For each of the clean surfaces P(E) is an increasing function of energy. Above 20 eV the interpretation of P(E) is more difficult because of noise. An important finding of this research is that the effect of oxygen coverage on the spectral ionization probability is different in the three cases.

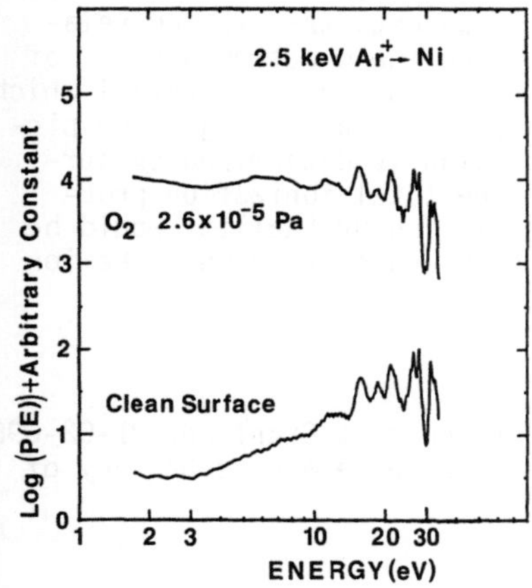

Fig.2 Spectral ionization probability P(E) for sputtered Ni for different oxygen coverages; note the change in slope. With P(E) αE^N, N = 0.54 for the clean surface and is near zero when oxygen is on the surface. This change is correlated with the suppression of the sharp d band in the filled electron density of states

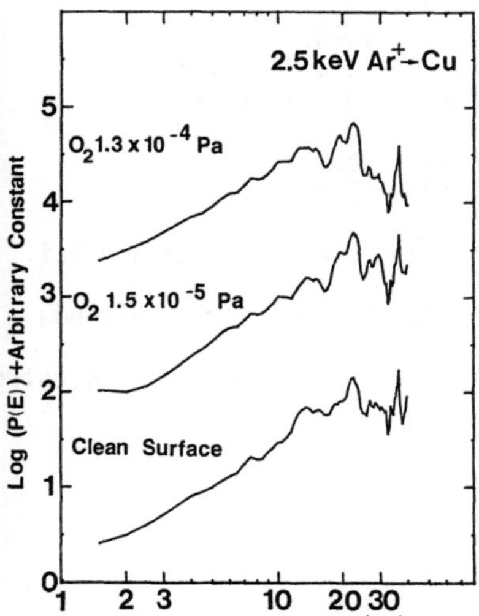

Fig.3 Spectral ionization probability P(E) for sputtered Cu; note the tendency towards linearity below 20 eV. The slopes are similar but decrease with increasing oxygen coverage. For this case also the d band is suppressed with oxygen coverage. The effect is less, presumably because of the smaller sticking probability of oxygen on Cu

CINI'S model for ion formation implies that the power dependence of the spectral ionization probability is related to the sharpness of the surface density of states [3], which, in fact, seems to be the case. Theoretical calculations find the filled density of states near the Fermi level to be greatest for Ti and least for Ni [11,12]. Furthermore, the correlation is consistent with experimental studies of oxygen effects on the density of states [13-16].

The total ionization probability involves two factors, both dependent on the chemical state of the surface. Although one, the energy dependence, seems to agree with CINI'S model, her model does not contain any quantity which can explain the other, the variation of the magnitude with the surface chemical structure--the "chemical effect." This suggests that the correlation between the spectral ionization probability and the electron density of states may not be directly related to ion formation, but to ion survival which is also expected to correlate with electron density of states [17]. The picture for secondary ion emission is that ion formation is determined by surface chemistry, giving rise to the magnitude of the total ionization probability. The energy dependence of the total ionization probability would be complicated due to the convolution of the energy dependences of both the ion survival probability and the ion formation probability.

Acknowledgement

Part of this research was supported by NASA under Contract/Grant No. 21-02-096 while the author was at the Institute for Physical Science and Technology of the University of Maryland.

References

1. G. Blaise, Radiation Effects 13, 235 (1973).
2. Z. Sroubek, Surface Sci. 44, 47 (1974).
3. M. Cini, Surface Sci. 54, 71 (1976).
4. A. Blandin, A. Nourtier and D. Hone, J. Physique 37, 369 (1977).
5. Z. Sroubek, J. Zavadil, F. Kubec and K. Zdansky, Surface Sci. 77, 603 (19
6. T.R. Lundquist, J. Vac. Sci. Technol. 15, 689 (1978).
7. T.R. Lundquist, to be published in Surface Sci. 1979.
8. V.I. Veksler and B.A. Tsipinyuk, Sov. Phys. JETP 33, 753 (1971).
9. E. Dennis and R.J. MacDonald, Radiation Effects 13, 243 (1972).
10. Z. Jurela, Radiation Effects 19, 175 (1973).
11. E.C. Snow and J.T. Waber, Acta Metal. 17, 623 (1969).
12. C.S. Wang and J. Callaway, Phys. Rev. B 9, 4897 (1974).
13. G.E. Becker and H.D. Hagstrum, Surface Sci. 30, 505 (1972).
14. K.T. Yu, W.E. Spicer, I. Kindau, P. Pianetta and S.F. Lin, Surface Sci. 57, 157 (1976).
15. D.E. Eastmen, Electron Spectroscopy, ed., D.A. Shirley, North-Holland, Amsterdam, 1972, p. 487.
16. A. Platau, L.I. Johansson, A.L. Hagstrom, S.E. Karlson and S.E.M. Hagstro Surface Sci. 63, 153 (1977).
17. J.C. Tully, Phys. Rev. B 16, 4324 (1977).

Physical Aspects of the Valence Model's Parameters

C. Plog and W. Gerhard
Bereich Neue Technologien
Dornier System GmbH
Postfach 1360, 7990 Friedrichshafen, Germany

1. Introduction

By evaluating experimental SIMS data of 15 metals oxidized in vacuum, an empirical formula ("Valence Model" [1], refer to Fig.1) was recently established which describes the absolute yield of secondary ions of the kind MeO_n^+ (n=0,1,2,...) and MeO_n^- (n=1,2,3,...). Similar to a Gaussian shape, this yield depends on the "fragment valence" K of the metal in the emitted fragment ion and the "lattice valences" G^+ and G^-. The maximum yields S^+_{max} and S^-_{max} of the secondary ions $MeO_n^{+/-}$, differ by a factor $m^{-2,4}$ which only depends on the mass, m, of the metal.

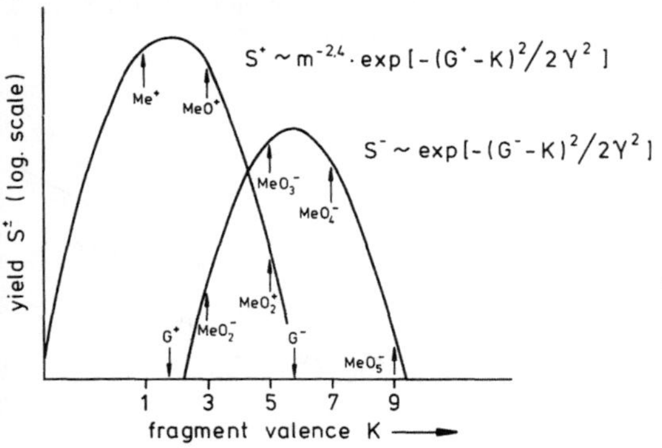

Fig.1 Yield $S^{+/-}$ of fragment ions $MeO_n^{+/-}$, from metal oxide specimens versus fragment valence K

As the Valence Model considers neither variations of the metal's valency nor multicomponent metal oxide specimens, two extensions were necessary. The influence of changing metal valencies on secondary ion intensities was investigated in the case of vanadium in full detail and was found to be describable by the dynamical Valence Model [2]. For multicomponent metal oxide specimens, it was recently shown with the aid of simultaneous SIMS and AES investigations that an extension in form of the "Extended Valence Model" [3] makes it possible to determine metal concentrations. This extension does not use internal standards.

We have successfully applied the Valence Model as well as its extensions to a great number of industrial specimens. Therefore, we were encouraged to try to elucidate its physical meaning by looking for correlations between the Model's parameters and more fundamental physical quantities.

2. Correlations

2.1 Mean Metal Valency in Metal Oxide Specimens

It was shown by simultaneous SIMS and AES investigations of pure single-component metal oxides [3] that in the case of transition metals, G^- is very closely related to the mean valency state G of the metal as present in the metal oxide. During oxygen exposure of clean metals, G^- therefore varies monotonically with G. In the case of vanadium, G^- changes from 0.6 to 5.5 (data taken from [2]). For the final state, a mean valency of 5.2 was found which is, within the experimental uncertainty, identical with the most likely valency of vanadium.

2.2 Sputtering Parameter Surface Binding Energy

During our efforts to elucidate the physical meaning of the Valence Model's parameter we found a remarkable relation between G^+ of metal oxide specimens and the metal sputtering yields S for pure metals bombarded perpendicularly with 1 keV Ar^+ ions [4] (refer to Fig.1). The difficulty left, however, is that the sputtering yields for metal oxides under our experimental conditions (3 keV, 70° Ar^+ bombardment) - as far as we know - are unknown for most cases of interest. However, if the experimental conditions are changed for all metals in the same manner, the sputtering yield will also change similarly. From this point of view this relation should hold in general.*

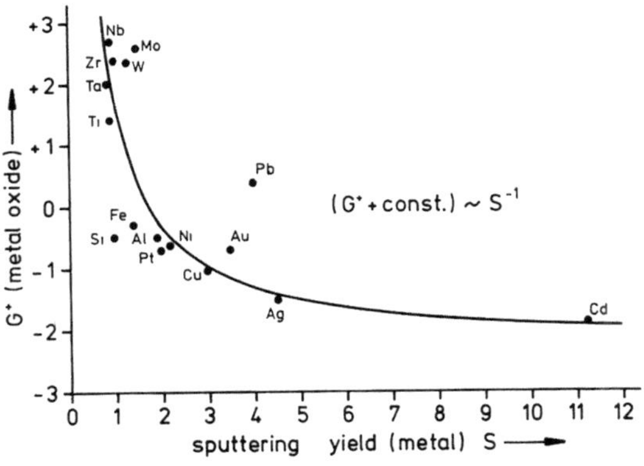

Fig.2 Lattice valence G^+ of metal oxide specimens versus metal sputtering yield

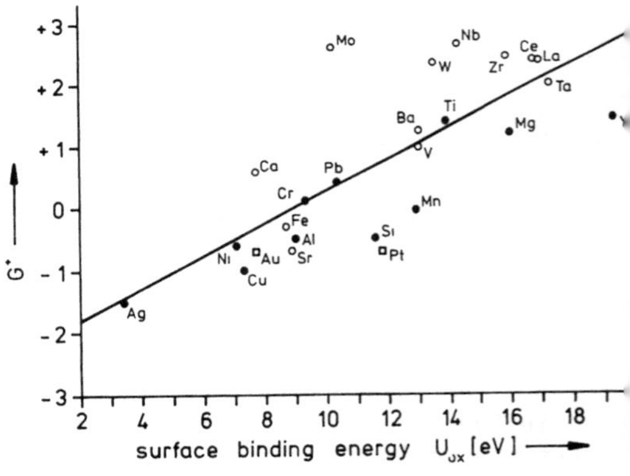

Fig.3 Lattice valence G^+ of metal oxide specimens versus surface binding energy U_{OX} of MeO

In order to confirm this result we looked for relations between G^+ and more fundamental quantities than S. The quantity U characterizes the surface fundamentally and is not influenced by the special bombardment conditions of the primary ions. According to [5], S is inversely proportional to U. Therefore, a linearity between G^+ and U (data taken for pure metal) should be expected, a relation which we actually found. But, as we deal with metal

*Note that for profile measurements this relation has the practical consequence of allowing depth determination by SIMS alone.

oxides the surface binding energies U_{ox} for MeO are needed--values of which are hardly given in the literature. A rough estimation of U_{ox}, however, is possible with the aid of available dissociation energies D (Me - Me) and D_{ox} (Me - O) [6]. One has to assume that, as was found for pure metals [7], the ratio U/D ≈ const ≈ 2 is also valid for metal oxides. The close connection between U, U_{ox} and D_{ox} is expressed also in the fact that a linear dependency of G^+ on all the three quantities were found.

2.3 Sputtering Parameter Incident Ion Mass

If, on the other hand, the material remains constant and only the bombardment conditions are changed by varying the incident ion mass (refer to Fig.4, data taken from [8]), the positive secondary ion yield ratios will not be effected as U_{ox} and therefore G^+ will remain constant. Only the total positive ion yield, represented by S^+_{max}, will change. Its behavior can then be described by the energy transfer function of binary collisions, if an effective collision mass is chosen which lies between the atomic masses of the oxygen and the metal.

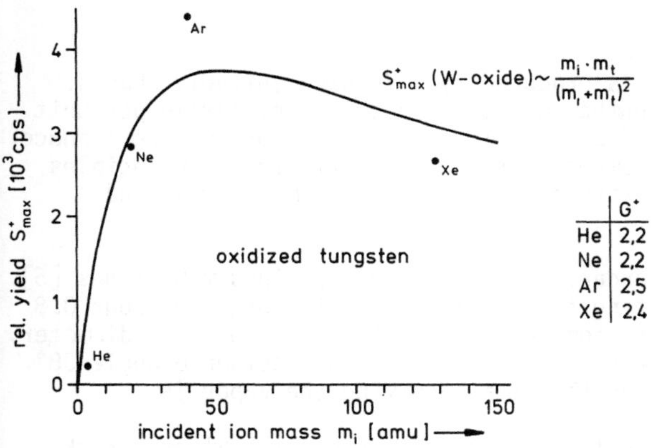

	G^+
He	2,2
Ne	2,2
Ar	2,5
Xe	2,4

Fig.4 Relative yield S^+_{max} versus incident ion mass m_i

References

1. C. Plog, L. Wiedmann, A. Benninghoven, Surf. Sci. 67, 565 (1977).
2. C. Plog, Thesis, Munster (1974).
3. C. Plog, DE-OS 2711793 (1977).
4. H. Oechsner, Appl. Phys. 8, 185 (1975).
5. P. Sigmund, Phys. Rev. 184, 383 (1969).
6. See for example: G. Herzberg, Spectra of Diatomic Molecules 2nd Ed., New York, 1950.
7. W. Gerhard, Z. Physik B22, 31 (1975).
8. A. Benninghoven, C. Plog, N. Treitz, Int. Journ. Mass. Spec. Ion Phys., 13, 415 (1974).

Negative Ion Emission from Surfaces Covered with Cesium and Bombarded by Noble Gas Ions

M. Bernheim, J. Rebière, G. Slodzian
Bât. 510 - Université Paris-Sud
91 405 Orsay-France

1. Introduction

It is well known [1-4] that either Cs^+ bombardment or Cs flooding of metallic surfaces enhances the yield of secondary negative ions considerably. In the present work, the samples are simultaneously flooded with Cs and bombarded with Ar^+ ions. This procedure has the advantage to offer a possibility of controlling the Cs coverage independently of the sputtering yield.

2. Experimental Set-up

The collimated cesium vapor jet is produced by a furnace suitably shaped. The Cs flux, J, defined as the number of atoms reaching the sample per unit area and unit time, can be varied by changing the temperature of the furnace. The J values reported here are computed from gas kinetic theory principles assuming a monoatomic vapor; no attempts to calibrate temperatures and fluxes have been made yet.

The collection system of secondary ions has been described many times [5]. The constant diaphragm was limiting the initial lateral energy to about 0.8 eV; the area from which secondary ions were collected was 200 m in diameter. The primary energy of Ar^+ ions was about 14 keV and the incidence angle 38°. The sputtered area was larger than the area receiving the vapor jet.

The sample was surrounded by a liquid nitrogen trap which allowed work with a residual pressure better than $5x10^{-9}$ torr in the vacuum chamber. In addition, secondary ion images of the surface could be observed on a channel plate. This last feature revealed to be extremely helpful for controlling the sample quality by selecting appropriate areas and adjusting the primary beam and the vapor jet.

The secondary ions were both mass and energy analyzed. The energy spectrum of secondary ions could be recorded by varying the 4 keV accelerating voltage with an amplitude of about 15 volts. Since the zero of the energy spectrum depends on the work function of the sample, the measurement of the displacement of the spectrum offered a way of determining variations of work functions; 0.1 eV could be easily detected.

3. Results

A. General

Figure 1 shows the typical variations of $I(M^-)$, negative currents, and W,

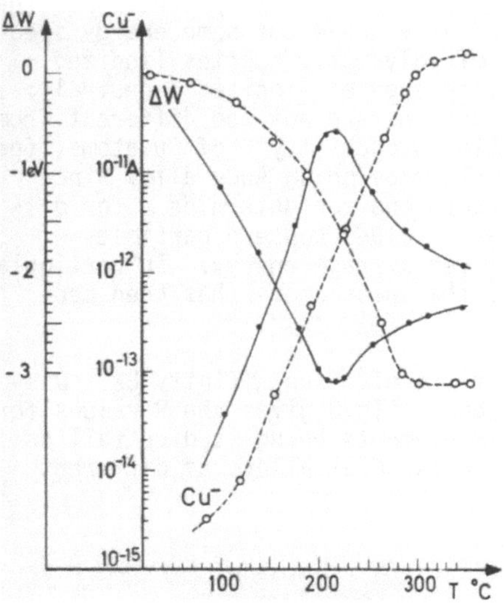

Fig.1 Work function and Cu^- secondary ion yield variations with the temperature of the reservoir. The continuous lines (corresponding to lower primary density) are describing the variations beyond the minimum of work function (and maximum of Cu^-)

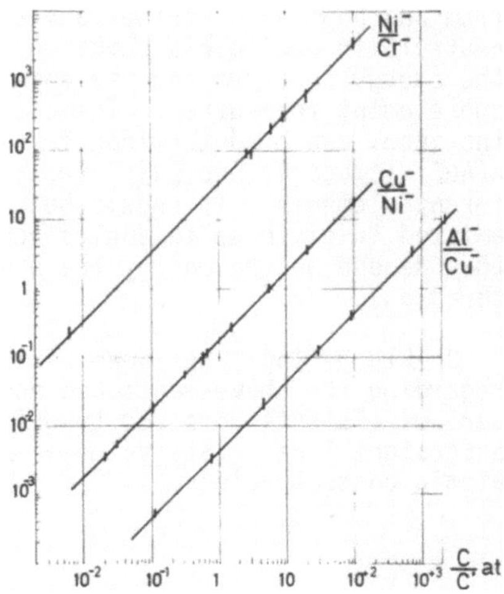

Fig.2

work function, observed on a metallic sample M (copper) when the Cs flux J is increased. Measurements were generally performed on a single crystalline grain. It can be seen that $I(M^-)$ is rising very sharply (it roughly obeys a J^4 law) and exhibits a smooth maximum. Simultaneously, W decreases (according a nearly linear law, $J^{0.85}$) and goes through a minimum. If, for bare copper, W is supposed to be 4.5 eV, the minimum is found at about 1.4 eV. All the samples which have been tested have shown similar features. However, there are significant differences; for instance, according to the element, the sharp increase of $I(M^-)$ may be described by different J^n exponent laws. It is also worthwhile to note that the spectra of initial energies may differ somewhat in shape from one element to another.

B. Alloys

It is of prime importance to know the relative yields of various elements. A first step towards that goal is to study alloys. Fig.2 summarizes the results obtained on three types of binary alloys. ($I(M^-)$ have been corrected for isotopic abundancies.) It can be seen that $I(M^-)$ is proportional to concentrations with a good approximation. Then, it is possible to deduce relative practical yields, R. Of course, R values may depend on the collection system, the energy pass band of the spectrometer, the flux J, but we could make sure that they were independent of J in the neighborhood of the maximum of $I(M^-)$. (The energy pass band was about 6 ev.)

An attempt was made to determine the probabilities of M^- production. The assumptions were: M^- ions and M° sputtered atoms have the same energy spectrum and angular distribution; the number of polyatomic species (ionized + neutral) is negligible compared to monoatomic species (ionized + neutral); the energy spectrum and the angular distributions are not too different from one element to another. Then, the ionization probability of Cu atoms, for instance, can be determined from measurements made on an AuCu alloy since $\eta(Au^-)$ is nearly one [4]. Thus, step-by-step, one can determine η for different elements. It should be noticed that values concern particles emitted in given solid angles and with a given average energy. In particular, they depend on the energy the pass band of the spectrometer has been centered on.

On Fig.3, log η has been plotted against the electron affinity E_A. Disregarding the above-mentioned assumptions on η, Fig.3 gives the R values for various elements compared to gold. The six elements being studied fall on a straight line; η(Si) value was obtained on two NiSi alloys at different atomic concentrations.

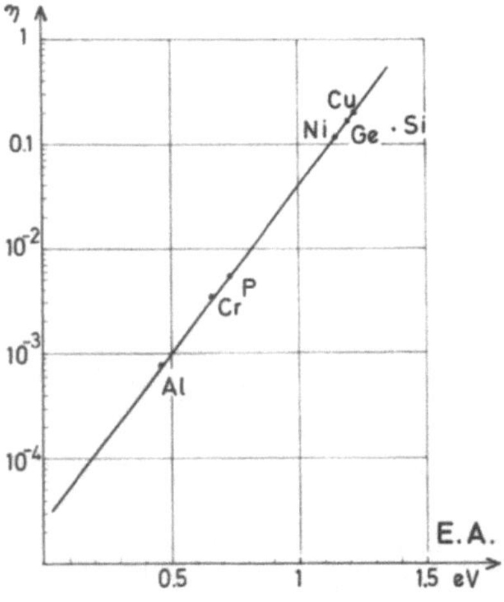

Fig.3

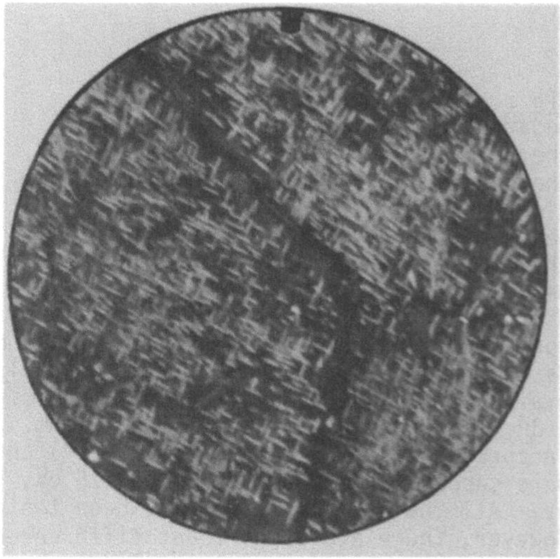

Fig.4 Cu^- micrograph of an aluminum copper 4 at% alloy obtained with cesium flooding (field of view 80µm)

C. Oxides

Cu_2O and NiO have been studied; the general behavior of $I(M^-)$ and W was similar to that of alloys. The ratios $I(M^-)/I(O^-)$, once corrected for isotopic abundances and concentrations, are respectively 0.15 and 0.17. Consequently, if one took $\eta(M^-)$ from Fig.3, $\eta(O^-)$ would be slightly greater than 1! Therefore, in this case, assumptions previously made might be wrong (for instance, if there are high yields of polyatomic species) or $\eta(M^-)$ might be smaller than the values measured on alloys (but surprisingly, $\eta(Cu^-)/\eta(Ni^-)$ is nearly the same).

4. Discussion

We will restrain the discussion to ionization processes. Two points are to be emphasized: the dependences of $I(M^-)$ on W (Fig.1) and $\eta(M^-)$ on E_A; E_A values are taken from Ref. [6].

It is known from literature [7] that W depends on Cs coverage and that a minimum is reached when the coverage is about 0.25. The sharp increase of $I(M^-)$ is mainly due to the lowering of W since it looks reasonable to assume that, at low coverages, the sputtering rates are not changed very much. A detailed account of the phenomena near the maximum of $I(M^-)$, and beyond it, is more difficult, especially in the dynamical equilibrium reached during sputtering.

Figure 3 strongly suggests that η is proportional to $\exp(E_A/\varepsilon)$ where ε is a parameter which would appear as kT in a thermal theory. However, in such a theory one would rather plot α against E_A, α being the "degree of ionization": $\alpha = \eta/(1 - \eta)$. More precisely, what is needed is $\alpha g_o/g_-$ where g_o and g_- are the degeneracy of the neutral atom and the negative ion ground states. If $\log(\alpha g_o/g_-)$ is plotted against E_A, the points do not fit a straight line as nicely as in Fig.3. Thus, let us look towards theories describing the electron sharing between the leaving particle and the surface. From Ref. [8] one can infer the expressions:

$$E_A < W: \qquad \eta = \frac{2}{\pi} \exp - \frac{|E_A - W|}{\varepsilon} ;$$

$$E_A > W: \qquad \eta = 1 - \frac{2}{\pi} \exp - \frac{|E_A - W|}{\varepsilon}$$

which cease to be valid near the Fermi level $E_A - W \simeq 0$. Taking into account the experimental uncertainties and the theoretical approximations, the factor $2/\pi$ may be replaced by 0.5 and the formulas tentatively extended to the whole range of E_A. Fig.3 yields the following values $W \simeq 1.34$ eV and $\varepsilon \cong 0.14$ eV.

5. Conclusion

The high yield of secondary negative emission induced by Cs and the relative easiness of calibration are of great analytical interest, but many problems have not been examined in this "summary." The reduced amount of data does not allow definite statements on ionization processes; however, the actual results lean our preference towards theories describing the electronic screening of particles leaving a surface.

References

1. V.E. Krohn, J. Appl. Phys., 33 3523 (1962).
2. M.K. Abdullayeva, A.K. Akukhanov and U.B. Shamsiya, Rad. Effects, 18 167 (1973).
3. H.A. Storms, J.D. Stein, K.F. Brown, Joint Japan, U.S. Seminar 1975.
4. M. Bernheim and G. Slodzian, J. de Phys. (Lettres), 38 L 325 (1977).
5. G. Slodzian, NBS Spec. Pub., 427 33-61 (1975).
6. H. Hotop and W.C. Lineberger, J. Phys. Chem. Ref. Data, 4 3, 539 (1975).
7. V.B. Voronin, A.G. Naumovets, A.G. Fedorus, Sov. Phys. JETP, L 15 370 (1970).
8. A. Blandin, A. Nourtier, D.W. Hone, J. de Phys. 37 369 (1976).

Angle-Resolved SIMS—A New Technique for the Determination of Surface Structure

B.J. Garrison, S.P. Holland, and N. Winograd
Department of Chemistry
The Pennsylvania State University
University Park, PA 16802

One of the central problems in surface science is the experimental determination of surface structure. Recently we have theoretically predicted that the angular distributions of the particles that eject during ion bombardment are sensitive to the original site of the adsorbate [1]. Furthermore, we found that by analyzing only the higher energy particles (kinetic energy $\gtrsim$ 10 eV) strong enhancement of the angular anisotropies occurs. Although the molecular dynamics calculations employed to predict this effect were for neutral ejected species, the ejected ions should exhibit similar angular distributions. For example, the polar deflection of a 20 eV particle due to the image force will be less than 2° [1]. Hence we feel that SIMS would be a viable technique with which to measure the angular distributions of ejected particles.

For the initial experiments of angle-resolved SIMS we have chosen to investigate a c(2x2) coverage of oxygen on Cu(001) and clean Ni(001). The angle-resolved SIMS experiments were performed using a previously described apparatus [2]. Angle selection was accomplished by placing a cylindrical shield with two small apertures around the crystal [3]. One aperture collimates the normally incident primary ion and the other selects a given polar angle for the ejected particles. Although this configuration is simply constructed it allows only one polar angle to be investigated. To maximize the structural information as indicated by the theoretical calculations we have chosen a polar angle of 45°. The crystal can be rotated to obtain a full 360° azimuthal scan. We estimate the angular resolution to be 8-10°. The energy selector collects ions with between approximately 10 and 50 eV of kinetic energy. For all the experiments the primary ion was Ar^+ at normal incidence with between 900 and 1500 eV of energy. The total dose of ions was kept below 10^{13} ions/cm^2 to avoid significantly altering the surface structure.

For a c(2x2) coverage of oxygen on Cu(001), the azimuthal angular distributions at a polar angle of 45° of both Cu^+ and O^- exhibit large anisotropies, changing by more than a factor of two between minimum and maximum values. The interesting feature is that the Cu^+ maximizes in the <100> directions while the O^- maximizes in the <110> directions, 45° out of phase. Comparing these results with the calculations indicates that the oxygen is in a fourfold bridge site.

The effect of the height of the adsorbate above the surface plane on the angular distributions is shown in Fig.1. In all cases the oxygen is in a fourfold bridge site on Cu(001). The patterns all have the maximum intensity of the oxygen in the <110> directions, however, the polar angle of the maximum intensity varies with height above the surface. The preliminary experimental results of oxygen on Cu(001) indicate that the adsorbate height is ∿1.2-1.5 Å above the surface. To be sure of the precise atomic position, however, SIMS measurements at other polar angles would be desirable.

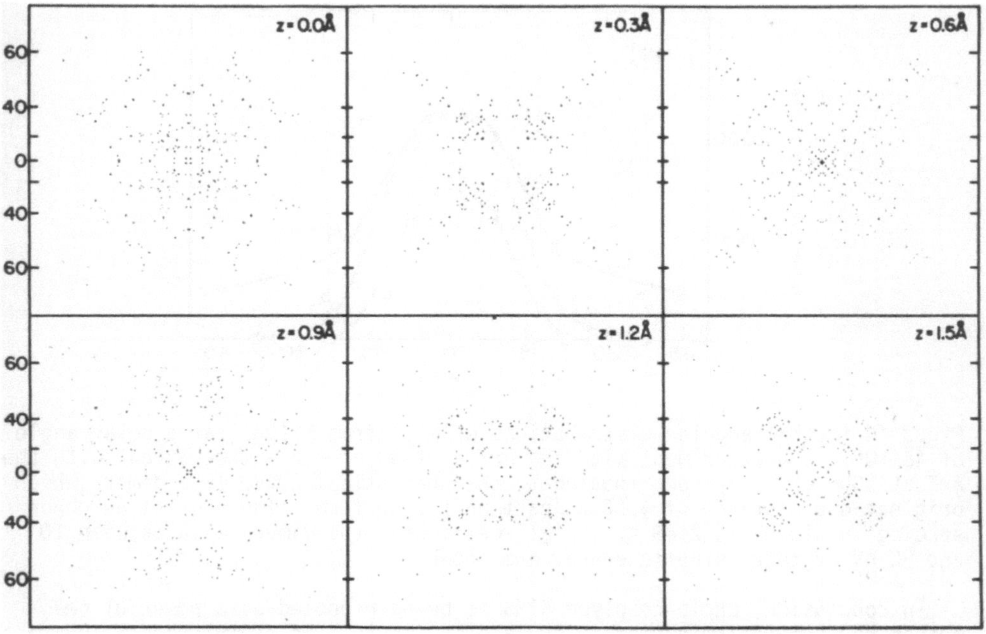

Fig.1 Calculated angular distributions of ejected oxygen atoms as a function of height (z) above the surface plane of Cu(001). In all cases the oxygen is in a fourfold bridge site at a c(2x2) coverage. Only the atoms with between 10 and 50 eV of kinetic energy are shown. The Ar^+ ion is at 600 eV for all these calculations. Each ejected atom is plotted on a flat-plate collector an arbitrary distance above the crystal. The numbers on the ordinate refer to the polar deflection angle given in degrees. The vertical and horizontal directions correspond to the <100> directions of the (001) face

The cluster ions also exhibit angular anisotropies. Shown in Fig.2 is the azimuthal angular distribution of Ni_2^+ from Ni(001) at a polar angle of 45°. The anisotropy is larger than a factor of 5, with the peak count rate being quite substantial (∿2000 cps) under "static" SIMS conditions. Also shown in Fig.2 is the calculated angular distribution of Ni_2. Excellent agreement is obtained.

For the angular distributions of the monomers we found that generally only one or two mechanisms of ejection were responsible for the majority of the anisotropy [1,4]. This appears to be true for the dimers as well. Shown in Fig.2 is the contribution to the theoretical curve for dimers that originated a distance of 4.98 Å apart on the surface. The mechanism that gives rise to this particular type of dimer causes the majority of the anisotropy. In fact, for 600 eV Ar^+ on Cu(001) this mechanism is responsible for *all* the dimers in this angle-energy regime. Recent calculations of oblique angle of incidence of the primary Ar^+ ion have shown that other mechanisms of dimer formation and thus other originating sites can be preferentially enhanced [5]. By adjusting the experimental conditions, such as angle of incidence of the primary ion and energy and angle of detection of the ejected dimers, one can hopefully be able to obtain structural information of alloy surfaces.

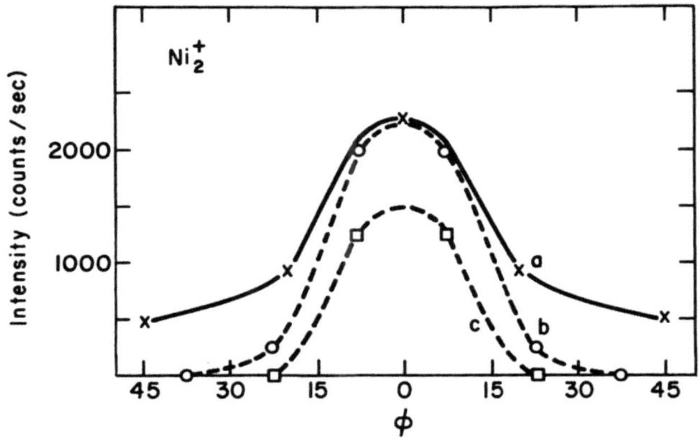

Fig.2 Azimuthal angular distributions of Ni_2^+ from Ni(001) at a polar angle of 45±10°. a - experiment with the Ar^+ at 1500 eV. b - theoretical with the Ar^+ at 1000 eV. c - contribution to the theoretical curve from dimers which originated a distance of 4.98 Å apart on the surface. The nearest neighbor spacing in nickel is 2.49 Å. In all cases only the dimers with between 10 and 50 eV of total kinetic energy are shown

In conclusion, angle-resolved SIMS is being proposed as a powerful new method for the determination of surface structure. An example of adsorbate site and height determination has been given. In addition, the angular distributions of Ni_2^+ dimers have been measured for the first time.

Acknowledgements

The authors wish to thank R.A. Gibbs for helping to construct many parts of the experimental apparatus and K.E. Foley, D.E. Harrison, Jr., S.Y. Tong, and W.N. Delgass for many stimulating discussions. We also acknowledge support by the National Science Foundation (Grant No. CHE 78-08728), and the U.S. Air Force Office of Scientific Research (Grant No. AF76-2974). Portions of the computations were supported by the National Resource for Computation in Chemistry under a grant from the National Science Foundation and the U.S. Department of Energy (Contract No. W-7405-ENG-48).

References

1. N. Winograd, B.J. Garrison, and D.E. Harrison, Jr., Phys. Rev. Lett., 41, 1120 (1978).
2. T. Fleisch, W.N. Delgass, and N. Winograd, Surf. Sci., 78, 141 (1978).
3. S.P. Holland, B.J. Garrison and N. Winograd, Phys. Rev. Lett., 43, 220 (1979).
4. B.J. Garrison, N. Winograd, and D.E. Harrison, Jr., Phys. Rev. B., 18, 6000 (1978).
5. K.E. Foley and B.J. Garrison, J. Chem. Phys., in press.

II. Quantitation

Factors Influencing Secondary Ion Yields

Vaughn R. Deline
Charles Evans & Associates
1670 S. Amphlett Blvd., Suite 120
San Mateo, CA 94402

Secondary ion mass spectrometry is the most sensitive analytical technique for surface and thin film elemental characterization. Each element of the periodic table has a certain probability for forming either a positive or negative ion as it is ejected from the surface of a bombarded solid. The large flux of ions typically produced ($\sim 10^8$-10^{12} ions/sec), coupled with the net efficiency of an ion probe for extracting, mass analyzing and detecting these ions ($\sim$0.01%) gives SIMS its exceptional sensitivities. It is the purpose of this report to provide a general overview of the physical processes governing secondary ionization and how these processes influence the quantitative interpretation of ion intensities.

Prior to the late 1960's, the majority of SIMS studies employed inert gas ion bombardment. During this period, the technique was plagued by inexplicable gross variations in ion yields and "matrix effects." A "matrix effect" is defined as a variation in an analytical signal for an impurity element not in direct proportion to concentration, but loosely associated with a change in the concentration of major constituents. Moreover, it was impossible to correlate ion yields (the ratio of the number of ions produced to the number of atoms of that element sputtered) with a simple physical parameter, thereby lending more confusion to the technique. Several workers noted that the introduction of oxygen into the vacuum system during inert gas ion bombardment dramatically enhanced and stabilized positive ion yields with respect to matrix variations. Moreover, Krohn [1] found that the use of cesium ion bombardment significantly increased the yield of negative secondary ions. In the late 60's, oxygen ion bombardment was exploited to take advantage of its enhancement effects for positive ion formation. From the stabilizing effect of this type of sputtering, Andersen and Hinthorne [2] were able to establish a relationship between positive secondary ion yield and ionization potential for each sputtered element. They found that when several elements were sputtered from the same matrix, the relative atomic ion yields approximately followed an inverse exponential dependence on ionization potential. However, they could not explain why the absolute atomic ion yield would change when the same element was sputtered from another material, i.e., the SIMS matrix effect.

In recent years several other successful procedures have been developed for quantitative SIMS analysis within a single matrix. However, no explanation was given for the SIMS matrix effect which would allow a prediction of ion yield variations for a given element sputtered from different matrices.

For the purpose of clarity, let us subdivide ion yield variations into three categories: a) different elements contained in a single matrix;

b) a single element sputtered from different matrices - the SIMS matrix effect; and c) different elements sputtered from different matrices. As previously stated, Andersen and Hinthorne [2] have shown that the ion yields of different elements within a given matrix (case a, above) exhibit an inverse exponential dependence of ion yield on ionization potential and, based on this observation, successful procedures have been developed for quantitative analysis within a single matrix. For many years the SIMS matrix effect (case b) has been the most mysterious of the ion yield phenomena. Both positive and negative ion yields were known to be greatly enhanced by the incorporation of oxygen or cesium, respectively, in the sputtered surface. But, even in the presence of these enhancing species, the ion yield of a given element is found to vary by many orders of magnitude from one matrix to another.

If we combine the basic observation that the presence of oxygen has a dramatic enhancing effect on positive ion yields and that different matrices sputter with different sputtering yields (the total number of atoms ejected per incident ion), an interesting and important relationship can be established. Under reactive ion bombardment, the primary ion species concentration, [P], at the substrate surface is inversely proportional to the sputtering yield, S [3], in the absence of preferential sputtering. This relationship can be written:

$$[P]_a \text{ (atom fraction) } \alpha \ (1 + S_a)^{-1} \quad (1)$$

(S_a = substrate atoms ejected/primary atom)

or, for a fixed sputtered area, A, and primary ion current density, J, (8)

$$[P]_v \text{ (volume concentration) } \alpha \ S_{\ell}^{-1} \ \alpha \ \frac{\rho A}{S_a J} \quad (2)$$

(S_{ℓ} = linear sputtering rate in Å/sec and ρ = density of the matrix)

The above relationships show that the near-surface concentration of the ion-yield-enhancing species, and hence the secondary ion yields, will vary inversely with the matrix sputtering yields. It is therefore evident that a matrix effect can be identified which is solely due to sputtering yield variations. In a recent study, Deline, et al. [4,5] have correlated substrate matrix effects with sputtering yield under oxygen and cesium ion bombardment. They found that indeed there is a strong inverse correlation between ion yield and sputtering yield, and suggested the surprising result that this appears to be the only cause of the substrate matrix effect. Using ion implanted standards (B, C, F, P, As and Sb) into several different matrices (C, Si, Ge, Sn and GaAs), it was found that a linear plot could be obtained when the useful ion yields of the implanted species were plotted vs. the reciprocal of the linear sputtering rate (Fig. 1 and 2). Other studies demonstrated that elements at high concentrations, such as Si in metal silicides, again follow this same relationship.

To determine how universally applicable this rather simple relationship is, Katz [6] examined over 20 different metals and compound semiconductors. This study employed ion implants of H, P and As into the various matrices. Again, the useful ion yields were found to vary as a power function of the reciprocal of the linear sputtering rate.

The results of these studies are rather startling. They indicate that

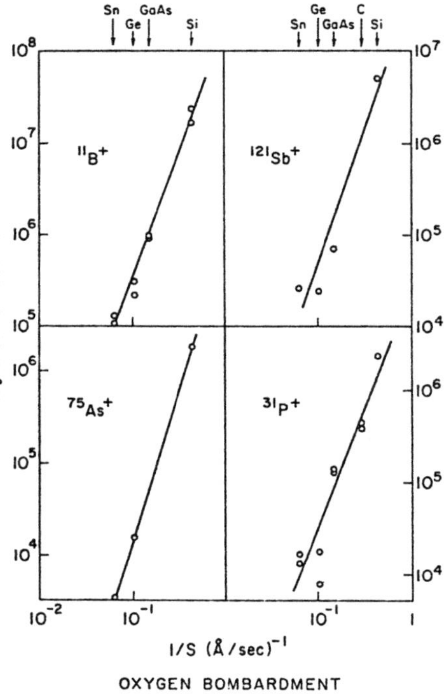

Fig.1 Integrated Positive ion counts for $^{11}B^+$, $^{121}Sb^+$, $^{75}As^+$, and $^{31}P^+$, vs. the reciprocal of the matrix sputtering rate, S_ℓ

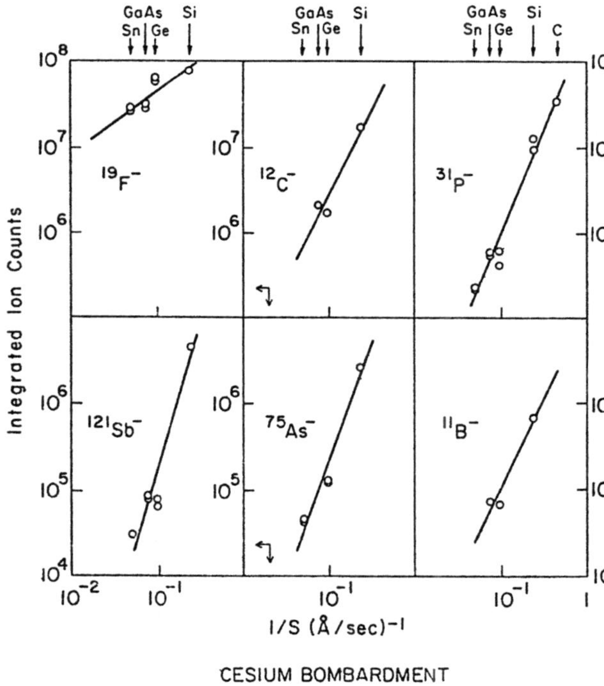

Fig.2 Integrated negative ion counts for $^{19}F^-$, $^{12}C^-$, $^{31}P^-$, $^{121}Sb^-$, $^{75}As^-$, and $^{11}B^-$, versus the reciprocal of the matrix sputtering rate S_ℓ

the substrate matrix effect (case b) is merely an artifact, which arises because the sputtering yield determines the near surface concentration of the enhancing species. That is, if we could compare ion yields in different substrates at the same oxygen or cesium concentration, we would find that there is no matrix effect.

The third subdivision of ion yields discussed earlier is that of different elements sputtered from different matrices. Secondary ion yields across the periodic table have been shown to vary by many orders of magnitude under oxygen or cesium ion bombardment [7]. Applying the relationships obtained for the first two subdivisions, it can be demonstrated that the ion yields of different elements from different matrices can be corrected for sputtering yield variations and the resulting corrected ion yield, follows a reasonable Saha type relationship (Fig.3). An understanding of these ion yield controlling factors can be used as a basic foundation for the quantitative interpretation of ion intensities. Each element has a certain probability for forming either a positive or negative ion. The efficiency of ionization can be predicted through the use of ionization potential (for positive ions), or electron affinity (for negative ions). Combining this knowledge of ionization efficiencies with an understanding of relative sputtering yield trends, one can easily predict the relative sensitivity of a given element within any homogeneous matrix.

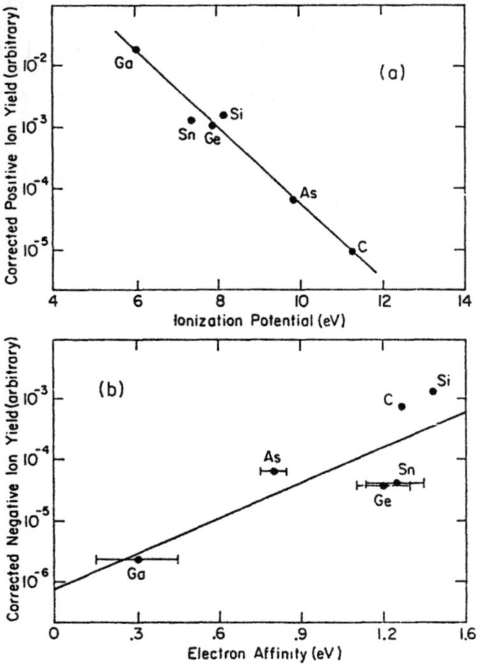

Fig.3 (a) Positive matrix ion yields corrected for sputtering rate as a function of ionization potential. (b) Negative matrix ion yields corrected for sputtering rate as a function of electron affinity

Heterogeneous matrices or interfaces between dissimilar materials require a more extensive interpretation. Since the primary enhancing species (O or Cs) is being implanted to a depth beyond the sputtering front, the ionization yield of an element from a region just below the interface can be controlled by the overlying matrix. This will result in complex ion intensity fluctuations as the beam sputters through the interface, which may be difficult to accurately correlate with the true elemental concentration.

These studies indicate that only two basic factors control secondary ion yields in any system:

a) the ionization potential or electron affinity of the sputtered atom,

and

b) the near-surface concentration of the enhancing species (oxygen for positive ion emission or cesium for negative ion emission). In simple sputtering situations (i.e., in the absence of differential sputtering effects, segregation, etc.), this near-surface concentration is inversely proportional to the substrate sputtering yield.

References

1. V. Krohn, J. Appl. Phys., 33, 3523 (1962).
2. C.A. Andersen and J.R. Hinthorne, Science, 175, 853 (1972).
3. J.E. Chelgren, W. Katz, V.R. Deline, C.A. Evans, Jr., R.J. Blattner and P. Williams, J. Vac. Sci. Technol., 16, 324 (1979).
4. V.R. Deline, W. Katz, C.A. Evans, Jr., and P. Williams, Appl. Phys. Lett. 33 (9), (1978).
5. V.R. Deline, C.A. Evans, Jr., and P. Williams, Appl. Phys. Lett. 33 (7), (1978).
6. W. Katz, Ph.D. Thesis, 1979.
7. H.A. Storms, K.F. Brown, and J.D. Stein, Anal. Chem., 49, 2029 (1977).
8. J.C.C. Tsai and J.M. Morabito, Surf. Sci., 44 247 (1974).

Instrumental Effects on Quantitative Analysis by Secondary Ion Mass Spectrometry

Dale E. Newbury
National Bureau of Standards
Washington, D.C. 20234

In developing methods for quantitative analysis by secondary ion mass spectrometry (SIMS), most attention has been paid to the problems of correcting for the large matrix (chemical) effects on secondary ion intensities in order to obtain accurate compositional values. [1,2] The possible influence of instrumental parameters on relative secondary ion intensities has often been neglected or else presumed to cancel through the measurement of selected ion intensities to serve as internal standards. Recent studies show that instrumental effects can actually exert a strong influence on relative as well as absolute secondary ion intensities, which can significantly affect the result in a quantitative analysis.

To assess the magnitude of instrumental effects on secondary ion intensities, a comparative SIMS study of selected glasses and steels has been carried out in cooperation with laboratories in the United States, Japan, and Europe. [3] To compare the secondary ion spectra of the glasses reported by the various investigators, the data were reduced to a common form through the calculation of relative sensitivity factors, $S_{X/M}$:

$$S_{X/M} = (i_X/C_X f_X)/(i_M/C_M f_M) \tag{1}$$

where (i) is the measured secondary ion intensity, (C) is the atomic concentration of the element in the sample, (f) is the isotopic abundance, and (X) and (M) denote any two elements. Ideally, (M) should be chosen as a major constituent of the matrix. For the analysis of the glasses, silicon was chosen for this role since it was present in the matrix at a concentration greater than 10 atom percent. The relative sensitivity factors for positive secondary ions derived from the spectra obtained by the various investigators are listed in Tables 1 and 2. Two major observations can be made from these tables. (1) The ratio of the maximum-to-minimum value reported for a given relative sensitivity factor ranges from 5 to 60. Note that this is a remarkable range of relative sensitivity. While we might expect that absolute sensitivity will vary strongly from instrument to instrument due to differences in primary ion conditions, secondary ion collection, and spectrometer transmission, the relative sensitivity should be similar. Such a range of relative sensitivity, however, indicates that the differences between SIMS instruments are not simply a matter of scaling. (2) With the exception of oxygen, which produces mainly negative secondary ions, low atomic number elements generally show less variation in the relative sensitivity factor from one instrument to another than do high atomic number elements.

Table 1 Relative sensitivity factors from NBS glass K-251

Analyst	O/Si	Al/Si	Ba/Si	Ta/Si	Pb/Si	Bi/Si
(1) Oishi	2.01E-3	2.63	2.83	0.396	0.261	0.0735
(2) Someno	1.83E-3	2.92	6.40	0.646	0.0951	0.0781
(3) Yaegishi	2.07E-3	10.4	17.2	0.246	4.75	2.51
(4) Tamura	2.11E-3	2.92	4.38	0.199	0.226	0.127
(5) Kobayashi	2.39E-3	2.55	3.54	0.338	0.305	0.0614
(6) Nishimura	--	6.31	8.01	0.139	2.43	1.20
(7) Iwamoto	2.03E-4	3.01	2.09	0.115	0.125	0.0651
(8) Fujino	2.50E-3	1.90	2.67	1.16	0.294	0.0901
(9) Tamaki	2.58E-3	2.17	4.80	0.873	1.10	0.302
(10) Suzuki	1.60E-3	2.96	2.43	0.122	0.256	0.0819
(11) Konishi	8.88E-3	5.66	11.6	--	3.17	--
(12) Reed	--	3.39	11.3	0.386	1.67	0.364
(13) Christie	2.01E-3	6.13	8.64	0.186	1.06	0.470
(14) Evans	--	3.92	14.5	0.613	3.08	0.386
(15) Storms	9.99E-4	2.53	4.68	0.452	0.666	0.186
(16) McHugh	1.02E-3	2.25	2.49	0.283	0.356	0.0782
(17) Garratt-Reed	4.99E-3	2.30	3.87	0.714	0.0830	0.0459
(18) Morrison	7.43E-4	5.69	10.7	0.191	2.18	--
(19) Satkiewicz	2.75E-3	4.78	3.49	0.115	0.0862	0.0647
(20) Newbury	1.53E-3	2.24	5.04	0.545	0.648	0.152
(21) Johnson	5.52E-3	1.88	2.36	1.00	0.208	0.131
(22) Hinthorne	1.78E-3	3.79	7.14	0.166	0.857	0.166
Max/Min	43.7	5.53	8.23	10.1	57.2	54.7

Table 2 Relative sensitivity factors from NBS glass K-309

Analyst	O/Si	Al/Si	Ca/Si	Fe/Si	Ba/Si
(1) Oishi	2.60E-3	4.64	9.96	1.78	5.84
(2) Someno	1.81E-3	4.21	7.87	0.725	6.41
(3) Yaegishi	2.06E-3	9.72	19.8	5.62	19.8
(4) Tamura	2.48E-3	4.23	11.3	2.16	9.13
(5) Kobayashi	2.65E-3	3.03	5.63	1.53	4.59
(6) Nishimura	1.07E-2	4.27	9.59	2.49	6.51
(7) Iwamoto	--	4.96	8.19	1.20	4.11
(8) Fujino	1.38E-3	2.62	4.03	1.42	3.43
(9) Tamaki	1.54E-3	2.39	4.74	1.65	7.57
(10) Suzuki	1.97E-3	3.38	5.54	1.21	2.84
(11) Konishi	1.85E-2	6.54	13.9	3.54	18.5
(12) Reed	--	2.82	4.86	1.82	7.31
(13) Christie	1.34E-3	5.07	11.4	2.84	14.2
(14) Evans	--	2.82	5.76	3.89	11.7
(15) Storms	1.10E-3	3.21	5.68	1.79	6.11
(16) McHugh	8.39E-4	3.25	7.80	2.07	6.43
(17) Garratt-Reed	9.67E-3	1.99	5.37	0.725	3.42
(18) Morrison	2.26E-3	4.15	7.24	4.02	5.93
(19) Satkiewicz	3.33E-3	4.10	14.2	0.856	9.13
(20) Newbury	1.44E-3	3.06	5.10	1.66	7.80
(21) Johnson	8.58E-3	2.16	3.18	0.668	2.06
(22) Hinthorne	9.77E-4	2.95	5.31	1.38	4.89
Max/Min	22.1	4.5	6.23	8.41	9.61

The data shown in Tables 1 and 2 were obtained with a variety of primary ion species, current densities, beam energies, and residual gas pressures, which undoubtedly accounts for some of the variation in the measured relative sensitivity factors. However, even when the data from a particular type of instrument operating under similar conditions are compared, the range of sensitivity factors is still found to be as high as 10. [3]

Further studies have been carried out to assess the impact of these variations in instrumental response on quantitative analysis procedures. The secondary ion intensities have been converted to compositional values with a matrix correction method based on the local thermal equilibrium (LTE) model. [4,5] In this particular LTE method, the secondary ion intensities observed for two different elements and the known concentrations for those elements are used in order to determine the characteristic properties of the environment in which the secondary ions are generated. This use of "internal standards" involving actual experimental data from the particular instrument and known compositional values to constrain the calculation might be expected to eliminate the influence of local instrumental factors. However, the errors in the compositions derived from the experimental spectra by the LTE model show a wide range similar to the range previously observed for the relative sensitivity, Table 3. When the relative sensitivity for an element is low, the LTE model underestimates the concentration (that is, the error factor, defined as F = C(true)/C(LTE) in Table 3, is high). The size of the errors is least for the light elements (Table 4) and greatest for the heavy elements. The instrumental bias against an element is reflected in the final LTE Analysis. The instrumental effects appear to be nearly as large as the matrix effects, at least for heavy elements, which may tend to mask the effectiveness with which a matrix correction method operates. It has, in fact, been noted previously that the LTE method has its greatest success with light elements and fails most noticeably for the heavy elements. [5] It appears that some of this failure must be ascribed to instrumental effects.

Given this situation, a method is needed to compensate for the strong instrumental influences. In principle, it should be possible to identify and characterize each of the instrumental effects, such as the secondary ion extraction characteristics, the transmission of the spectrometer, the influence of primary ion species, energy, and current density, etc. Quantitative descriptions of each effect could then be incorporated in the quantitative analysis procedure to correct the measured secondary ion intensities to eliminate the instrumental effects and obtain the secondary ion intensities actually emitted from the sample. These emitted intensities would then be appropriate input to a matrix correction procedure such as the LTE method. However, considering the variation observed in relative sensitivity among instruments of the same nominal design, such a characterization procedure would have to be carried out on each individual instrument.

A more pragmatic approach could be based on the direct use of relative sensitivity factors measured on the individual instrument from known samples such as glasses. It has already been demonstrated in the blind analysis of glasses by the relative sensitivity method that a reasonably high level of accuracy can be achieve [6]. What the relative sensitivity factor method lacks, however, is the flexibility to reliably follow large changes in the character of the matrix. A hybrid procedure combining the relative sensitivity factor method with a physical model in a fashion similar to the procedure used in quantitative x-ray microanalysis may provide the answer. In such a procedure, relative sensitivity factors would first be measured for all elements of interest. These sensitivity factors would then be modified to a

Table 3 Comparison of LTE errors and measured relative sensitivity

Laboratory	$S_{Pb/Si}$	F_{Pb}	$S_{Ta/Si}$	F_{Ta}
1	0.261	5.9	0.396	6.1
2	0.095	21.9	0.646	4.9
3	4.75	1.1	0.246	4.9
4	0.226	8.0	0.199	14.2
5	0.305	5.3	0.338	7.7
6	2.43	1.2	0.139	34.3
7	0.125	10.2	0.115	18.1
8	0.294	4.9	1.16	2.0
9	1.10	1.8	0.873	3.6
10	0.256	5.4	0.122	18.8
11	3.17	1.01	--	--
12	1.67	1.7	0.386	11.0
13	1.06	2.4	0.186	21.8
14	3.08	1.1	0.613	7.5
15	0.67	2.8	0.452	6.6
16	0.36	3.8	0.283	8.0
17	0.083	19.9	0.714	3.7
18	2.18	1.4	0.191	22.3
19	0.0862	20.1	0.115	23.9
20	0.65	3.0	0.545	5.5
21	0.21	6.4	1.00	2.2
22	0.86	2.7	0.166	21

Material: NBS glass K-251

Table 4 Comparison of LTE errors and measured relative sensitivity

Laboratory	$S_{Al/Si}{}^{(1)}$	F_{Al}	$S_{Al/Si}{}^{(2)}$	F_{Al}
1	4.6	0.68	4.92	0.43
2	4.2	0.69	3.48	0.61
3	9.7	1.39	3.64	0.58
4	4.2	0.76	4.24	0.50
5	3.0	0.95	4.11	0.52
6	4.3	0.75	4.88	0.43
7	5.0	0.63	7.00	0.30
8	2.6	0.77	3.73	0.57
9	2.4	1.22	2.82	0.75
10	3.4	0.80	6.11	0.35
11	6.5	1.12	4.15	0.51
12	2.8	1.26	2.69	0.79
13	5.1	0.83	5.62	0.38
14	2.8	1.7	2.60	0.82
15	3.2	1.2	3.42	0.62
16	3.2	1.0	4.46	0.48
17	2.0	0.95	3.64	0.58
18	4.2	1.3	4.49	0.47
19	4.1	0.81	7.71	0.27
20	3.1	1.05	2.94	0.72
21	2.2	0.67	4.02	0.53
22	3.0	0.76	4.00	0.53

(1) K-309 (2) K-251

value appropriate to a new matrix:

$$S'_{X/M} = S_{X/M} \cdot m_1 \cdot m_2 \cdots \tag{2}$$

where (m_i) is a matrix correction factor [7]. The (m_i) values could be determined experimentally, from a physical model, or through a combination of both. Specific experimental factors which could be measured in the secondary ion spectrum include the ratios of elemental and oxide ions, single- and double-charged ions, and atomic and molecular oxygen. The original measurement of the sensitivity factor would automatically incorporate the local instrument characteristics, and the (m_i) factors would improve the flexibility of the approach to the analysis of samples of substantially different character.

References

1. J.M. Schroeer, in Secondary Ion Mass Spectrometry, ed. K.F.J. Heinrich and D.E. Newbury, National Bureau of Standards Special Publication 427 (Washington, 1975) 121.
2. C.A. Andersen and J.R. Hinthorne, Science, 175, 853 (1972).
3. D.E. Newbury, Report on the United States - Japan Cooperative Analysis of Glasses by Secondary Ion Mass Spectrometry, Second US-Japan Joint Seminar on SIMS, Osaka (Osaka, Oct. 23-27, 1978). (Available from Prof. M. Someno, Faculty of Engineering, Tokyo Inst. of Technology, 2-12-1, Ohokayama Meguro-ku, Tokyo, 152 Japan.)
4. C.A. Andersen and J.R. Hinthorne, Analyt. Chem., 45, 1421 (1973).
5. C.A. Andersen, NBS SP 427, ibid, 79.
6. D.E. Newbury and K.F.J. Heinrich, Mikrochem. Acta, in press.
7. J.A. McHugh, NBS SP 427, ibid., 129.

A Quantitative Model for the Effects on Secondary Ion Emission of Gaseous Absorption at Solid Surfaces Under Noble Gas Ion Bombardment

J.N. COLES
Research School of Earth Sciences,
Institute of Advanced Studies,
Australian National University,
Canberra, A.C.T. 2600, Australia

The origin of SIMS "fingerprint" mass spectra [1,2], in which the yields of molecular ions ($B_mA_n^q$) of (B)ulk and (A)dsorbate atoms maximize at different degrees of coverage (θ) of adsorbate gas, is explained in a new mathematical model based on the concept of "complex sputtering centres"[3,4]. From statistical reaction kinetics [4] the fractional abundance (N_{mn}) of the species B_mA_n in all charge states in the "sputtered assemblage" immediately outside the surface depends on the product of the abundances of the constituent atoms raised to the (integer) powers m and n respectively, multiplied by some velocity-averaged constant for that particular molecule, such that we have:

$$N_{mn} = \{N_B(\theta)\}^m.\{N_A(\theta)\}^n.(\text{constant for the molecule } B_mA_n) \ . \tag{1}$$

An empirical equation for the probability P(x) of sputtering an atom of a given species from a depth x inside the solid relative to that, P(o), of sputtering the same species from the surface, averaged over all velocities of emission, is proposed by the author in the form:

$$P(x)/P(o) = \exp(\frac{-x}{\nu d}) \ ; \tag{2}$$

where ν is a dimensionless constant for the target, taken to be independent of θ, and is the same for all species of atoms in the absence of preferential sputtering; while $d_{[hkl]}$ is the average interatomic spacing normal to the surface. This equation is supported by an independent computer simulation study [5] and by its use one can derive a "matrix correction" giving the free-atom abundances of species A and B in the sputtered assemblage, ie:

$$N_A = [1 - \exp(\frac{-\theta}{\nu})] \ ; \qquad (3) \quad \text{and} \qquad N_B = \exp(\frac{-\theta}{\nu}) \ ; \tag{4}$$

which should be corrected for molecular bonding, strictly, before substituting into (1).

The θ-dependence of the degree of ionization of the q-fold-charged ion $B_mA_n^q$, according to the JURELA/LTE model [6,7] comes from the change in local work function, $\emptyset(\theta)$, which, due to sub-surface incorporation of adsorbate atoms, is not normally thought to be sufficient to cause the large ion yield changes that are observed. Theoretically [8,9], however, assuming no sub-surface incorporation, $\Delta\emptyset$ is expected to be larger and given by [9]:

$$\Delta\emptyset = \psi\theta/(1+\chi\theta) \ , \tag{5}$$

where ψ and χ are constants (fitting parameters) depending on the mutual-depolarization, electrostatic dipole moment, and polarizability of the adsorbate atoms on the surface. It is proposed in this model that sub-surface incorporation of adsorbate gas atoms is strongly inhibited, especially at sputtering centres, by the momentum propagated back towards the surface if it is under ion bombardment, and therefore that (5) may be closely valid.

Thus in the simplest treatment of this model one deduces from the fore-

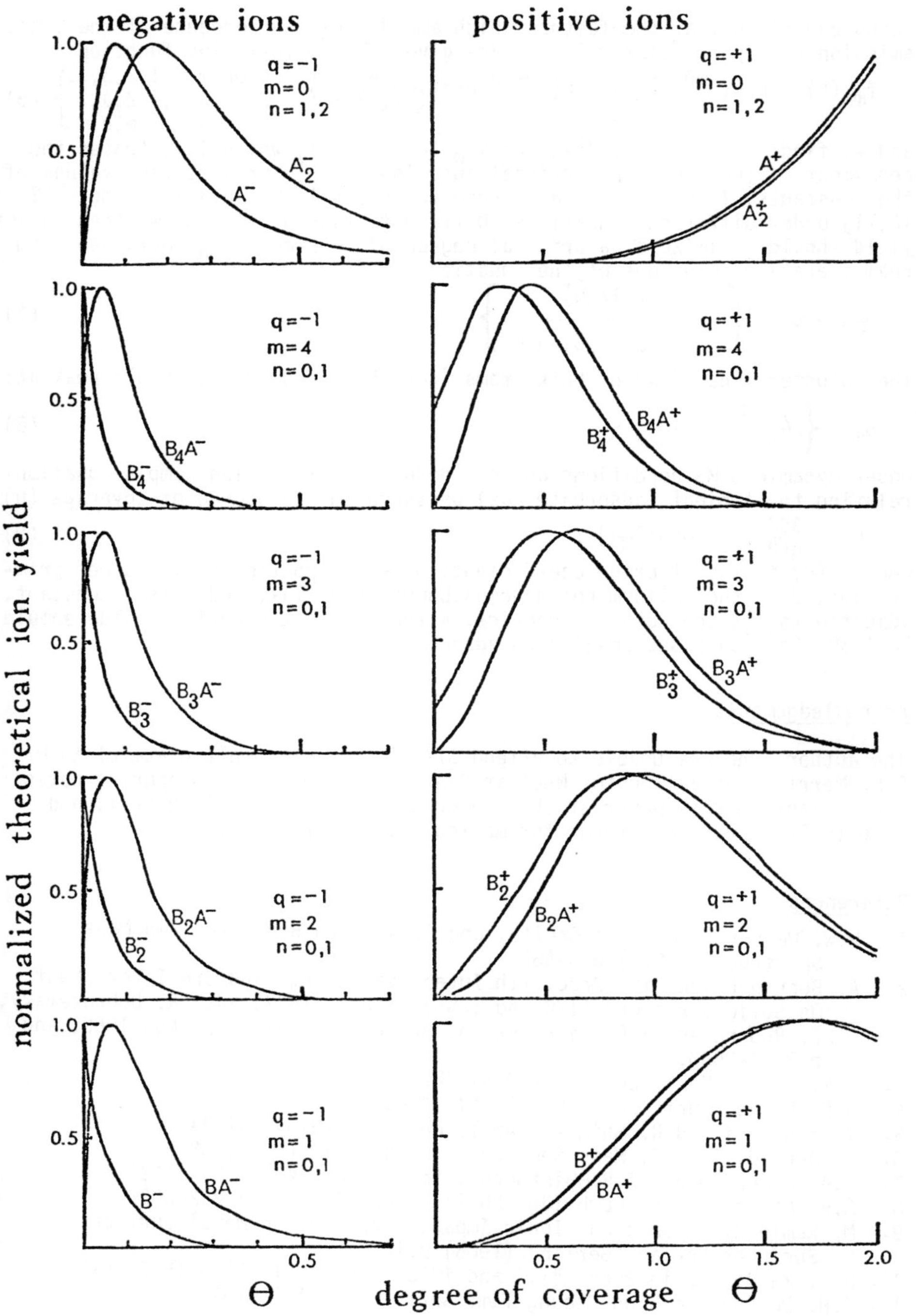

Fig. 1 Behaviour of f_{mn} (Θ) for the values ν=0.5, χ=1, ψ=7eV, and kT_e=1eV, where T_e is the effective electron temperature, and for increments in m,n,q

going equations that the factors which embody the θ-dependence of the ionic emission of a particular molecule are given in the following function:

$$f_{mn}(\theta) = \exp(\frac{-m\theta}{\nu})\,[1 - \exp(\frac{-\theta}{\nu})]^n \exp\left\{\frac{q\psi\theta}{(1+\chi\theta)kT_e}\right\} \quad \left\{\begin{array}{l}\text{for } m=0,1,2,3,..\\ n=0,1,2,3,..\\ q=\pm 1, \text{ etc.}\end{array}\right\} \qquad (6)$$

and we find: $\nu \simeq 1$, $\chi \simeq 1$, $\psi \lesssim 7$eV, and $kT_e \lesssim 1$eV. Fig.1, which illustrates the behaviour of the yield-proportional function $f_{mn}(\theta)$ for specified values of the constants, is similar to data reported recently by several authors [2, 10,11] under different conditions. Differentiation of f(θ) shows that an ion yield should maximize at a critical degree of coverage, $\hat{\theta}_{mn}$, obtained from real iterative solutions of the equation:

$$\hat{\theta}_{mn} = \nu \ln\left\{1 + \frac{nkT_e(1+\chi\hat{\theta})}{mkT_e(1+\chi\hat{\theta}) - q\nu\psi}\right\}. \qquad (7)$$

The "cluster ions" (B_m^+) of bulk atoms (for which n=0, q=+1) should peak at:

$$\hat{\theta}_m = \left\{\sqrt{\frac{q\nu\psi}{mkT_e}} - 1\right\} \Big/ \chi \qquad (8)$$

Under dynamic SIMS conditions one can deduce the following simple equation relating the ambient (adsorbate gas) pressure (p) to degree of coverage (θ):

$$p = \frac{ASJ}{\eta(\theta)}[1 - \exp(\frac{-\theta}{\nu})] \; ; \qquad (9)$$

where $\eta(\theta)$ is the sticking coefficient; S is the sputtering yield per primary ion; J is the uniform homogeneous bombarding flux; and A is a constant, adjusted to fit the data in practice, although its theoretical value should be 1.38×10^{-21}[torr sec cm^2]for O_2 adsorbed at 300K.

Acknowledgements

The author, who was unable to attend SIMS-II, is extremely grateful to Dr. G.L. Merrill, of Vallecitos Nuclear Center, for carefully preparing notes and presenting this paper at the conference on the author's behalf, and also to Dr. Howard A. Storms for making this arrangement.

References

1. H.W. Werner, H.A.M. De Grefte, and J. Van Den Berg, Adv. in Mass Spectrom. 6 (1974) 673-682
2. A. Benninghoven, in "Proc. 7th Inter. Vac. Cong. and 3rd Inter. Conf. On Solid Surfaces" (IVC and ICSS) Vienna (1977) Eds: R. Dobrozemsky, F. Rüdenauer, F.P. Viehböck, A. Breth, p.723-730, p.1063-1066, and p.2577-2580.
3. J.N. Coles, Surf. Sci. 55 (1976) 721-724.
4. J.N. Coles, Surf. Sci. 79, (1979) 549-574.
5. T. Ishitani and R. Shimizu, Appl. Phys. 6 (1975) 241-248.
6. Z. Jurela, Int. J. Mass Spectrom Ion Phys. 12 (1973) 33-51.
7. C.A. Andersen and J.R. Hinthorne, Anal. Chem. 45 (1973) 1421.
8. A.R. Miller, Proc. Cambridge Philos. Soc. 42 (1946) 292-303.
9. M. Kaminsky "Atomic and Ionic Impact Phenomena On Metal Surfaces" Springer-Verlag (Berlin) (1965) p.16.
10. R.J. MacDonald in Proc. "IVC and ICSS" Vienna (1977) p.1513-1516.
11. P.H. Dawson and Wing-Cheung Tam, Surf. Sci. 81 (1979) 464-478.

The Application of Ion Implantation to Quantitative SIMS Analysis

D.P. Leta and G.H. Morrison
Department of Chemistry
Cornell University
Ithaca, NY 14853

Abstract

The applicability of SIMS analysis for the micro and depth characterization of semiconducting matrices has been widely realized in recent years. Efforts for quantification have been hindered, however, by the absence of a workable theoretical model of the sputtering ionization process and more critically by the unavailability of homogeneously distributed standard materials for elements present in trace concentrations. The use of quantitative ion implantation and its particular applicability to single crystal semiconductors offers a solution to this difficulty.

Ion implant standards will be discussed in terms of their unique properties for SIMS analysis. The advantages of this method for use as both external and internal standards will be presented as well as the limitations inherent in the technique. The methods of concentration calculations both before implantation and after SIMS analysis will be given and the error factors discussed.

The use of such standards has allowed us to systematically study the effects of matrix composition on ionization probabilities for a wide range of elements. Correlations of matrix properties to the ion yield will be presented for Si, Ge, GaP, GaAs and InP substrates.

Energy Filtering and Quantitative SIMS Analysis of Silicates for Major and Trace Elements

N. Shimizu
Department of Earth and Planetary Sciences
Massachusetts Institute of Technology
Cambridge, Massachusetts 02139 U.S.A.

The potential capabilities and usefulness of SIMS analysis in geochemistry including localized analysis of trace elements or of isotopic compositions of trace elements in natural minerals far outweigh existing difficulties. Systematic studies are needed to understand the sputtering/ionization process in polycomponent targets in order to establish procedures for quantitative analysis.

The abundance of molecular ions (oxides, dimers, etc.) makes secondary ion mass spectra of geologic samples extremely complex and mass overlaps produced by molecular ions generally preclude simple conversion of intensity to concentration.

The energy filtering technique has been developed to suppress intensities of molecular ions relative to single-atom ions, based on the difference in their energy distributions. The molecular ion intensity decreases with increasing kinetic energy much more rapidly than single-atom ions and the high-energy ion population is therefore almost free from molecular ions. This technique is not only useful in obtaining interference-free mass spectra for SIMS instruments with relatively low mass resolution but also instructive in providing insight into the sputtering process in silicates. The determinations of the secondary ion energy distribution curves of major constituent ions in various mineral species suggest: (1) the energy distribution curve for Al determined for an Al plate agrees well with JURELA'S [1] measurements; (2) most single-atom ions show the energy distribution which can be approximated by $N = AE^{-n}$ $(1<n<2)$ in the energy range E>10eV; (3) alkali elements show energy distribution similar to molecular ions, suggesting that they are predominantly low energy ions; (4) the shape of the energy distribution curves for individual elements remained unchanged in various silicate structures.

Interference-free secondary ion intensities obtained in the range E>100eV were used to calculate relative intensities between elements (against Si in silicates), which were then compared with the atomic abundance ratios of the elements in the sample. Many elements (both major and trace elements) show linear relationships between relative intensity and atomic abundance ratios of the elements in the sample. In the case of trace elements, linear relationships were obtained between relative intensity and concentration, providing simple working curves. This technique has been used for various trace elements in silicates [2,3].

Difficulty arises where relative intensity of an element varies as a function of concentration of another element. For instance, Ca/Si intensity ratio increased with increasing Fe content in pyroxenes ranging from $CaMgSi_2O_6$ to $CaFeSi_2O_6$ despite constant atomic abundance of Ca [4]. A similar "enhanced" ion production has been observed in olivines $(Mg, Fe)_2SiO_4$ [4] and is currently under investigation [5].

The fact that the shape of the energy distribution curves of Ca remain unchanged among the pyroxenes indicates that the "enhanced" ion production may be caused by enhanced ionization. Understanding of the enhanced ion production is crucial because many natural silicate minerals form solid solutions involving Ca-Mg-Fe substitutions. The observations require re-evaluation of ionization models, e.g., the LTE model [6], to incorporate polycomponent dissociation equilibrium, or extension of sputtering models (e.g., [7]) to polycomponent systems. Our calculations suggest that the ANDERSEN-SIGMUND model of ternary collision [7] could explain enhanced Ca ion production as a function of Fe concentration in pyroxenes.

Despite the difficulty of establishing a general procedure, replicate analyses of well documented standard samples of limited composition ranges have yielded reasonable precision ($2\sigma<5\%$) of relative intensity measurement. Chemical compositions calculated from a set of empirical equations and a simple normalization agree with other analytical methods to within 5-10%.

References

1. Z. Jurela, Rad. Effects, 19, 175 (1973).
2. N. Shimizu, Earth Planet. Sci. Lett. 39, 398 (1978).
3. N. Shimizu and C.J. Allegre, Contrib. Mineral. Petrol. 67, 41 (1978).
4. N. Shimizu, M.P. Semet and C.J. Allegre, Geochim. Cosmochim. Acta 42, 1321 (1978).
5. I. Steele and I. Hutcheon, Abst. 14th Microbeam Anal. Conf. (1979).
6. C.A. Andersen and J.R. Hinthorne, Anal. Chem. 45, 1421 (1973).
7. N. Andersen and P. Sigmund, Kong. Dansk, Vidensk. Selsk. Mat.-fys. Medd. 39, 1 (1974).

Quantitative Analysis of Doped GaAs by Quadrupole SIMS

W. Gerigk and M. Maier
Basislabor of SFB 56 "Festkörperelektronik"
Technical University
D-5100 Aachen, FRG

Multilayer heteroepitaxial structures consisting of binary and ternary or quaternary III-V compound semiconductors are of continued interest for optoelectronic applications. For the determination of dopant profiles in such structures, SIMS is principally suited because of its excellent detection sensitivity for trace impurities. However, calibration of SIMS signals, which is usually done with standards, is impractical in this case because of the changing matrix. In this paper the basic applicability of the LTE model, which presents an alternative method for the calibration of SIMS signals [1], was investigated for some common dopants in gallium arsenide, like zinc, chromium, silicon and tin, using a quadrupole SIMS apparatus.

The measurements were performed with an Atomika DIDA secondary ion mass spectrometer [2]. The quadrupole mass filter was operated in the mass independent transmission mode [2] (relative transmission:10 %) and the bandpass of the energy filter preceding the quadrupole mass filter adjusted for optimum peak shape [3]. By variation of the target potential the energy spectrum of each type of secondary ion was shifted to position the energy bandpass at maximum intensity [4].

12 keV O_2^+ primary ions were used to achieve a high secondary ion yield. Additional oxygen could be bled into the target chamber. To obtain manageable matrix signals the primary ion current was reduced by a predetermined factor in the case of measuring the Ga^+ signal. To guarantee the required linear relationship between secondary ion intensity and primary ion current the primary ion beam had to be focussed because otherwise residual gas adsorption impeded fulfillment of this requirement. The current density of the focussed beam amounted to 10 mA/cm^2.

For each impurity a different homogeneously doped bulk sample was used, which was polished and presputtered. Only the zinc doped sample was a vapour grown epitaxial layer. The dopant concentrations were measured by atomic absorption spectrometry, with exception of the zinc concentration, which was obtained from Hall measurements. For the determination of the background, which was considerable for the zinc and silicon lines, the respective signals of the three samples not containing that particular dopant were averaged.

In Fig. 1 the dependence of some secondary ion signals on oxygen pressure in the specimen chamber is shown for unfocussed (7 µA/cm^2) and focussed (7 mA/cm^2) primary ion beam. The missing curves for the Sn^+ and Si^+ dopant signals for focussed ion beam can be obtained from those shown in Fig. 1 for unfocussed beam by shifting them to higher pressures. In doing so the ratio between sputtering rate and rate of adsorption is kept constant. To obtain

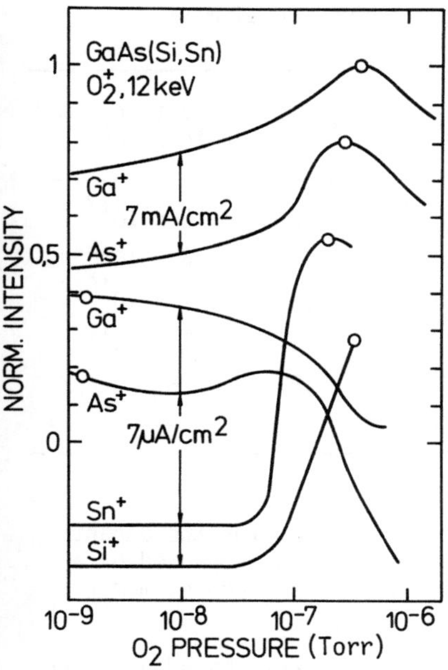

Fig.1 Dependence of secondary ion intensities upon oxygen pressure in specimen chamber. Intensities are normalized to maximum (o). Ordinate refers to uppermost curve (Ga^+, 7 mA/cm^2)

stable and reproducible secondary ion signals for quantitative analysis it is important to work within the flat region of the curves. Therefore no additional oxygen was bled in. The background pressure amounted to approx.2×10^{-9} Torr and was due to leakage from the ion source.

Relative concentrations of the elements were calculated using expression (9) of [5]. Gallium was used as reference element because arsenic showed large oxide secondary ion intensities. The only fitting parameter of this equation, T, was determined by minimization of chi-square [6] in

$$X^2(T) = \sum_i (c_{cal,i}(T) - c_{act,i})^2 / [(\Delta c_{cal,i}(T))^2 + (\Delta c_{act,i})^2] .$$

$c_{cal,i}(T)$ is the concentration of element i calculated by means of the above mentioned expression from [5] with T as parameter; $c_{act,i}$ is the "actual" concentration determined by AAS or Hall measurements or known from the composition of the matrix, as in the case of As; Δc_i are the corresponding standard deviations.

Table 1 shows the calculated concentrations which were obtained from the elemental secondary ion signals without making oxide corrections, in comparison with the "actual" ones. The value of the parameter T amounted to (4500 $\pm$ 200) K. In calculating this value the dopant zinc was not taken into account. Similar results were obtained in two further independent analyses. Surprisingly, application of oxide corrections had a negative effect on the accuracy of the analysis. From the fourth column of Table 1 it may be seen that the calculated concentrations are within a factor of two of the "actual" values. The disagreement with the zinc data derived from Hall measurements may be due to the fact that electrical methods tend to underestimate the impurity concentration at high values.

Table 1 SIMS analysis of doped GaAs with Ga as reference element

Element	Concentrations [atoms/cm^3] "Actual"	Concentrations [atoms/cm^3] Calculated	Ratio	Detection Limit [atoms/cm^3]
Si	$(3.1 \pm 0.3) \times 10^{18}$	$(2.7 \pm 0.8) \times 10^{18}$	1.2	2×10^{17}
Cr	$(8.8 \pm 0.9) \times 10^{16}$	$(8.9 \pm 0.8) \times 10^{16}$	1.0	6×10^{15}
Zn	$(3.6 \pm 0.5) \times 10^{19}$	$(2.0 \pm 0.9) \times 10^{20}$	5.6	2×10^{18}
Sn	$(1.4 \pm 0.1) \times 10^{18}$	$(9.9 \pm 1.6) \times 10^{17}$	1.4	1×10^{16}
As	2.2×10^{22}	$(4.2 \pm 2.3) \times 10^{22}$	1.9	---
Ga	2.2×10^{22}	2.2×10^{22}	1	---

Additionally shown in Table 1 are the detection limits for the dopants analysed in this work. The relatively poor value for zinc is caused by a high background from the nearby gallium peak, that for silicon by memory effects in the apparatus.

This study thus shows that a simple correction procedure based upon the LTE model, using only one fitting parameter and disregarding oxide ion signals, allows the quantitative determination of the dopants zinc, chromium, silicon and tin in gallium arsenide with a quadrupole SIMS apparatus. It may be expected that choice of a reference element like arsenic with low signal intensity would allow the use of an unfocussed primary ion beam. This would make the profiling of dopants in optoelectronic structures feasible.

References

1. C.A. Andersen and J.R. Hinthorne, Anal. Chem. 45, 1421 (1973).
2. K. Wittmaack, J. Maul, and F. Schulz, Proc. 6th Int. Conf. El. and Ion Beam Soc. and Techn., ed. R. Bakish (The ECS, Princeton 1974) p. 164.
3. K. Wittmaack, Proc. 7th Int. Vac. Congr. and 3rd Int. Conf. on Solid Surface, eds. R. Dobrozemsky, F. Rüdenauer, F.P. Viehböck, A. Breth F. Berger & Söhne, Wien 1977) p. 2573.
4. K. Wittmaack, Surf. Sci. 53, 626 (1975).
5. A.E. Morgan and H.W. Werner, Anal. Chem. 48, 699 (1976).
6. A.H. Bowker, G.J. Liebermann, "Engineering Statistics," (Prentice-Hall, New Jersey 1972).

Trace Element Analysis of Silicates by Ion Microprobe

C. Meyer, Jr.
SN7-NASA Johnson Space Center
Houston, Texas 77058

Quantitative ion microprobe analysis of silicates for trace element concentrations requires the close intercomparison of unknowns with standards using identical instrumental conditions and an internal standard such as silicon. Data from various silicate standards can be compared by relative sensitivities of elements calculated with respect to silicon [1].

$$R.S. = \frac{El^+/Si^+}{\%\ El/\%\ Si}$$

Peak heights are corrected for deadtime and isotopic percentage and concentrations are in atom percent. Fig.1 summarizes relative sensitivity for various elements in silicates as measured on the JSC-ARL-IMMA. A negative oxygen primary beam at 16.5 kV was used to sputter a 20 micron spot with 1 nanoamperes samples current. Positive secondaries were measured with 20 mil slit,α = 515, β = 515, and 15 kV. Tuning, spot size and β-aperture all cause variations in relative sensitivities [2] and were kept constant for these measurements. KREEP glass has been determined by isotope dilution and is our primary standard [3]. CHODOS glass is a set of 3 glasses obtained from A. CHODOS at Cal-Tech [4]. W-1 glass was prepared by RUCKLIDGE [5] from the USGS standard rock powder [6]. Reproducibility of these measurements on a given standard range from ± 10% for Li and Ba to ± 30% for K. In silicates, the range of relative sensitivities for a given element are found to be within a factor of two, allowing standards with high concentration to be used in analysis of trace elements in an unknown. At trace concentration, molecular ion interferences are the biggest cause for errors.

A sensitive test for matrix effects on sputtered ion intensities among various minerals is the variation in Ca^{++}/Ca^+ ratio. Table 1 illustrates a wide variation in this ratio even in a feldspar matrix. Consequently, it is expected that the relative sensitivity of elements with ionization potentials different from Si will also vary with matrix. The greatest variation in a silicate was for wollastonite. SHIMIZU, et al., [7] have reported systematic matrix effects for pyroxenes.

To look for orientation effects, crystals of kyanite, andalusite, and sillimanite (Al_2SiO_5 were mounted in different crystallographic orientations. Once a steady state was established, there was no difference in the Al^+/Si^+ or Si^{+2}/Si^{+1} ratios of these samples. Glasses prepared from Lake Co. plagioclase also gave results similar to crystalline material. Apparently the primary ion beam establishes a damaged boundary layer at the surface of the sample from which sputtered atoms are ejected. Atoms sputtered have no

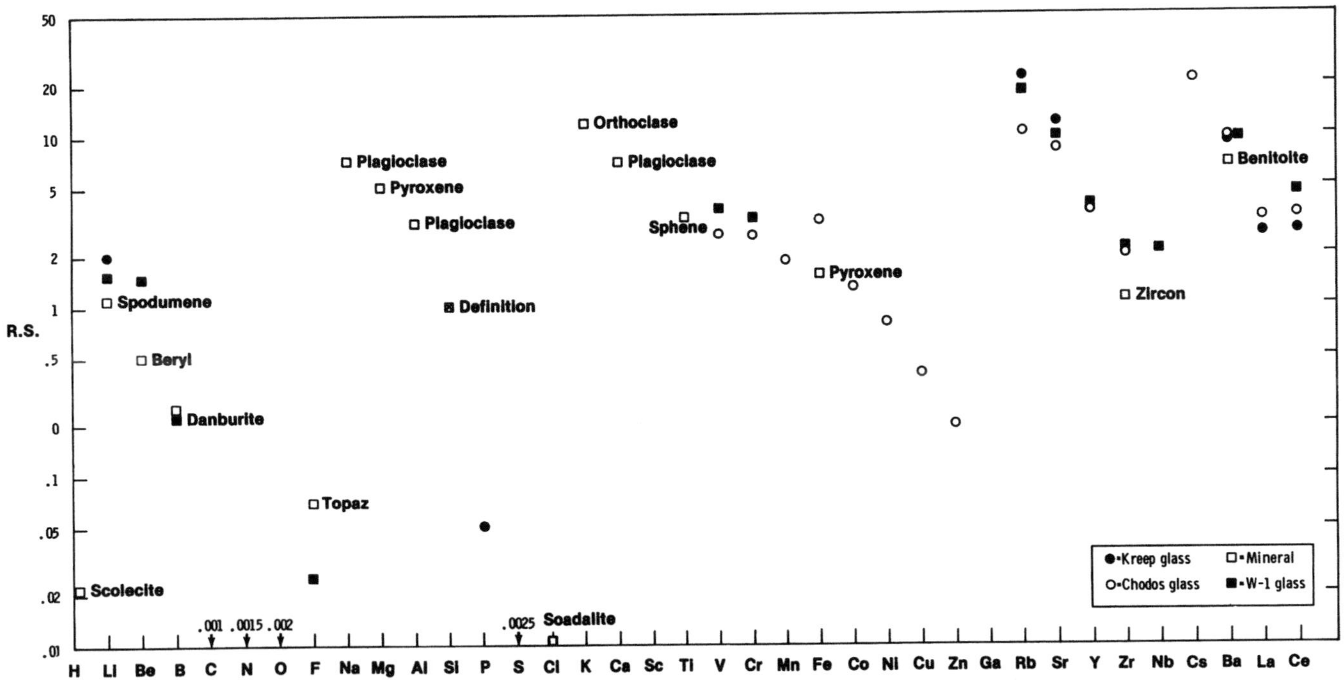

Fig.1 Relative sensitivities of elements in silicates. Note Si is 1. These sensitivities vary from laboratory to laboratory and should only be used as a guide for planning experiments

memory of crystallinity of surface. For this reason, glass standards are satisfactory to analyze minerals.

Table 1 Ion probe intensity ratios

Feldspars		(Ca/Si)*	Ca^{+2}/Ca^{+1}	Si^{+2}/Si^{+1}
Anorthoclase	(K,Na) $AlSi_3O_8$	.014	.072	$7x10^{-4}$
Orthoclase	$KalSi_3O_8$	.00045	.060	7.7
Adularia	$KalSi_3O_8$	.0002*	.065	---
Albite plag.	$NaAlSi_3O_8$	.0023	.081	7.7
Oligoclase plag.	(Na,Ca) $Al_{1-2}Si_{2-3}O_8$	.073	.042	8.4
Nain plag.	(Na,Ca) $Al_{1-2}Si_{2-3}O_8$	.183	.039	7.4
New Mexico plag.	(Na,Ca) $Al_{1-2}Si_{2-3}O_8$	.201	.041	7.7
Lake County plag.	(Na,Ca) $Al_{1-2}Si_{2-3}O_8$	.283	.041	7.4
Crystal Bay plag.	(Na,Ca) $Al_{1-2}Si_{2-3}O_8$	.329	.039	7.0
Hakone plag.	(Na,Ca) $Al_{1-2}Si_{2-3}O_8$	.463	.035	6.7
Pacaya plag.	(Na,Ca) $Al_{1-2}Si_{2-3}O_8$	.434	.034	7.0
Miyake plag.	(Na,Ca) $Al_{1-2}Si_{2-3}O_8$	.467	.033	6.7
15415 plag.	$CaAl_2Si_2O_8$	.478	.031	7.0
Other Silicates				
Danburite	$CaB_2Si_2O_8$	.434	.024	7.4
Sphene	$CaTiSiO_5$	.942	.025	8.0
Wollastonie	$CaSiO_3$	.869	.005	16.1
Scolecite	$CaAl_2Si_3O_{10}3H_2O$	.305	.042	6.7
KREEP Glass Standard 12033,97,1A		.241	.05	7.0
Beryl	$Be_2Al_2Si_6O_{12}$	$2x10^{-4}$*	.07	7.4
Zircon	$ZrSiO_4$	$6x10^{-5}$*	.05	4.0
Benitoite	$BaTiSi_3O_9$	$3x10^{-4}$*	.055	11.0
Chodos Glass Standard X,V,W		.116	.05	5.9
Non-Silicates				
Apatite	$CaPO_4$		.007	
Fluorite	CaF_2		.005	
Gypsum	$CaSO_42H_2O$		.002	
Perovskite	$CaTiO_3$		.0014	

*Atom percents estimated from Ca^+/Si^+ ratios using 15415 plagioclase as the standard

References

1. J.A. McHugh (1974) Secondary ion mass spectrometry. In Methods and Phenomena: Methods of Surface Analysis. Eds., Wolsky and Czanderna, Elsevier
2. J.G. Bradley, D. Jerome, and C.A. Evans, Jr., (1975) A comparison of mass spectra from three ion probes. In Secondary Ion Mass Spectroscopy. NBS Spec. Publ. 427, 69-77.
3. C. Meyer, Jr. (1978) Ion microprobe analysis of aluminous lunar glasses. Proc. Lunar Planet. Sci. Conf. 9th, p. 1551
4. A. Chodos (1975) personal communication. Also known as "Probe Society Glasses X,V,W."
5. J.C. Rucklidge, F. Gibb, J. Fawcett, and F. Gasparrini (1970) Rapid rock analysis by electron probe. Geochimica et Cosmochimica Acta 34, 243
6. F.J. Flanagan (1976) 1972 Compilation of data on USGS standards. Geol. Survey Prof. Paper 840, p. 171
7. N. Shimizu, M.P. Semet, and C.J. Allegre (1978) Geochemical applications of quantitative ion-microprobe analysis. Geochimica et Cosmochimica Acta 42, 1321

Imaging of Element Distributions by Ion Microprobe

J. H. Schilling*

1. Introduction

Imaging of surfaces by secondary ion mass spectrometry (SIMS) is a widely used technique. However, such images only record sputtered ion intensities. As has been shown previously for semiconductors [1]and metallic samples[2] the distributions of these intensities do not necessarily correspond to the element concentrations. This has been attributed to matrix and topographical effects. Recent measurements on geological samples also exhibit these artifacts [3].

Image processing is a promising approach for correcting such effects [2]. For this purpose, digitized images have to be obtained from the distribution of sputtered ion intensities. They can be converted to elemental distributions by applying quantifying algorithms (like CARISMA[6] or schemes based on reference materials) to each picture element. Image analysis can also be used to extract other information inherent in such images. Digitized images can either be obtained by digitizing photographic images[4] or directly by collecting images on a raster across the sample in computer-supported measurements. To utilize the high dynamic range of SIMS analysis, this second method of image collection was used in the present study for geological samples. A set of image processing programs was designed and applied to this data.

2. Measurements

An ARL ion microprobe supplemented with a Tracor-Northern multi-channel analyzer and controlled by a PDP 11/20 computer was used for the measurements. Geological samples were polished and gold coated prior to imaging. Primary ions of $^{16}O^-$ at 15 KeV were used for the experiments. Images were collected for all relevant mass peaks present in the spectrum of a sample. Each image consisted of 1000 to 4000 picture elements covering a sample area up to 400 x 400 µm at a smallest raster step size of .4 µm. Depending on the desired sensitivity, measuring times per picture element ranged from 10^{-2} to 10 seconds. Fig. 1 shows a typical sample area in an optical photograph and the corresponding area in sputtered ion intensities of $^{28}Si^+$ ions.

3. Image Processing

The first task of the image processing routines is to reconstruct and display the images. Routines for line printer images (fast display), gray

* The author is with the National Physical Research Laboratory, South Africa. At present, he pursues an associateship of the National Research Council at the Johnson Space Center, NASA, Houston, Texas 77058.

scale images (Fig.1), three dimensional plots, and contour maps (Fig.2) are available. Contrast enhancement is possible by scaling the intensities logarithmically (enhancement of low intensities) or by flat histogram scaling (presents the most detailed information [5]). Corrective routines can be applied to delete outliers from the images or to smooth noisy images. The images can be deblurred at each image point by subtracting a portion of its Laplacian. Portions of the original image can be enlarged. Statistical properties to describe the images are available: mean deviation, histogram, autocorrelation, and fast Fourier transform.

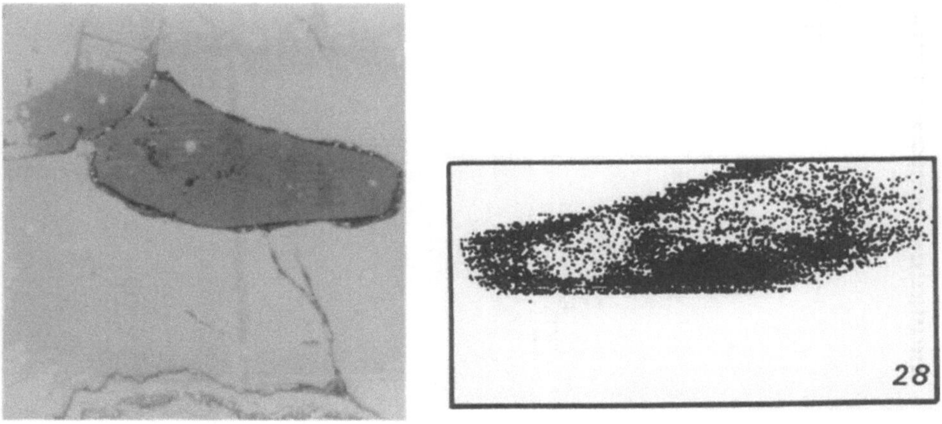

Fig.1 Optical photograph of sample area of an amphibole boardening an apatite inclusion in a magnetite matrix and sputtered $^{28}Si^{+}$ distribution of the same area

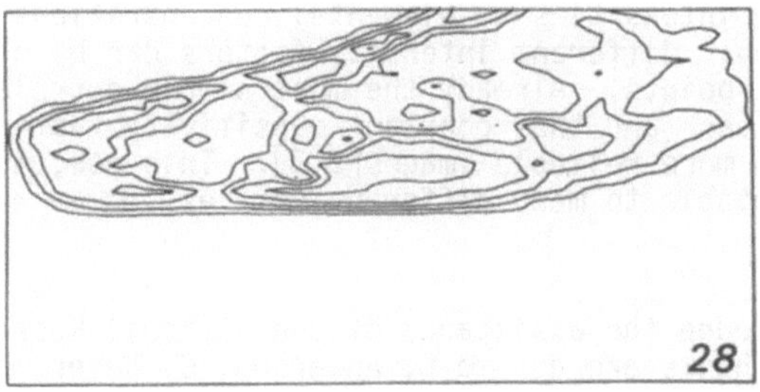

Fig.2 Contour map of $^{28}Si^{+}$ distribution in Fig. 1

For the description of the relations of images from different chemical elements, three programs can be used: cross correlation, correlation pattern, and classification. The correlation-pattern program gives a diagram of intensities of element A against element B (Fig.3), showing how many picture elements of a specific combination can be found. Very different applications of this program are possible: correlation, anticorrelation, or independence of one element from another can be detected. Clustering of points in different areas of the diagram indicates changing composition in different parts of the sample. The pattern of two isotopes of an element gives the isotope ratio and if very high count rates are involved, allows determination of the counting deadtime of the instrument.

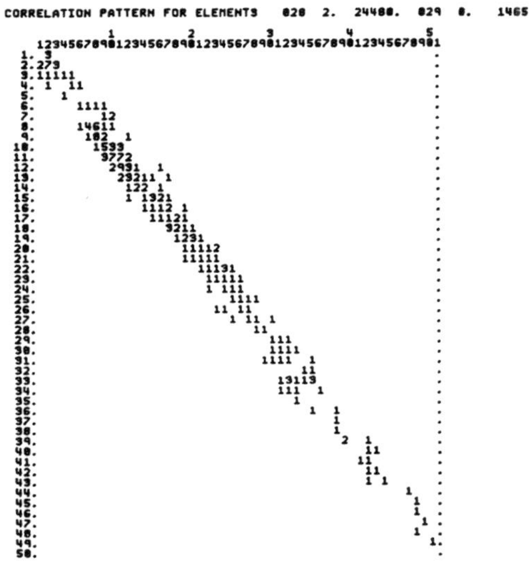

Fig.3 Correlation pattern $^{28}Si^+$ against $^{29}Si^+$ from area in Fig. 1

The classification program provides an unsupervised cluster analysis which can consider up to 10 different images. It determines sample areas of differe composition and/or changing ion yield. It results in a map of such areas and a table of means for all elements and all classes.

Converting the intensities to elemental concentrations is based on the classification map: different intensity factors can be used for different classes of sample points. Already the most simple correlation procedure using only one class (and thus constant sensitivity factors across the whole sample) results in more reliable images[2,3]. This new, multiclass quantificat procedure is adaptable to many different samples.

Acknowledgments

I want to acknowledge the assistance of the National Research Council towards this research. Thanks are due to D. Anderson, C. Meyer, and D. Phinney for many discussions and their collegial support.

References

1. J.H. Schilling, P.A. Büger, Appl. Phys. 15 115 (1978).
2. J.H. Schilling, P.A. Büger, Int. J. Mass Spect. Ion Physics 27, 283 (1978)
3. J.H. Schilling, Proc. Microbeam Analysis Society Conf., San Antonio, (1979), p. 345.
4. J.D. Fasset, J.R. Roth, G.H. Morrison, Anal. Chem. 49 2322 (1977).
5. A. Rosenfeld, A.C. Kak, "Digital Picture Processing," Academic Press, New York (1976).
6. C.A. Anderson, Nat. Bur. Stand. (U.S.) Special Publ. 79 427 (1975).

Secondary Ion Emission from Titanium Alloys Under Argon and Oxygen Bombardment

J.M. Schroeer, A. Dely, and L.L. Deal
Department of Physics
Illinois State University
Normal, Illinois 61761, USA

1. Introduction

Recently, a number of researchers have shown that in many cases a simplified thermal equilibrium (LTE) model adequately describes the emission of secondary ions from solids [1,2,3]. The simplified LTE model predicts that the probability of a sputtered atom leaving the surface as an ion is given by

$$R^+ = \text{const } (B^+/B^o) \exp(-I/kT), \qquad (1)$$

where B^+ and B^o are the partion functions of the sputtered ion and atom, respectively, I is the ionization energy of the sputtered atom, and T is the apparent surface plasma temperature.

The experiments described here intend to test whether secondary ion emission from titanium alloys can be described by a simplified LTE model when the alloys are bombarded by argon or oxygen in the presence of varying amounts of ambient oxygen. We will also test for correlation between T in the LTE model and the ratio of metal oxide ion to metal ion, $[TiO^+]/[Ti^+]$.

2. Procedure

Titanium alloys (NBS 642: 6.7% Mn; NBS 646: 3.4% Cr, 2.1% Fe, 1.1% Mo; NBS 654a: 6.3% Aℓ, 3.9% V) were bombarded with 5.5 keV Ar^+ or O_2^+ in the CAMECA IMS-300 ion microscope [4] at Cornell University in the presence of different partial pressures of oxygen. The spot size of the beam was 125 µm, and it was rastered over an area 500 µm square. Only ions emitted from the central area 200 µm in diameter were detected. Secondary ions with energies between zero and 20 eV passed through the mass spectrometer and were registered by pulse counting. The background pressure in the target region was between 1 and 2 x 10^{-7} torr. During argon bombardment there was an additional partial pressure of argon in the target region of approximately 4 x 10^{-7} torr. The total bombarding ion current and the partial pressure of oxygen in the target region are tabulated in Table 1.

3. Results

Equation (1) was compared with the experimental results as follows. The detector outputs were corrected for isotopic abundance and for elemental concentration, were then multiplied by the square root of the ionic mass to correct for detector sensitivity, and finally were normalized to titanium. As in [2] and [3], we include an empirical oxide correction: the ion yields of the metals and their oxides as corrected above were added together to obtain the experimental values for R^+. For a given set of bombarding conditions the data from all three alloys were combined.

Log $[R^+(B^+/B^o)]$ was then plotted against I. The slope of the best straight line through the data points gives T in (1). Since the ratio B^+/B^o does not vary much with temperature, all partition functions were evaluated at 7000K and were taken from BOLTON [5]. The plotted values of $R^+(B^+/B^o)$ were all found to lie within a factor of two of the best straight lines drawn through the data points. See [3] for examples.

Table 1 lists the bombarding conditions and the observed values of T and of the ratio $[TiO^+]/[Ti^+]$. In Fig.1, T is plotted against the ratio $[TiO^+]/[Ti^+]$. The amount of oxygen on the surface increases from the lower left to the upper right in the figure. Runs C, I, and N represent essentially saturation of the surface with oxygen.

If we had used the correction factors of RUDAT and MORRISON [6] when correcting for detector sensitivity, the experimental values of T would all have been lower. Otherwise, our conclusions would be unaffected.

Table 1 Bombarding conditions and experimental results

Date	Bombarding Species	Bombarding current [nA]	Oxygen partial pressure [10^{-7} torr]	T[K]	$\frac{[TiO^+]}{[Ti^+]}$	Run
1977	O_2^+	60-240	0	6 300	0.38	A
		15-60	7	9 700	1.4	B
		15-30	80	11 200	2.0	C
1979	O_2^+	1000	0	5 590	0.22	D
		1000	0.5	5 200	0.21	E
		100	3	6 060	0.30	F
		50	10	8 370	1.01	G
		19	7	10 030	1.06	H
		20	60	9 700	1.60	I
1977	Ar^+	300	0	4 190	0.17	J
		300	2	4 690	0.24	K
		50	2	4 760	0.28	L
		20	2	5 430	0.29	M
		21	60	14 000	0.78	N

4. Discussion

As shown in Fig. 1, one can draw a straight line through all the data points except the one for Run N, which represents a surface bombarded with oxygen in the presence of large amounts of oxygen. Thus it appears that except for conditions similar to Run N, Fig. 1 can be used to deduce T in Eq. (1) within ±10% from the ratio $[TiO^+]/[Ti^+]$. An uncertainty of ±10% in T results in an uncertainty of about 10% in R^+ in Eq. (1).

Normally, when using Eq. (1) in the mass analysis of a sample, T and the constant can be found if the concentration of two elements in the sample are known. If T can be determined from the metal ion to metal oxide ion ratio of some element in the sample, then the concentration of only one element in the sample need be known to evaluate the constant.

It is suspected that a different line in Fig.1 would result if a different mass spectrometer were used. We intend to examine what happens to Fig.1, if the titanium is contained in an iron matrix rather than the present titanium matrix.

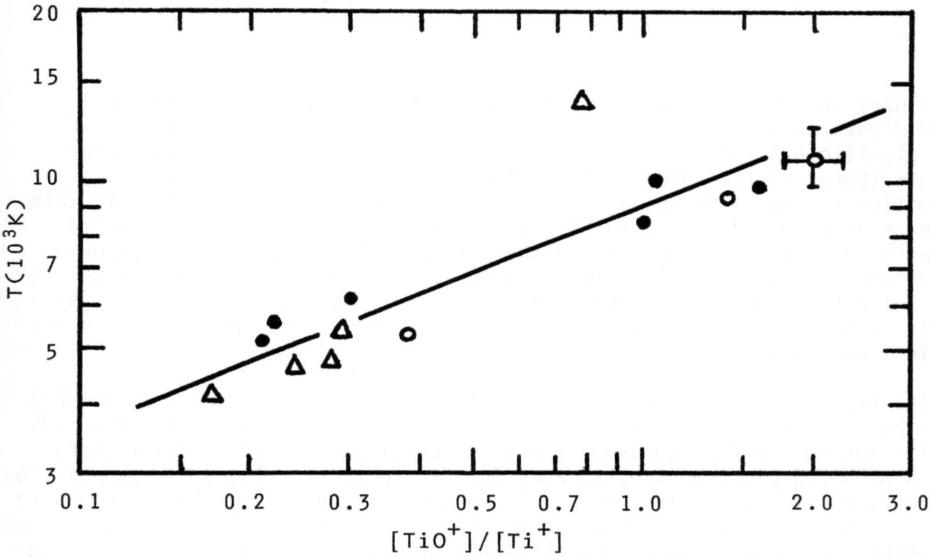

Fig. 1 The LTE temperature T plotted against the ratio of the secondary ion yields of TiO^+ and Ti^+. o Bombardment by O_2^+ in 1977, • bombardment by O_2^+ in 1979, and Δ bombardment by Ar^+ in 1979. The error bars indicate an uncertainty of ±10% in T and in the ion ratio

It appears that for the particular titanium alloys and bombarding conditions used here, a simplified LTE model, Eq. (1), satisfactorily describes the ionization of sputtered atoms. Similar results were previously obtained for iron alloys [3].

Acknowledgements

We thank Prof. G.H. Morrison of the Chemistry Department at Cornell University for making the ion microscope available to us. The research was supported through released time, Summer Research Appointments and a Research Grant provided by Illinois State University. A.D. and L.L.D. were undergraduate students at Illinois State University.

References

1. D.S. Simons, C.A. Evans, and J. Baker, Anal. Chem. 48, 1341 (1976)
2. A.E. Morgan and H.W. Werner, Anal. Chem. 48, 699 (1976)
3. J.M. Schroeer, J. Vac. Sci. Technol. 14, 343 (1977)
4. G.H. Morrison and G. Slodzium, Anal. Chem. 47, 932A (1975)
5. C.T. Bolton, Astrophys. J. 161, 1187 (1970)
6. M.A. Rudat and G.H. Morrison, Int. J. Mass Spectr. Ion Phys., 27, 249 (1978)

Effect of Alloying in Secondary Ion Emission from AgPd and CrNi Systems

M.L. Yu and W. Reuter
IBM T.J. Watson Research Center
Yorktown Heights, NY 10598

Secondary ion emission during sputtering depends very much on the local atomic environments of the sputtered atoms. For example, sputtering with O_2^+ significantly enhances positive ion formation [1]. This enhancement is attributed to the incorporation of oxygen atoms in the target. The formation of the oxide bonds intensifies the secondary ion yields. In this experiment we moved a step further and tried to see whether the presence of a second metallic component in the sample could influence this oxygen enhancement effect. We also tried to look for any corresponding changes in the oxide bonds with X-ray photoemission (XPS). We examined two alloy systems: CrNi and AgPd. Both Cr and Ni form strong oxides, while both Ag and Pd form weak ones.

The experiment was performed in an all-metal ultrahigh vacuum system [2] with base pressure in the low 10^{-9} torr region. The system included a SIMS system equipped with a quadrupole mass spectrometer and an energy filter, and a HP 5950B electron spectrometer for XPS analysis. The O_2^+ beam was produced by a differentially pumped cold cathode ion gun. The chamber pressure rose to about $1x10^{-8}$ torr when the gun was in operation. The O_2^+ beam, focuse to about 100 μm at 1 μA, was rastered to sputter an area of 1.5mm square at normal incidence. Electronic gating limited data acquisition to within the center 40% of the crater. We also integrated the ion yields over the ion energy distribution (IED). The crater sizes were measured to determine the sputtering coefficients. By normalizing to the same concentration and sputtering coefficient, we obtained the relative ionization coefficient α^+ of the different alloy components as a function of composition.

Figure 1 shows the values of α^+ for Ag and Pd in various alloys normalized to the values for the pure elements. α^+ of Ag was enhanced by over an order of magnitude in the $AgPd_3$ alloy. On the other hand, α^+ of Pd was suppressed by alloying with Ag. The Ag^+ IED was also narrower in the alloys while the Pd^+ IED was substantially broadened. As shown in Table 1, the biggest change was about a factor of two in the full width at half maximum (FWHM) of the IEDs.

Table 1 FWHM (eV) of IEDs obtained with 15 keV O_2^+

	Ag	Ag_3Pd	AgPd	$AgPd_3$	Pd
Ag^+	5.8	4.3	3.3	3.0	
Pd^+		18.6	15.8	11.8	9.7

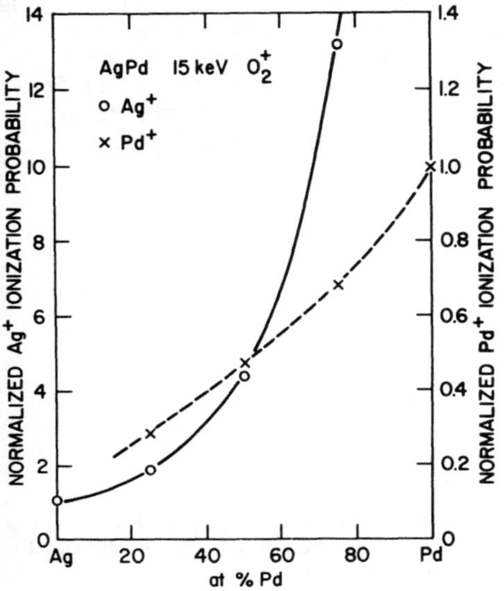

Fig.1 Normalized Ag^+ and Pd^+ ionization probabilities as a function of alloy composition

The Cr^+ and Ni^+ ionization coefficients showed a different trend. Fig.2 shows the variation of α^+ with the CrNi alloy composition. The emission of Ni^+ was enhanced strongly by the presence of Cr while the effect of Ni on the Cr^+ yield was relatively small.

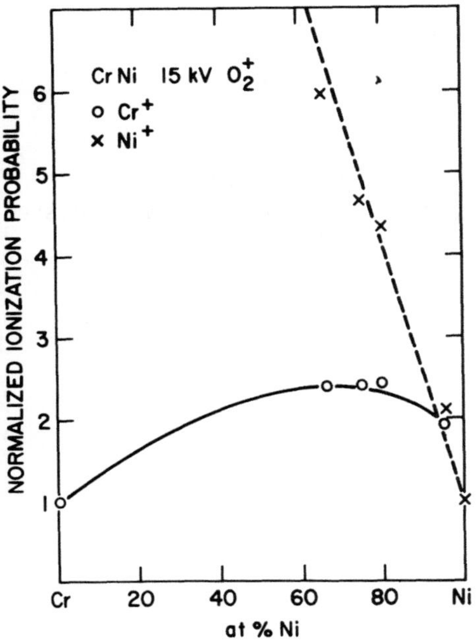

Fig.2 Normalized Cr^+ and Ni^+ ionization probabilities as a function of alloy composition

XPS showed that the oxidation of Ni was different in different CrNi alloys. We monitored the Ni and Cr 2p level binding energies after each O_2^+ bombardment. Fig.3 shows the Ni $2p_{3/2}$ lines. We noticed that for pure Ni, oxidation was mostly in the form of NiO [3]. But with the introduction of Cr, the

Ni peak broadened and shifted to higher binding energies. This indicates the formation of a mixture of oxides and a gradual shift to higher oxidation states, e.g., Ni_2O_3. The enhancement of Ni^+ seems to be linked to these higher oxidation states. The Cr 2p line did not show appreciable shift in energy with alloy composition. This again correlates well with the relatively small change in the ionization coefficient of Cr with alloying.

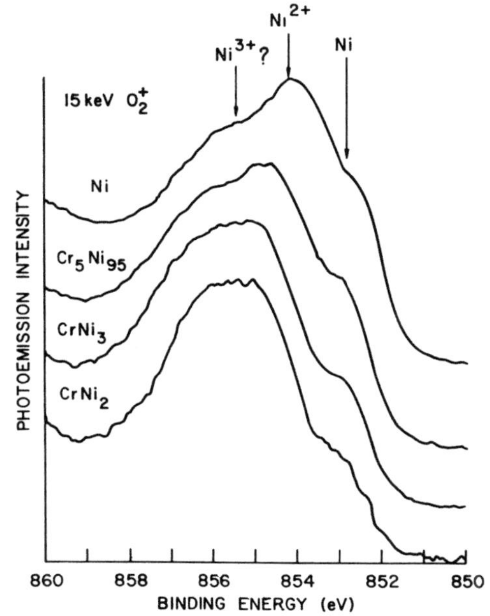

Fig.3 Nickel $2p_{3/2}$ photoemission peaks of O_2^+ bombarded CrNi alloys

References

1. C.A. Andersen, Int. J. Mass Spect. Ion Phys. 2, 61 (1969).
2. M. Frisch, W. Reuter and K. Wittmaack, Rev. Sci. Instr. (to be published).
3. K.S. Kim and W. Winograd, Surf. Sci. 43, 625 (1974).

III. Semiconductors

Quantitation of SIMS for Semiconductor Processing Technology

T. W. Sigmon
Stanford Electronics Laboratories
Stanford, California 94305

The quantitation of the depth and concentration scales of SIMS measurements for semiconductor device processing technology is discussed. Using ion implantation and MeV ^{4}He backscattering, one is able to provide absolute depth and concentration scales for many important impurities in materials such as silicon dioxide, silicon and gallium arsenide. This paper will discuss how the use of ion implanation, Rutherford backscattering and SIMS can provide measurement capabilities for semiconductor processing, heretofore unavailable. Recent results and earlier examples will be presented.

With the availability of high quality SIMS equipment a new tool has been provided for the semiconductor materials analyst's arsenal of weapons for looking at detailed impurity concentrations in semiconductor materials. With the push towards smaller structures in both silicon and gallium arsenide integrated circuit technology, a more careful knowledge of the detailed in-depth impurity profiles is needed. Since secondary ion mass spectrometry can provide information concerning atomic concentration profiles vs depth over many orders of magnitude, this technique can be useful in materials where electrical measurements are not applicable, such as oxides, semi-insulating layers, etc. The technique also can be used to verify the relation between the atomic concentration profile and the measured electron or hole concentration profiles in semiconducting materials. The combination of SIMS, ion implantation and MeV ion backscattering can provide a unique quantitation of atomic profiles in semiconductor material systems. This in turn makes possible detailed analysis of new semiconductor processing techniques such as those being studied for <u>V</u>ery <u>L</u>arge <u>S</u>cale <u>I</u>ntegrated <u>C</u>ircuit applications.

The fitting of ion implantation profiles in silicon, silicon dioxide and gallium arsenide by the LSS range statistics [1] has allowed controlled standards to be generated for most impurities of interest in these substrates. Since the theoretical distribution of the ion implanted impurity is well characterized by the implantation energy and total ion dose selected, it is only necessary then to fit the ion yield vs sputtering time to a theoretical Pearson IV distribution to obtain a quantitative profile of the impurity in the material. Also, sputter ion yields can be obtained for a given impurity in a matrix thus enabling a variety of distributions of this impurity to be measured once the sputtering rate has been calibrated. An important advantage of SIMS over electrical measurements is that the impurity profile may be measured without post-implantation annealing, thereby allowing theoretical predictions to be examined. Also impurities, such as fluorine in silicon, which are not electrically active may be measured [2].

For many cases, when the mass of the impurity elements are heavier than the mass of the substrate materials, MeV ^{4}He ion backscattering can also be used in conjunction with SIMS measurements to establish excellent profiling capabilities in a material. Ion backscattering allows determination of both layer thickness and impurity concentrations. [3] The measurement of thin film thicknesses in multilayered structures can be useful for calibration of the sputter yield as a function of depth.

The combination of these three techniques has been quite successful in the past few years for semiconductor materials and process evaluation [4-6]. The following discussion will briefly illustrate some of the cases where the techniques have been successfully combined to yield quantitative profiles unmeasurable by other techniques.

The first such example [7] will illustrate the use of SIMS measurements combined with ion implantation and MeV ^{4}He ion backscattering to assess the effect of a laser beam upon an implanted impurity in a silicon crystal. In Fig.1 we plot the number of ^{75}As atoms vs depth for an implanted profile in silicon. For this case, 3 x 10^{14} As/cm^2 has been implanted at 100 keV into silicon that was initially held at room temperature. The wafer was then annealed with a free running ruby laser with an on-target energy level of 15-30 J/cm^2 and a pulse length of approximately 1 msec. In the figure we show SIMS results taken from the as-implanted profiles, 8, 16, 32 and 64 pulses from the laser with the substrate held at room temperature, along with a thermal anneal at 1000°C for 30 minutes.

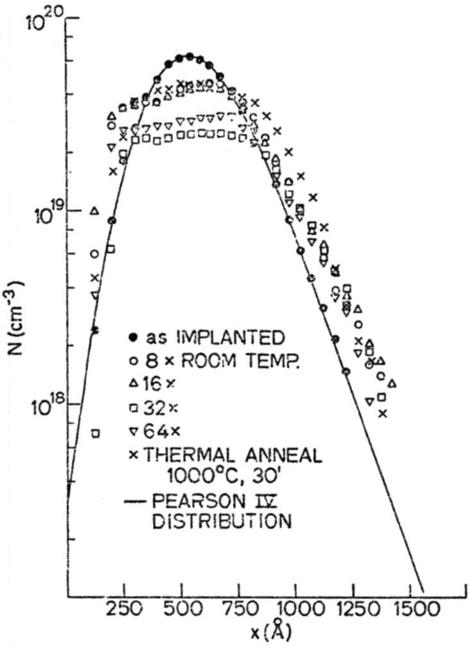

Fig. 1 Arsenic concentration profiles obtained from SIMS (Normalized with RBS data) for (●) as-implanted (o, △, □, ▽) pulse laser annealed and (x) thermally annealed Si samples. The Pearson IV (-) distribution is also plotted for comparison

Also the Pearson IV distribution fit to the implanted arsenic profile is shown for comparison. For calibration of the total arsenic dose both before and after anneal, MeV ^4He backscattering has been employed. Normalization of the Pearson IV distribution function to this measured dose is then used to produce the atom concentration vs depth scale. One can see that the distribution of arsenic atoms can be measured quite accurately in the range from 10^{20} to 10^{18} cm^{-3} using the SIMS. These measurements were performed using an early SIMS. Presently much better results can be obtained with new equipment, such as the Cameca IMS-3f.

In Fig.2 we show atom concentration (cm^{-3}) vs depth (μm) for boron implanted into silicon [8]. The boron profile consists of an ^{11}B implant to a dose of 10^{15} B/cm^2 at 34 keV, followed by a ^{10}B implant to a dose of 4×10^{12} $^{10}B/cm^2$ at 130 keV. The dark line represents a composite profile which is the sum of the two individual profiles. Again concentration and depth scales have been obtained from the range distribution and for ^{11}B as published by Gibbons, et al. [1]. It can be seen that the composite profile has a shoulder occuring at approximately 0.37 μm. The purpose of this experiment was to explore problem encountered when using BF_2 implantation into silicon from a pre-analysis implantation machine [9]. Here we see an example of the sensitivity of the SIMS, with concentrations varying from 10^{19} to almost 10^{14} cm^{-3} for ^{11}B, and similar concentrations for ^{10}B. Also the ability to separate both mass 10 and 11 is illustrated by this example.

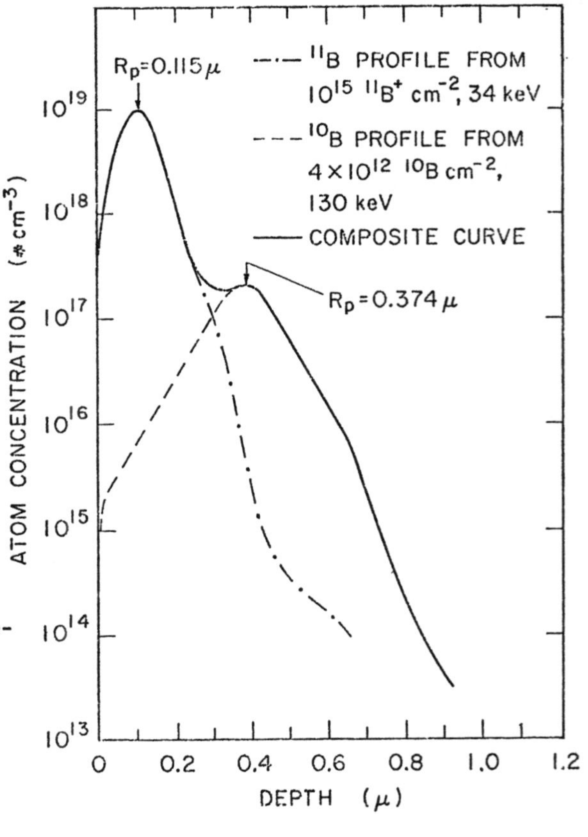

Fig. 2 ^{11}B and ^{10}B concentration versus depth profiles resulting from a 10^{15} $^{11}B/cm^2$ 34 keV and 4×10^{12} $^{10}B/cm^2$ implant. The solid line represents the composite profile

We have also investigated the use of standards created by ion implantation for compound semiconductors. In Fig. 3 we show data taken on impurities in a GaAs substrate as a function of depth. On the vertical scale is plotted reltive intensity of the impurity. The first atomic species of interest is ion implanted selenium. By using the LSS range distribution theory we are able to determine the peak concentration of the implanted Se in the GaAs; this then yields atomic concentration vs depth for this impurity. We have also measured Cr as a function of depth in the near surface region of the GaAs [10]. The Cr concentration of 1 x 10^{17} atoms/cm^3 shown in Fig. 3 has been obtained by using a Cr implantation into GaAs for a SIMS standard. The resulting depth and concentration (ion yield) scales were then obtained from the LSS theory. Also shown for comparison is the chromium concentration vs depth in the GaAs substrate after various annealing steps. It can be seen from these measurements that these calibration procedures will allow detailed measurement of the movement of important impurities in presend-day compound semiconductors, thereby contributing to the advance of the state-of-the-art in fabrication of devices in these materials.

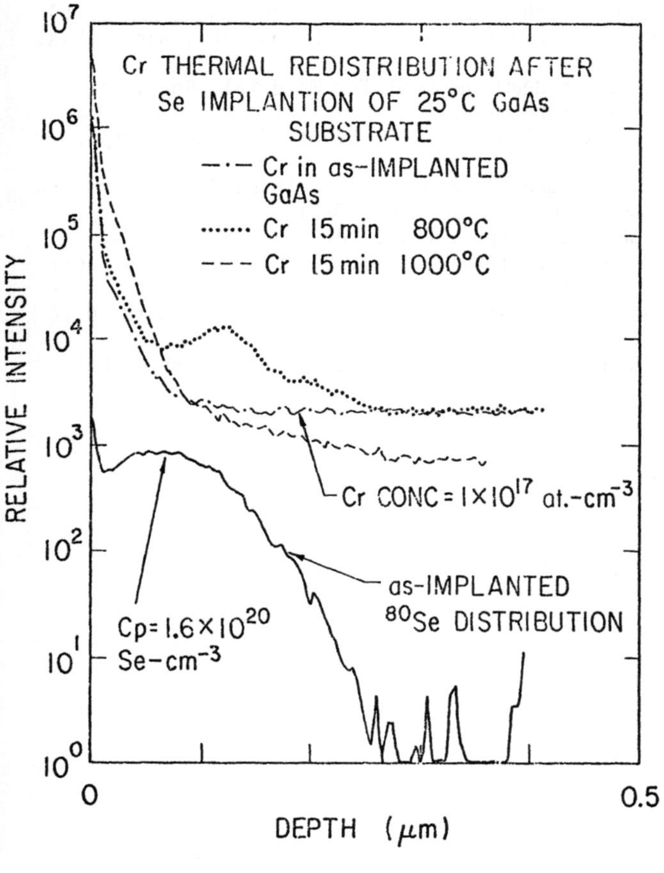

Fig. 3 SIMS profiles of implanted Se in GaAs at 150 keV to a dose of 1 x 10^{15} at/cm^2 with the substrate maintained at room temperature. Also plotted are the SIMS profiles for Cr as function of thermal processing of the substrate

The use of secondary ion mass spectrometry, ion implantation and MeV ion backscattering can be combined to provide a very powerful quantitative analytical tool for the semiconductor processing industry. Examples of the measurement of distributions of important impurities in silicon and GaAs have been discussed. Absolute concentration and depth scales can be established by a combination of the above-mentioned three techniques. It is the belief of the author that continued complementary use of these three techniques will provide new data useful for submicron, very high speed device processing and modeling.

Acknowledgements

We are pleased to acknowledge the continued encouragement and financial support of Dr. R. Reynolds, DARPA (Contract No. MDA903-73-0290), the helpful comments by Dr. C.A. Evans, Jr., and the SIMS profiles provided by Charles Evans & Associates.

References

1. J.F. Gibbons, W.S. Johnson, and S.W. Mylroie, Projected Range Statistics in Semiconductors (Dowden, Hutchinson and Ross, Stroudsburg, PA, 1975).
2. M.Y. Tsai, B.G. Streetman, P. Williams and C.A. Evans, Jr., Appl. Phys. Lett. 32, 144 (1978).
3. J.W. Mayer and E. Rimini, Ion Beam Handbook for Materials Analysis, Sec.2, (Academic Press, New York, 1977).
4. A. Lidow, J.F. Gibbons, V.R. Deline and C.A. Evans, Jr., Appl. Phys. Lett. 32, 15 (1978).
5. A. Gat, J.F. Gibbons, T.J. Magee, J. Peng, V.R. Deline, P. Williams and C.A. Evans, Jr., Appl. Phys. Lett. 32, 276 (1978).
6. C.A. Evans, Jr. and P. Williams, Appl. Phys. Lett. 30, 559 (1977).
7. J.F. Gibbons, J.L. Regolini, A. Lietoila, T.W. Sigmon and others, Jour. Appl. Phys. (to be published).
8. T.W. Sigmon, V.R. Deline, C.A. Evans, Jr., and W.M. Katz, submitted to Jour. Electrochem.
9. C.A. Evans, Jr., V.R. Deline, T.W. Sigmon, A. Lidow, Appl. Phys. Lett. (to be published).

Problems Encountered in Depth Profiling of Nitrogen and Oxygen in Silicon by Means of Secondary Ion Mass Spectrometry

W. Wach and K. Wittmaack

Gesellschaft für Strahlen- und Umweltforschung mbH
Physikalisch-Technische Abteilung
D-8042 Neuherberg, FR Germany

Secondary Ion Mass Spectrometry (SIMS) has become the technique most frequently used for compositional in-depth analysis in the 1 nm to 10 µm regime. It has been noticed, however, that one may encounter implications and problems in SIMS depth profiling studies which are not due to insufficient signal heights. For example, the damage introduced in the sample by primary ion impact can alter the original composition profile significantly. Effects due to collisional displacement [1,2] and radiation-enhanced diffusion [3,4] have been observed. In order to keep bombardment-induced atomic mixing as small as possible, the primary ion energy should not exceed ~1 keV [5] . Radiation-enhanced diffusion can occur not only in the bulk but also at the surface of the ion-bombarded sample. Accordingly, adsorbed gases may diffuse into the sample [6] . These findings are important in SIMS experiments because (i) saturation of positive secondary ion yields requires incorporation of oxygen rather than mere adsorption [7] and (ii) uptake of residual gases will result in unwanted background levels of the respective atomic ions (e.g. H^-, C^-, O^-) [8] and, equally important, of molecular ions (e.g. $^{30}Si^1H^+ = {}^{31}P^+$) [9]. In this contribution we report on a systematic study of background generation in depth profiling of nitrogen and oxygen in silicon.

The experiments were performed in the DIDA ion microprobe described elsewhere [10]. The base pressure in the unbaked target chamber was 5×10^{-7} Pa (4×10^{-9} Torr). Mass-analysed focussed Ar^+ beams at energies between 2.5 and 10 keV were raster scanned across the silicon samples, with beam currents ranging from 0.2 to 1 µA. The scan speed was typically 1 frame/s. The widths of the square-shaped areas covered by the normally incident beam were varied between 0.4 and 2.5 mm. An electronic aperture [11] was employed to record only secondary ions emitted from the central part of the sputtered area (aperture size 10 to 20 %).

Examples of depth profiles of 5 keV N in Si (10 keV N_2^+ implantation) are presented in Fig.1. Si_2N^+ secondary ions were used to evaluate the nitrogen profile. These ions constitute the most intense nitrogen-carrying species in the positive spectrum [12]. In the present context, the most important result of Fig.1 is that the background observed after having passed through the implanted profile depends upon the average current density of the probing argon beam. In order to investigate the quantitative aspects of background production in depth profiling of nitrogen and oxygen in Si, we have measured the intensity of Si_2N^+ and Si_2O^+ under various experimental conditions. Si_2O^+ was used for comparison with Si_2N^+ because both molecules contain the same number of Si atoms.

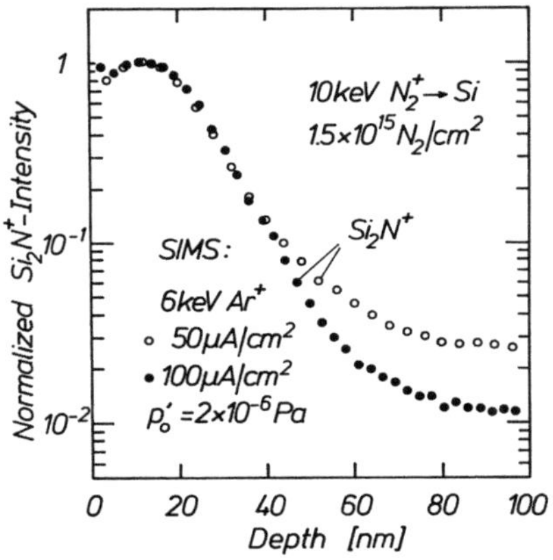

Fig.1 Normalized depth profiles of nitrogen in silicon recorded at two different current densities of the probing argon beam. The peak concentration was approximately 10^{21} N atoms/cm^3

Figure 2 shows the variations in the steady state signal height, $\Delta I(Si_2N^+)$ and $\Delta I(SiO_2^+)$, introduced by increasing the nitrogen partial pressure $\Delta p(N_2)$. ΔI has been obtained by subtracting from the measured signal, $I = I(\Delta p)$, the signal $I_0' = I(p_0')$ observed at the base "working" pressure p_0' (Ar^+ beam on), prior to bleeding in nitrogen. $\Delta I(Si_2N^+)$ increases linearly with increasing pressure of N_2. The (delayed) increase of $\Delta I(Si_2O^+)$ is probably due to liberation of oxygen and/or water from the walls, which can occur by exchange reactions with N_2.

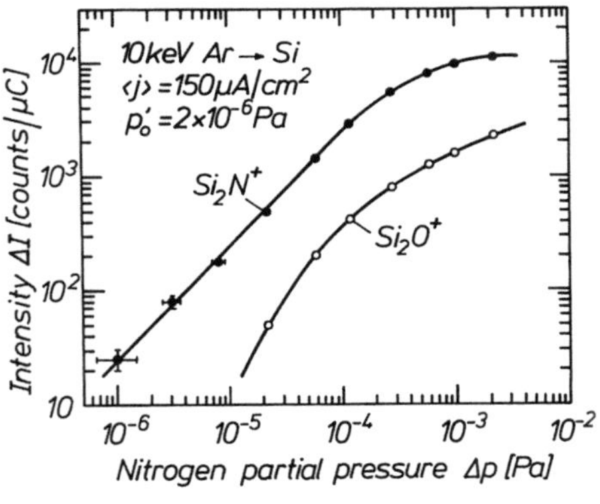

Fig.2 Enhancement of the background intensities due to raising the nitrogen partial pressure

The effect of the average current density $\langle j\rangle$ and the argon energy on the background intensities at base pressure is depicted in Figs. 3(a) and (b), respectively. $\langle j\rangle$-variations have been achieved either by varying the scanned area A at a fixed current i or by varying i at a fixed A. The data in Fig.3 have been corrected for the decrease in spectrometer sensitivity with increasing area [10]. According to Fig.3(a) the background intensities at base pressure depend only upon the average current density, a result which is supported by the observation that variations in scan speed between 0.5 and 50 frames/s do not affect the background level. Moreover, Fig.3(b) shows that variations in primary ion energy introduce only a small effect.

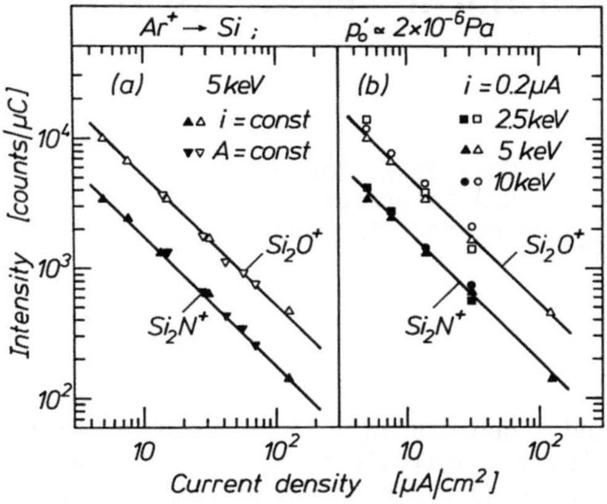

Fig.3 Background intensities at base pressure versus the primary ion current density. Parameters are (a) the beam current i and the sputtered area A, (b) the argon energy

These results indicate that the integrated background intensities resulting from the uptake of ambient gases are *proportional to the residual gas pressure, the area viewed by the spectrometer and the time required to sputter through a certain layer*. Accordingly, the current density (and the sputtering yield) should be as high as possible in order to minimize the background intensity. For various reasons, however, the average current density in raster scanning depth profiling cannot exceed certain limits. (i) The maximum target current is usually space-charge limited and decreases rapidly with decreasing energy [13]. (ii) The width of the sputtered area must be sufficiently large compared to the beam diameter in order to avoid crater edge effects [11]. (iii) The analysed area may not be reduced below a certain limit set by requirements of counting statistics. Therefore, the only satisfactory way to lower background intensities of the type discussed here is to reduce the residual gas pressure (and to choose a proper projectile mass to maximize the sputtering yield). Detection limits better than 10^{18} nitrogen atoms/cm^3 can be achieved only at pressure-to-current density ratios$<10^{-4}$Pa/Acm^{-2}.

References

1. F. Schulz, K. Wittmaack, J. Maul, Rad. Effects 18, 211 (1973).
2. J.A. McHugh, Rad. Effects, 21, 209 (1974).
3. P. Blank, K. Wittmaack, Rad. Effects Lett.
4. B.Y. Tsaur, Z.L. Laiu, J.W. Mayer, Appl. Phys. Lett., 34, 168 (1979).
5. H.H. Andersen, Appl. Phys., 18, 131 (1979).
6. W. Wach and K. Wittmaack, Nucl. Instrum. Meth., 149, 259 (1978).
7. K. Wittmaack, Surface Sci., 68, 118 (1977).
8. P. Williams, R.K. Lewis, C.A. Evans, Jr., P.R. Hanley, Anal. Chem., 49, 1399 (1977).
9. K. Wittmaack, Appl. Phys. Lett., 29, 552 (1976).
10. K. Wittmaack, Adv. Mass Spectrometry VII, 758 (1977).
11. K. Wittmaack, Appl. Phys., 12, 149 (1977).
12. A. Benninghoven, W. Sichtermann, S. Storp, Thin Sol. Films, 28, 59 (1975).
13. K. Wittmaack, Nucl. Instr. Meth., 143, 1 (1977).

Depth Profiling of Phosphorus in Silicon Using Cesium Bombardment Negative SIMS

Charles W. Magee
RCA Laboratories
Princeton, NJ 08540

1. Introduction

One of the most often used n-dopants in silicon solid state technology is phosphorus. Unfortunately, its analysis by SIMS at the trace level ($\lesssim 10^{17}$ at/cm^3) has been limited by sample contamination from the vacuum system. Any hydrogen-containing species which condense onto the sample will create ^{30}SiH sputtered molecular ions which interfere with the detection of ^{31}P if one cannot separate the species through use of high mass resolution. Previous efforts [1,2] at SIMS profiling of P in Si with oxygen bombardment had detection limits of approximately 1×10^{18} at/cm^3. The SIMS instrument at RCA Laboratories [3] maintains a pressure of 10^{-10} torr in the sample chamber, but cannot form a primary ion beam of oxygen ions. An oxygen jet is used on most samples which require an oxide surface layer to enhance positive secondary ion yields. However, this technique cannot be used when analyzing for P in Si because the oxygen unavoidably will contain enough H_2O to produce a significant ^{30}SiH signal. To circumvent this problem, advantage was taken of the high negative ion yield of P under Cs^+ bombardment [4], and a surface ionization Cs^+ primary ion source was developed [5]. Thus, taking advantage of the clean sample environment of the instrument and using Cs bombardment negative SIMS, it was hoped that P in Si could be depth profiled at trace concentrations.

2. Results and Discussion

Since the residual pressure in the sample chamber has been shown to be the determining factor in P analysis, an experiment was performed to quantitatively determine the limits imposed on ^{31}P detection due to ^{30}SiH interference. A p-type silicon wafer of (100) orientation with a resistivity of 5-10 Ω-cm was implanted with 15 keV protons to a maximum concentration of 3.5×10^{21} at/cm^3 (as determined by referencing to a standard sample of protons implanted into silicon [6]). The sample was then depth-profiled with the SIMS instrument using Cs^+ bombardment and detection of both $^1H^-$ and $^{30}Si^1H^-$. By previously calibrating the instrument sensitivity for P in Si (from P-implanted samples) the intensity of the ^{30}SiH signal could be converted into an "equivalent phosphorus" signal.

A comparison of the H^- and SiH^- profiles Fig.1 shows that the hydrogen detection limit for phosphorus is directly related to the hydrogen concentration in the silicon. Furthermore, it indicates that in order to depth profile P in Si to 5×10^{16} at/cm^3, the hydrogen content must be $<5 \times 10^{18}$ at/cm^3. Thus, at a sputtering rate of 5 Å/sec, the partial pressure of hydrogen-containing species must be $<5 \times 10^{-10}$ torr.

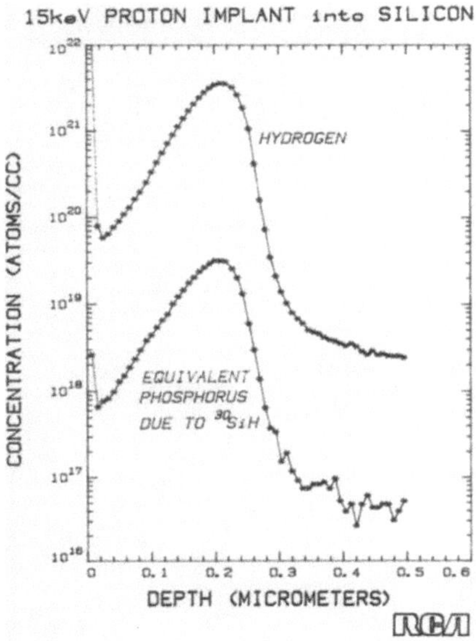

Fig.1 H^- and $^{30}SiH^-$depth profiles of a 15 keV proton implant in Si using Cs^+ bombardment. The $^{30}SiH^-$ profile intensity has been adjusted to represent the equivalent concentration of $^{31}P^-$. Sample chamber pressure: 4 x 10^{-10} torr. No background has been subtracted from either profile

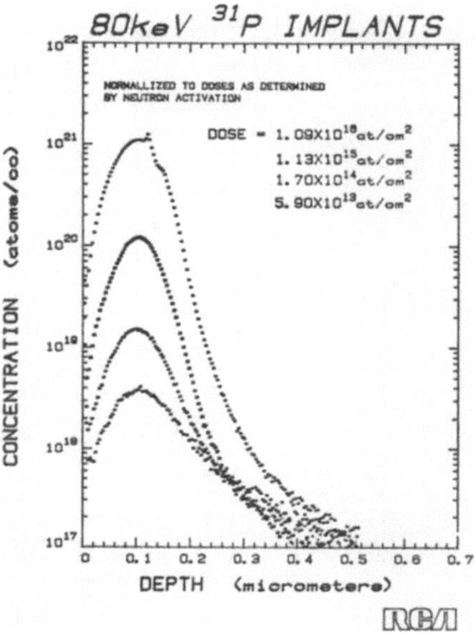

Fig.2 $^{31}P^-$ depth profiles of 80 keV phosphorus implants [7] into silicon. Profiles were normalized to the doses as determined by neutron activation analysis

To test the phosphorus depth profiling performance of the instrument, 80 keV P implants into silicon [7] were analyzed. The nominal doses were 1 x 10^{13}, 1 x 10^{14}, 1 x 10^{15} and 1 x 10^{16} at/cm^2. The actual doses were determined by neutron activation analysis. Profiles of these samples Fig.2 show good sensitivity, depth resolution and dynamic range.

Figure 3 shows the depth profile of another 80 keV P implant into silicon, but the sample was also implanted with B at 80 keV. Both doses were 1 x 10^{14} at/cm^2. Under these same experimental conditions used to obtain the profiles shown in Fig.2, and by monitoring the molecular ion $^{11}B^{28}Si^-$ at mass 39, simultaneous B and P depth profiling in Si was demonstrated. Fig.3 shows sensitivity for both dopants in the low 10^{16} at/cm^3 range. Fig.4 shows how this capability can be used to determine electrical junction depths (depth at which the concentrations of both n- and p-type dopants are equal) directly from the SIMS profile once a calibration sample has been analyzed. Angle-lapping and staining techniques were also used to determine the junction depth of this sample and the two measurements agreed to better than 10%.

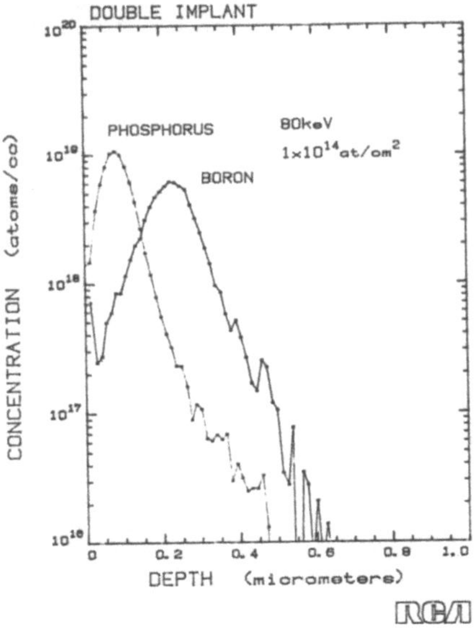

Fig.3 Depth profile of a ^{11}B and ^{31}P double implant into silicon. Both B and P were 80 keV, 1 x 10^{14} at/cm^2. Detected ions: $^{11}B^{28}Si^-$ and $^{31}P^-$

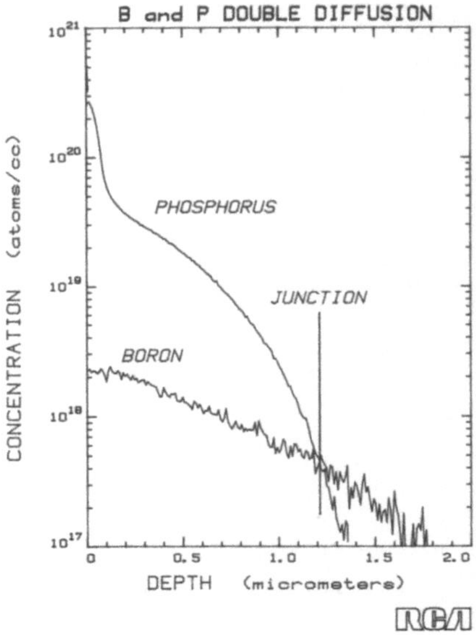

Fig.4 Example of using simultaneous B and P depth profiling to determine electrical junction depths. Concentration scales were established by analyzing the double implant standard shown in Fig.3

3. Conclusions

The use of Cs^+ ion bombardment in conjunction with negative secondary ion detection has been shown to be capable of sub-part-per-million sensitivity for P in Si in a depth profiling mode of operation. Furthermore, by monitoring the ion $^{11}B^{28}Si^-$ together with $^{31}P^-$, both B and P can be analyzed simultaneously. Thus, electrical junction depths can be determined directly from the SIMS depth profiles.

References

1. J. M. Morabito and J. C. C. Tsai, Surf. Sci. 33, 422 (1972).
2. R. D. Fralick, J. Vac. Sci. Technol. 13, 388 (1976).
3. C. W. Magee, W. L. Harrington and R. E. Honig, Rev. Sci. Instrum. 49, 477 (1978).
4. H. A. Storms, K. F. Brown and J. D. Stein, Anal. Chem. 49, 2023 (1977).
5. C. W. Magee, J. Electrochem. Soc. 126, 660 (1979).
6. J. F. Ziegler, et.al. Nucl. Instrum. Methods 149, 19 (1978).
7. R. E. Dobrott, private communication, (1979).

Chromium and Iron Determination in GaAs Epitaxial Layers

A.M. Huber, G. Morillot, P. Merenda, N.T. Linh
Laboratoire Central de Recherches
THOMSON-CSF
91401 Orsay, France

The characteristics of gallium arsenide metal semiconductor field effect transistors (MESFET's) are mainly determined by the interface region between the epitaxial layer and the Cr-doped GaAs semi-insulating substrate. A number of adverse effects associated with this interface are eliminated and device performances are improved by the insertion of a high resistivity buffer layer between the substrate and the active layer. It has also been suggested that more reliable results can be obtained by doping the buffer layer with a low concentration of deep level elements such as chromium or iron for compensating the residual donor or acceptor levels.

For a rapid optimization of the growth conditions of such device material, we have carried out a quantitative study of these elements in the epitaxial layers by the secondary ion mass spectrometry (SIMS).

SIMS has already been successfully applied to the determination of the concentration profiles of several impurities in gallium arsenide. For interface studies one needs a detection limit of about 5×10^{14} atoms/cm^3. Unfortunately, the detection limit of shallow donors and acceptors is not sufficiently low despite the recently developed Cs^+ ion bombardment technique. Nevertheless, the quantitative analysis of the three most important deep level elements, Cr, Fe, O, is already possible with a detection limit of 5×10^{14} atoms/cm^3.

Quantitative analysis is performed with the well known CAMECA IMS 300. Standard samples in which chromium and iron concentrations are determined by spark source mass spectrometry (SSMS) are first measured. High quality single crystals have been selected for the standards: their etch pit density was less than 10^{14} cm^{-2} and the homogeneity of chromium and iron was carefully checked by SIMS. The conditions of analysis are as follows: 3µA O_2^+ primary ion beam accelerated by a 5.5 keV potential scanned over a surface area of 1.5x1.5mm, the ions collected for the analysis come only from the center of the eroded crater which has a surface area of ∿300µm in diameter. The SIMS instrument maintains a pressure of 1×10^{-7} torr. For the analysis, $^{52}Cr^+$, and for the internal ion intensity reference, $^{75}As^+$, are respectively used. The chromium relative signal measured by SIMS is plotted versus chromium concentration determined by SSMS in four different Cr doped GaAs samples (Fig.1). Each point on the curve represents the average of six measurements taken on a standard sample. The unknown Cr concentration in GaAs can be determined from this curve. For the iron determination only one crystal containing iron was used for reference, due to the lack of samples. We made the assumption that the concentration diagram is linear as with $^{52}Cr^+$. The measurement accuracy is limited by SSMS standards measurements (±20%).

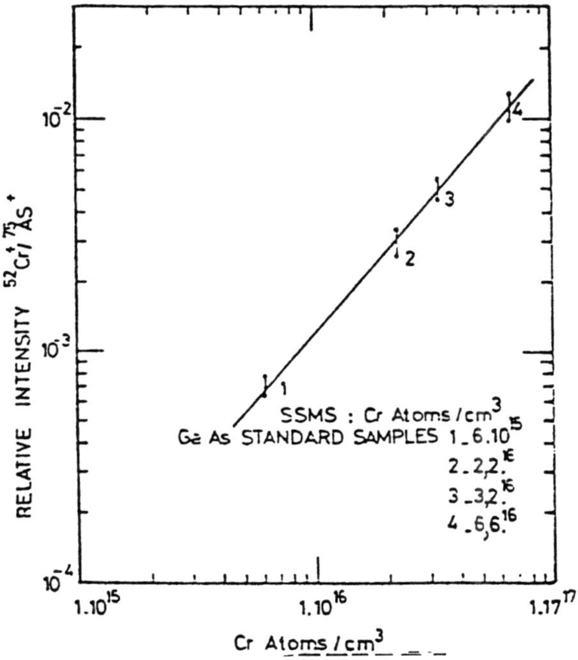

Fig.1 Relative $^{52}Cr^{+}/^{75}As^{+}$ signal versus Cr concentration in GaAs determined with SSMS

Chromium profiles were determined in a large number of epitaxial layers (~60 samples), doped or undoped, with and without buffer layer, prepared by various methods. Fig.2 shows typical profiles of Cr in two vapor phase

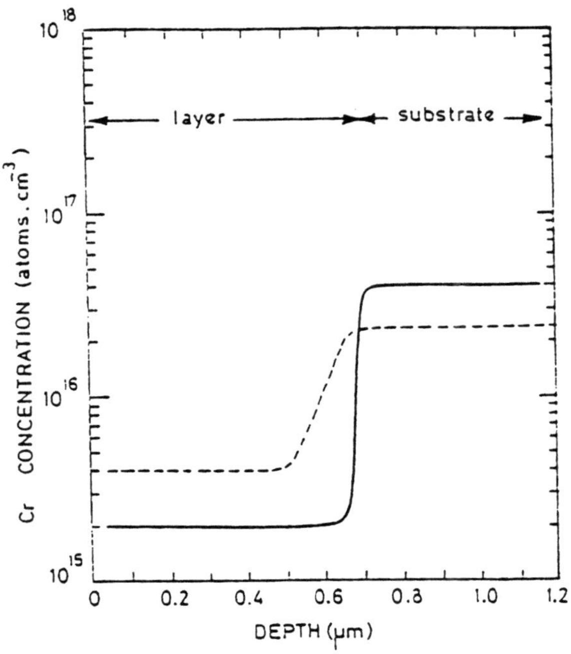

Fig.2 Cr profiles, in GaAs epitaxial layers. VPE in Ga/$AsCl_3/H_2$ system (----) in TMG system (——)

epitaxial layers (VPE). The dashed line curve is a Cr profile of an undoped layer grown in a $Ga/AsCl_3/H_2$ process. The second layer (continuous line) was prepared by trimethyl gallium process (TMG). From the results we can draw the following conclusions:

- if the diffusion coefficient of Cr is taken to be: $4.3\times10^3 \exp(-3.4/KT)$, the tail of outdiffusion in both VPE systems is in fairly good agreement with this data. Note the low chromium outdiffusion from the substrate into the layer. (∼1000 Å for a 1μm thick layer in the $Ga/AsCl_3/H_2$ system.)
- However, a background chromium level, probably existing in the growth system, contributes to the compensation of buffer layers. It can be detected even in layers grown on N^+ substrates in which very low Cr level is found. Expitaxial gallium arsenide can be controllably doped with chromium or iron.

Figures 3 and 4 present, respectively, the SIMS results obtained in a TMG system grown chromium doped layer and on an iron doped layer grown in a Ga/ $AsCl_3H_2$ system.

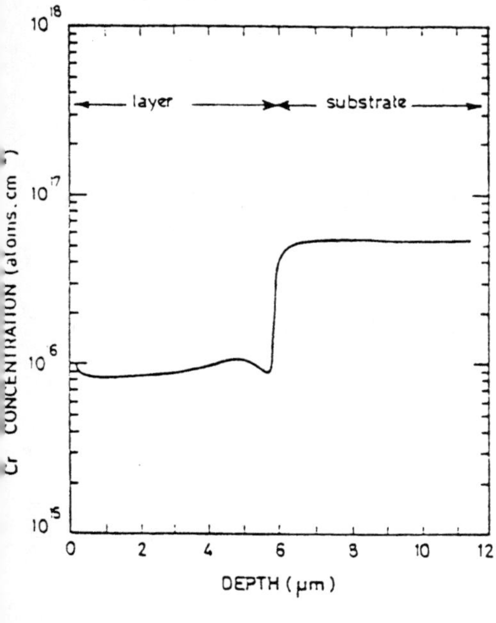

Fig.3 Cr profile in GaAs epitaxial layer doped with this element (VPE-TMG system)

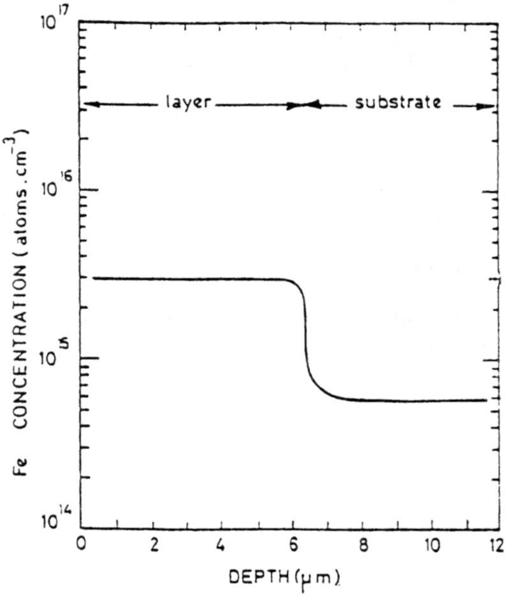

Fig.4 Fe profile in GaAs epitaxial layer doped with this element (VPE $Ga/AsCl_3/H_2$ system)

The chromium profile of a layer grown by molecular beam epitaxy (MBE) (Fig.5) shows a Cr accumulation at the interface and a zone where the Cr concentration is depleted. This peak is always much higher when the substrate is not cleaned by in situ ion sputtering. The profile of this figure shows the situation after one hour of sputtering. A simple model of diffusion, like that used for VPE, would not explain the high Cr accumulation at the interface.

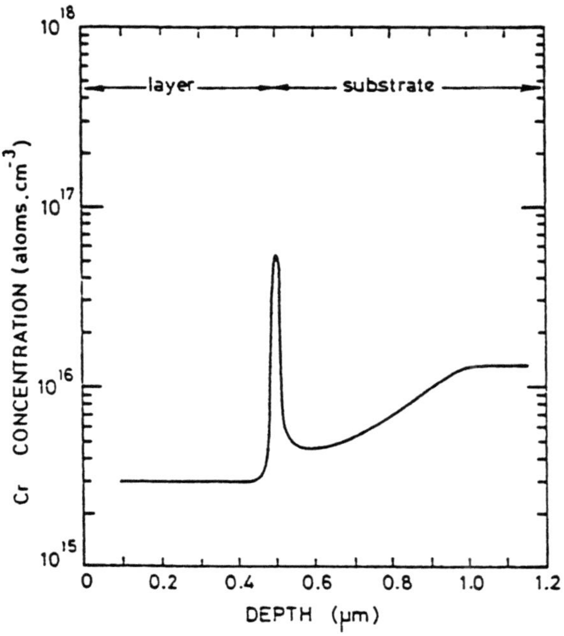

Fig.5 Cr profile in GaAs epitaxial layer. MBE

At present SIMS is the only quantitative chemical method allowing such an accurate study of the deep level elements in GaAs epitaxial layers.

Acknowledgement

The authors wish to express their thanks to Mr. Duchemin for providing VPE samples and valuable discussion.

Thermal Redistribution of Cr in GaAs Due to Damage, Stress and Concentration Gradients

C. A. Evans, Jr., and V.R. Deline
Charles Evans & Associates
1670 S. Amphlett Blvd., Suite 120
San Mateo, CA 94402

T.W. Sigmon
Stanford Electronics Laboratory
Stanford University
Stanford, CA 94305

Abstract

The recent availability of a high sensitivity secondary ion mass spectrometer, the CAMECA IMS-3f, has permitted the direct observation of Cr and its movement in GaAs. The high sensitivity (3×10^{15} at-cm^{-3}) has fostered a variety of studies into the role and redistribution of Cr added to GaAs to render it semi-insulating. The results of these studies provide an insight into phenomena heretofore not understood.

1. Cr has been observed to getter into residual damage left after annealing of Se ion implanted SI GaAs ($\geq 1 \times 10^{14}$ Se-cm^{-2}) [1].

2. The near-surface stress caused by thermal expansion mismatch between the annealing encapsulant and the GaAs can getter Cr to the extent that the Cr level is depleted for significant distances into the bulk substrate [1].

3. ASBECK, et al., [2], have quantitatively correlated the depletion of Cr with the de-compensation of residual donors, the resultant electrical profile and the performance of a GaAs FET device [2].

4. Pre-growth gettering of Cr into back-surface damage reduces the Cr outdiffusion into as-grown VPE GaAs layers and results in a higher quality epi-layer and materials performance [3].

5. The Cr from the substrate outdiffuses into LPE layers on SI GaAs substrates during only moderate thermal processing (850°C for 30 minutes) and can influence the devices grown on that epi-layer [4].

This presentation will provide an unified overview of these data and more recent results on the role and redistribution of Cr in GaAs.

Acknowledgement

This research was supported in part by Defense Advance Research Projects Agency Contract DARPA MDA 903-78-C-0290 (R. Reynolds).

References

1. C.A. Evans, Jr., V.R. Deline, T.W. Sigmon and A. Lidow, App. Phy. Lett., Accepted for Publication (August, 1979 issue).
2. "Chromium Redistribution In Annealed Semi-Insulating GaAs," P. Asbeck, J. Tandon, D. Siu, R. Fairman, B. Welch, C.A. Evans, Jr., and V.R. Deline, Presented at the Device Research Conference; to be published.
3. "Back Surface Gettering and Cr Outdiffusion in VPE GaAs Layers," T.J. Magee, J. Peng, J.D. Hong, C.A. Evans, Jr., V.R. Deline and R.M. Malbon, App. Phy. Lett., Submitted for Publication.
4. "Outdiffusion of Cr from SI GaAs into LPE GaAs Layers," J. Gladstone, V.R. Deline, C.A. Evans, Jr., and T.W. Sigmon, manuscript in preparation.

On-Line Sputter Rate Measurements During SIMS, AES Depth Profiling

J. Kempf
IBM Deutschland GmbH
Sindelfingen, Germany

In high-sensitivity depth profiling techniques like SIMS, AES, ESCA, ISS, etc., the continuous measurement of sputter rate or sputtered depth during sputtering becomes increasingly important. Established on-line techniques for sputter rate measurements based on optical and electro-mechanical sensing are less suited because of the extensive calibration and sample preparation, the low depth resolution and the restricted availability with respect to different material. In this paper a simple electro-optical technique is used as an on-line method for measuring the sputtered depth during SIMS/AES depth profiling. Its optical arrangement is shown schematically in Fig.1.

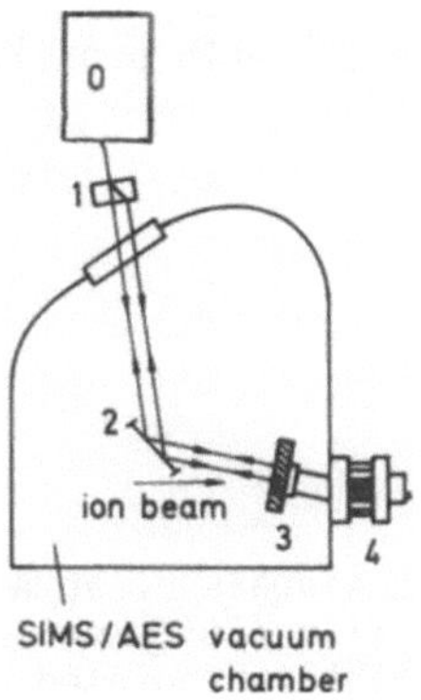

Fig.1 Experimental arrangement used in measuring sputtered depth during SIMS/AES depth profiling. 0: High resolution laser interferometer, 1: Calcite crystal, 2: Plane mirror, 3: Specimen surface, 4: Specimen manipulator

The method is based on two laser beams which are reflected from the sputtered and the unsputtered area, respectively. The phase difference ϕ_M between the two reflected beams changes with changing of the sputtered depth σ due to

$$\phi_M = 4\pi\frac{\sigma}{\lambda} \qquad (1)$$

where λ = the wavelength of the photon employed (λ_{HeNe} = 632.8 nm)

ϕ_M is measured with a high resolution electro-optical signal evaluation technique described in [1] (details to be published in the near future). This technique offers measurements of extremely high depth resolution (<1 nm) and shows no disadvantages mentioned above.

Figure 2 shows a typical plot of the sputtered depth σ vs. sputtering time measured during the Ar^+ bombardment of homogeneous Si. The relative sputt rate $(d\sigma/dt)/i_{abs}$ evaluated from $\sigma(t)$ in the beam current range between 0. and 2 µA was found to be constant within 1.5%. The depth resolution quote here is limited to about 1 nm by sample vibration. The intrinsic depth resolution using the electro-optical technique is better than .1 nm.

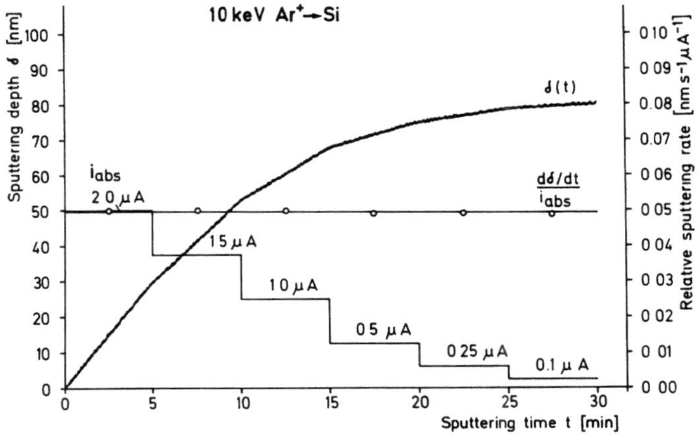

Fig.2 Sputtering depth plotted vs. sputtering time. A raster scanned Ar^+ beam of 10 keV impa energy at different abs lute beam current i_{abs} used for sputtering hom geneous Si

The advantage of the electro-optical method used in depth profiling of inhomogeneous material is illustrated in Fig.3.

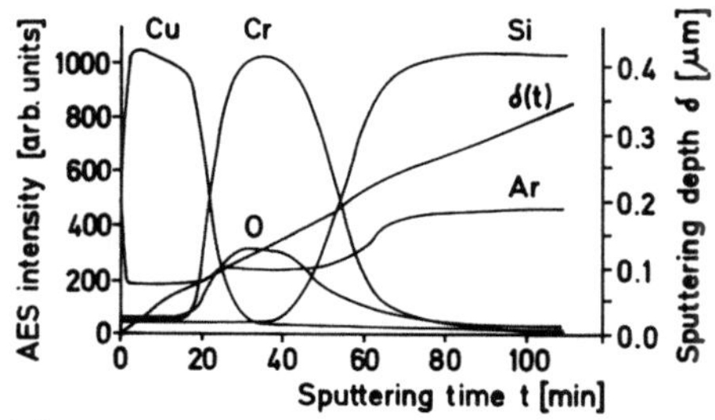

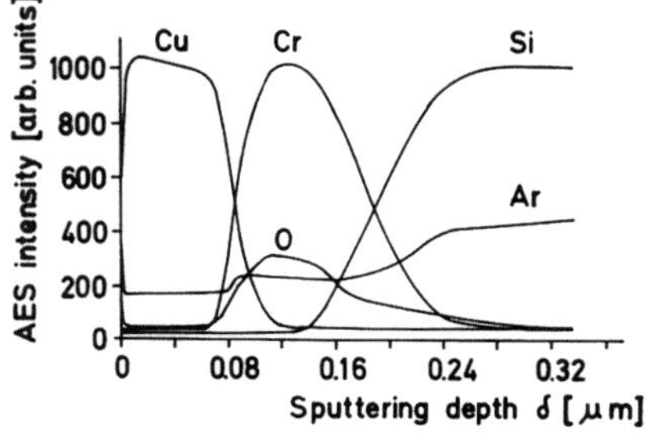

Fig.3a AES signals of an Ar^+ bombard multilayer film (Cu(90 nm thick) and Cr(90 nm thick) evaporated on Si substrate) vs. sputtering time

Fig.3b Corrected AES depth profiling obtained by using the $\sigma(t)$ curve in Fig.3a

Fig.3a shows an uncorrected AES depth profile of the multilayer film by assuming a constant sputter rate (film thickness ratio of Cr and Cu is 1.5 The corrected AES-depth profiling in Fig.3b was obtained by using the elec optical measured $\sigma(t)$ in Fig.3a.

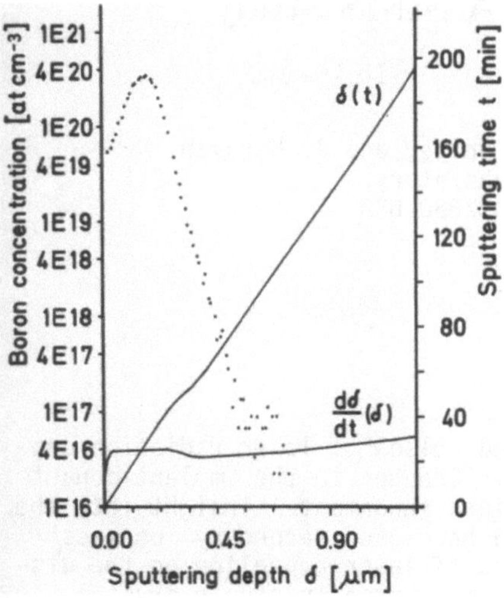

Fig.4 SIMS depth profiling of 40 keV ^{11}B implanted and annealed Si(100) correlated with the electro-optical measured depth profiling of the sputter rate dσ/dt

In Fig.4 the electro-optically determined depth profile of the sputter rate dσ/dt was correlated with a SIMS depth profile of 40 keV ^{11}B implanted and annealed Si(100).

The sputter rate up to 1300 nm sputtered depth shows variations due to surface effects and lattice distortions caused by incorporated B in Si ($\gtrsim$ 3 x 10^{18} at/cm^{-3}). The increase with the sputtering depth likely is an instrumental effect which is not well understood up to now.

The electro-optical method used above represents an excellent technique for high sensitivity and accurate measurements of sputtered depth during the sputtering process. The new technique is of great importance for SIMS/AES depth profiling and will be useful for fundamental studies of the sputtering process.

References

1. G. Makosch, B. Solf, IBM pat. appl., disclosure no. 877197, doc. no. 2851750.

Laser Induced Redistribution of Ion Implanted and Surface Deposited B in Silicon: A SIMS Study*

W.H. Christie, R.J. Warmack, C.W. White, and J. Narayan
Oak Ridge National Laboratory
Oak Ridge, Tennessee 37830 USA

1. Introduction

Recently, it has been shown that high powered pulses of laser radiation can be used to anneal ion implanted silicon [1]. Changes in the implant dopant profile as a result of laser annealing provides fundamental insight into the laser annealing mechanism. Consequently, we have used secondary ion mass spectrometry (SIMS) to investigate the effect of laser annealing on the distribution of ion implanted and surface deposited boron in single crystal silicon.

2. Experimental

To facilitate SIMS depth profiling, the samples were cut into chips approximately 2 mm on a side. The individual specimens were ultrasonically solvent cleaned, affixed to a flat conducting sample mount and overcoated with 20-30 nm of spectrographic grade carbon. The primary beam used was O_2^+ at 16.0 keV impact energy, 15×10^{-9} A beam current and a 5 µm beam diameter. The beam was raster scanned (50x40 µm) and $^{11}B^+$ sputtered ions were detected from the central 10-15% of the rastered area using an electronic aperturing technique. Depth and concentration scales were established by profiling ion implantation standards prepared at 35 keV energy at doses of 10^{14}, 10^{15} and 10^{16} ^{11}B atoms cm^{-2}. Laser annealing was carried out in air using the output of a Q-switched ruby laser (λ = 0.694 µm, 60 ns pulse duration).

3. Results and Discussion

The result of laser annealing (1.6 J cm^{-2}, 60 ns) silicon implanted to different boron dose levels in the range 10^{14} to 10^{16} cm^{-2} is shown in Fig.1. The as-implanted profiles are shown to establish the depth and concentration scales. The extent of boron migration and the shape of the final profile is almost the same for crystals implanted to 10^{15} and 10^{16} cm^{-2}, but is somewhat less at a dose of 10^{14} cm^{-2}. The increased redistribution at doses of 10^{15} cm^{-2} and greater is believed related to more efficient absorption of laser light due to increased damage and/or dopant concentration in these crystals. The absorption coefficient at the ruby wavelength is a strong function of the damage in silicon crystals, being about an order of magnitude higher in

---

*Research sponsored by the U.S. Department of Energy under contract #W7405-eng-26 with the Union Carbide Corporation.

amorphous silicon as compared to single crystal silicon. In addition, the dopant concentration should increase the absorption coefficient due to free carrier absorption [2]. Integration of the profile peak areas for the as-implanted samples and the corresponding laser annealed samples indicates that no appreciable boron is lost from the silicon during the laser annealing process.

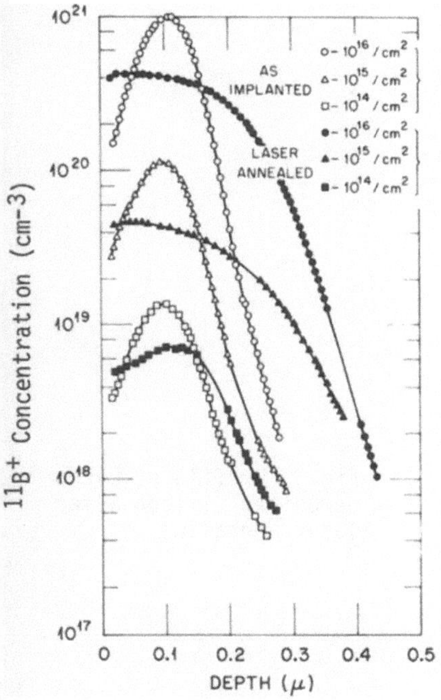

FIG 1. BORON PROFILES IN SILICON IMPLANTED TO DIFFERENT DOSES

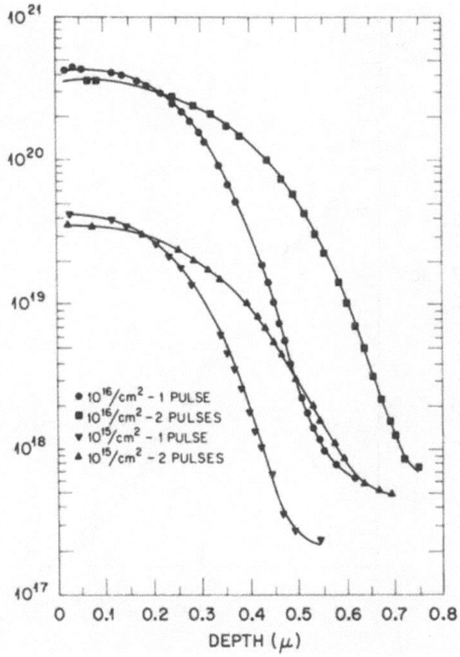

FIG 2. BORON PROFILES IN SILICON AFTER ANNEALING WITH SUCCESSIVE LASER PULSES

Figure 2 shows the relationship of the dopant profile on the number of successive laser pulses. This is illustrated for silicon implanted with boron to doses of 10^{15} and 10^{16} cm^{-2} and annealed with one and two successive laser pulses (1.7 J cm^{-2}, 60 ns). Most of the redistribution takes place during the first pulse, but additional redistribution is observed during the second pulse. A third pulse produced no significant further broadening of the profile. These results show that the crystal can be remelted after the first laser pulse. This may occur due to increased free carrier absorption by the implanted dopant. Calculations [3] show that for pure single crystal silicon the absorption is too low for melting to occur under these conditions.

Another application involves the use of laser irradiation to form large area p-n junctions in silicon by a process of laser induced diffusion [4]. For this application, the dopant is deposited on the surface as a film approximately 100 Å thick by evaporation. The deposited surface is then irradiated with pulsed laser light and during the time the near surface region is melted, the deposited dopant diffuses into the crystal and is made electrically active by being incorporated into substitutional lattice sites. The profile of boron obtained after laser irradiation of boron deposited silicon is given in Fig.3

and shows the penetration of boron to a depth of several thousand angstroms. This appears to be a very attractive method for fabricating shallow p-n junctions because the depth of melting and subsequent dopant diffusion can be carefully controlled by the laser energy density.

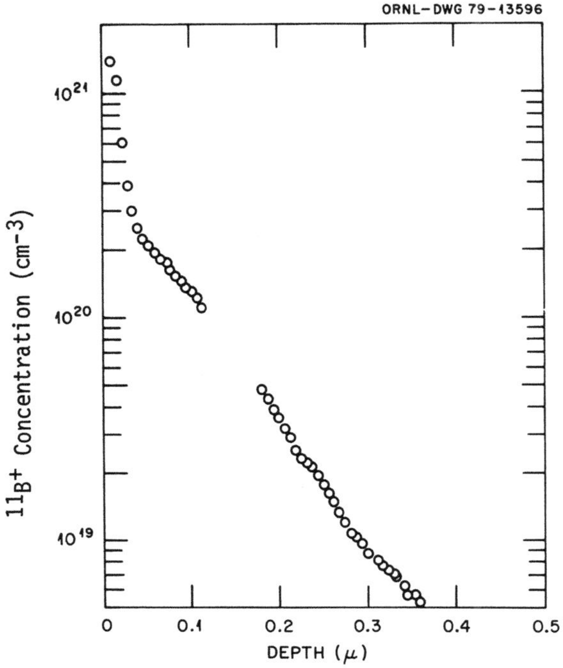

Fig.3 Profile of boron deposited silicon after laser annealing

4. Conclusions

SIMS investigations have shown that pulsed laser annealing of ion implanted and surface deposited silicon leads to significant dopant redistribution. Comparison of calculated profiles to SIMS determined profiles provides strong evidence that the laser annealing mechanism involves melting of the crystal, diffusion of dopants in liquid silicon, followed by epitaxial regrowth [5].

References

1. R.T. Young, C.W. White, G.J. Clark, J. Narayan, W.H. Christie, M. Murakami, P.W. King and S.D. Kramer, Appl. Phys. Lett. 32, 139 (1978).
2. A.A. Grinberg, R.F. Mekhtiev, S.M. Ryvkin, V.M. Salmanov and I.D. Yaroshetskii, Sov. Phys.-Solid State 9, 1085 (1967).
3. R.F. Wood, private communication.
4. J. Narayan, R.T. Young, R.F. Wood and W.H. Christie, Appl. Phys. Lett., 33, 338 (1978).
5. J.C. Wang, R.F. Wood, and P.P. Pronko, Appl. Phys. Lett. 33, 455 (1978).

SIMS Identification of Impurity Segregation to Grain Boundaries in Cast Multigrained Silicon

L.L. Kazmerski, P.J. Ireland and T.F. Ciszek
Photovoltaics Branch
Solar Energy Research Institute
Golden, CO 80401, USA

1. Introduction and Experimental Details

The application of secondary ion mass spectroscopy (SIMS) and complementary Auger electron spectroscopy (AES) to the comparative compositional analysis of grain and grain boundary regions in multigrained silicon is presented. The method incorporates in-situ ultrahigh vacuum fracturing of the material with the surface sensitive analytical techniques to delineate differences in elemental composition and chemistry between the inter- and intra-grain region. By this analysis, the first direct physical evidence for the localization of impurities at the grain boundaries in silicon grown by casting [1] and the related directional solidification technique [2] is provided.

The multigrained silicon used in this study was obtained from three different sources. Two of these (termed Si-A and Si-B) were produced by a "conventional" casting process in which the silicon was molten when poured into its shaping crucible held slightly below the melting point of silicon. The third sample type (termed Si-C) was produced by the directional solidification process.This differs from casting in that solid silicon was loaded into the crucible and subsequently heated to the molten phase. The cooling and cooling rate were precisely controlled to provide optimum grain size (1-10 mm diameters) and structure. Both carbon and Al_2O_3-based crucibles were used to form multigrained silicon ingots which were then sliced into thin (<1 mm thick) sheets.

The SIMS and complementary surface investigations were performed in a Physical Electronics Industries Model 590 scanning Auger microprobe-based system. The SIMS utilized a differentially-pumped ion gun with a focused beam diameter of 140 µm. An Extranuclear Laboratories quadrupole mass analyzer (QMA) provided 1-1000 AMU elemental and molecular species analysis. The minimum beam diameter of the SAM was measured to be 1600Å. The samples were inserted using an introduction/transfer stage which preserved the UHV chamber conditions, minimizing contamination to this volume. A sample fracture stage provided for the in-situ ultrahigh vacuum exposure and comparison of both intragrain and grain boundary areas. A typical fracture is shown in Fig. 1 in which both the grain (labeled b) and grain boundary (labeled a) are observable and available for side-by-side analysis under identical conditions. By the UHV fracturing, potential contamination to the inner sample surfaces from sorbed species is minimized, thus ensuring that the resulting data were not artifacts of the experiment. The chamber environment was continuously monitored during the experiment with a separate QMA having residual gas analysis capability.

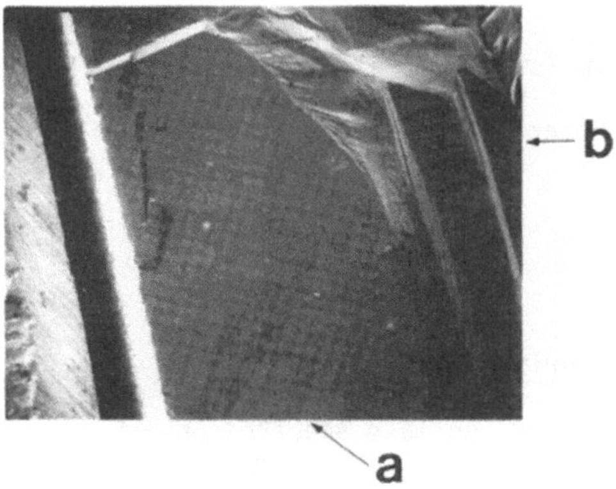

Fig.1 SED Image of Fractured Sample: (a) Grain Boundary; (b) Grain

2. Surface Analysis Results

This investigation emphasizes the necessity of using two complementary surface analysis techniques for the unambiguous solution of the impurity segregation problem. Each can contribute its inherent diagnostic quality (e.g., submicron spatial resolution, non-destructive analysis, monolayer surface sensitivity, quantifiable data and nearly uniform sensitivity for SAM; increased sensitivity, usually 100 times better than Auger, isotope and molecular fragment identification, hydrogen and trace analysis for SIMS).

Auger elemental mapping of the fractured Si is utilized to give preliminary evidence for the localization of impurities at the grain boundaries. The Auger maps indicate some build-up of impurities (including Ni, Al, C and O) at the grain boundaries, with none detectable in the grain interiors. High resolution AES analysis indicates that the impurity localization varies over the grain, with both alloys and separate phases present. SIMS was performed on these same regions without moving or otherwise disturbing the sample. Fig. 2 presents a typical SIMS spectrum for a region fractured through a grain (i.e. region b in Fig. 1). A controlled oxygen leak ($\sim 10^{-7}$ torr) was used to enhance secondary ion yields, providing increased sensitivity for trace impurity detection. The SIMS spectrum indicates the primary presence of Si; the only impurity is the intentional dopant. The oxides result from the oxygen leak, and Na and K from the inevitable inclusion of the top and/or bottom surfaces of the thin samples (previously exposed to atmosphere) in the SIMS analysis. In contrast, many impurities are observed in fractures at the grain boundary, indicated in the SIMS spectrum of Fig. 3. These data indicate the presence of Ti, Cu, B and Mg in addition to the C, Ni and Al detected by AES. None of these impurities appear in the corresponding analysis of the grain region.

SIMS and AES depth compositional profiles (perpendicular to grain boundaries) are presented to confirm the localization effects of these impurities. Differences in impurity localization along the grain boundary and from grain-to-grain in the same sample are documented. Preliminary data on the diffusion of P into this multigrained material are presented. Differ-

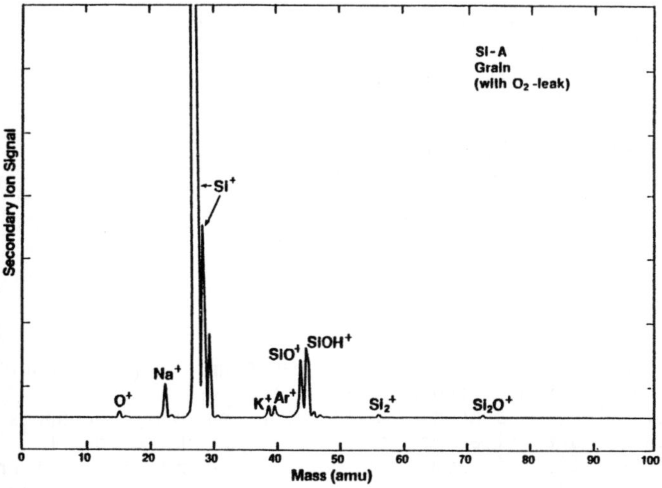

Fig.2 Positive SIMS spectrum of fractured grain region

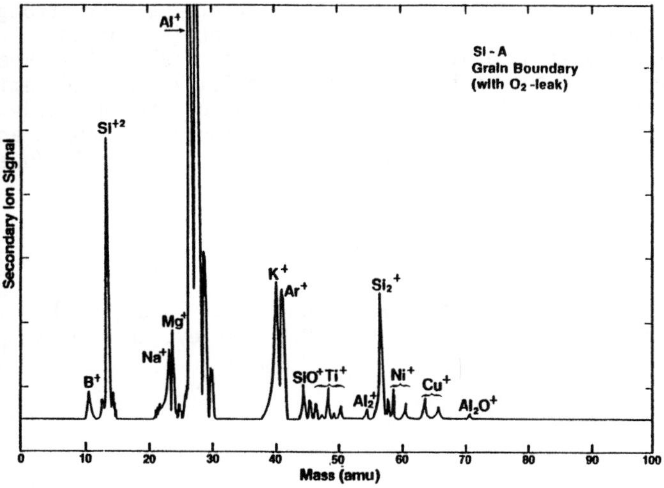

Fig.3 Positive SIMS spectrum of grain boundary

ences and similarities in the diffusion coefficients as determined by SIMS between the grain and grain boundary regions are discussed. Differences in the electrical activity of the grain boundaries in a given sample are correlated with the segregation evidence provided by the SIMS results.

References

1. J. Lindmayer, Proc. 13th IEEE Photovoltaics Spec. Conf., Washington, D.C. (IEEE, New York, 1978) pp 1096-1099.
2. T.F. Ciszek, G.H. Schwuttke and K.H. Yang, J. Cryst. Growth 46, 527 (1979).

SIMS Studies in Compound Semiconductors*

J.E. Baker and P. Williams
Materials Research Laboratory
University of Illinois
Urbana, IL 61801

D.J. Wolford, J.D. Oberstar and B.G. Streetman
Coordinated Science Laboratory
University of Illinois
Urbana, IL 61801

Abstract

In recent years, the SIMS technique has approached significantly closer to the desired role as a universal microanalytical technique with part-per-million sensitivity. Initial successes with boron and other p-dopants in silicon demonstrated the analytical power of the technique for depth profiling of dopant distributions in semiconductors. With the introduction of the cesium ion source [1,2], electronegative elements became accessible to analysis. Recent work has been directed towards extending the applicability of the technique to those elemental species which are neither strongly electronegative or electropositive and exploiting the capability of the technique to perform multi-element analysis in a given material. This paper will review recent progress in two such areas involving III-V compound semiconductors: analysis of nitrogen in gallium arsenide and gallium arsenide phosphide and studies of the redistribution of both the implanted dopant and the bulk compensating dopant during annealing of Be^+ ion-implanted, Fe-doped indium phosphide.

References

1. H.A. Storms, K.F. Brown and J.D. Stein, Anal. Chem. 49, 2023 (1977).
2. P. Williams, R.K. Lewis, C.A. Evans, Jr., and P.R. Hanley, Anal. Chem., 49, 1399 (1977).

*Work supported in part by the National Science Foundation MRL Grant DMR-77-23999, Joint Service (U.S. Army, U.S. Navy, U.S. Air Force) under contracts DAAG-29-78-C-0016 and N00014-79-C-0424.

Characterization for Composition and Uniformity of MCVD Glass Film by Secondary Ion Mass Spectrometry (SIMS)

D.L. Malm
Bell Laboratories
Murray Hill, New Jersey 07974

G.W. Tasker
Department of Materials Science & Engineering
Massachusetts Institute of Technology
Cambridge, Massachusetts 02138

1. Introduction

Inorganic oxide glasses coated with organic polymers for optical fiber applications have been studied vigorously for the past few years. Among the most important but least understood aspects of these materials are the structure and chemistry at the interfaces between the dissimilar phases. The first part of a study [1] proposed to further the understanding of the physio-chemical nature of the glass-polymer interface involves examination of the surfaces of high purity glasses of known chemical composition. The B_2O_3-SiO_2 system was chosen for this purpose for its simplicity and because films of high purity can be fabricated via modified chemical vapor deposition (MCVD) [2]. Some modification of the MCVD process is required during the final phase of deposition in order to provide surfaces acceptable for study. To insure that the composition and uniformity of the MCVD film is not subsequently altered from that predicted from the vapor composition used for film deposition, a composition-depth profile is desirable. Secondary ion mass spectroscopy is used in this study to monitor changes in the Si-B ratio, thus providing a measure of the quality of the MCVD films. For comparison, Si/B ratios were also measured for boro-silicate glasses of similar composition fabricated by RF plasma fusion. The silicon and boron contents of the respective melts were determined by atomic absorption spectroscopy and the uniformity of the boule was checked by X-ray fluorescence. These samples facilitated calibration of the SIMS measurements, thus enabling quantitative comparisons to be made of the near-surface composition of glasses fabricated by other techniques. In this work we report those results and the experimental details of the SIMS technique.

The SIMS apparatus used in this experiment was constructed at B.T.L.; a detailed description is given in the literature [3]. The ion beam conditions used for both sputtering and analysis were 1.9 keV, ^{40}Ar at a nominal current density of 10 $\mu A/cm^2$ on a conducting substrate. Charge neutralization at the glass surface was supplied by a heated filament positioned near the sample surface.

2. Results and Discussion

The SIMS composition-depth profiles for both RF plasma glasses and MCVD film are plotted in Figs.1 and 2, respectively. The Si/B ratio measurements used to compute the data are expressed in terms of B_2O_3 content and plotted as concentration vs. depth (Å). The points used to define the curve represent the average values computed from data taken from at least three undisturbed spots at the surface. The error bars mark the limit of the maximum spread.

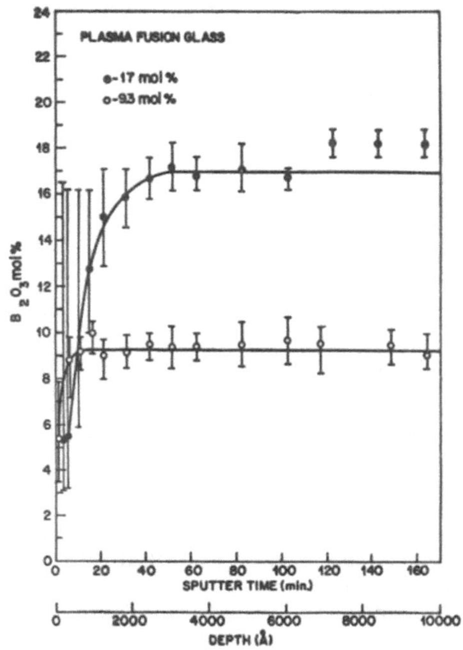

Fig.1 SIMS, composition-depth profile of B_2O_3 concentration for R.F. Plasma Fusion glass, BSS-2 (9.3 mol%) and BSS-3 (17 mol%)

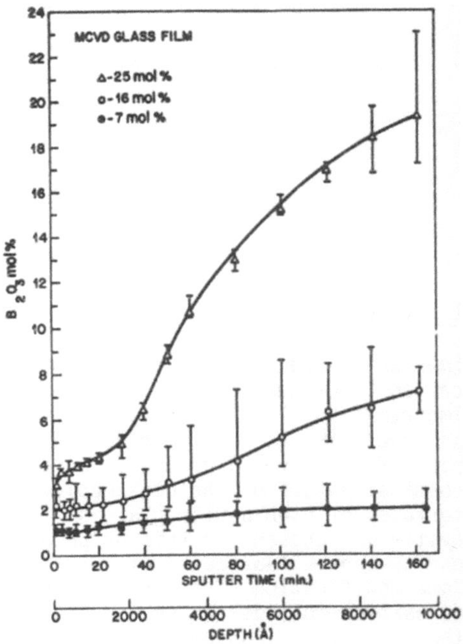

Fig.2 SIMS, composition-depth profile of B_2O_3 concentration for MCVD glass film BS-2 (7 mol%), BS-4 (16.0 mol%) and BS-5 (25 mol%)

The nominal sputter rate of 59 Å/min for borosilicate glass was determined independently by optical interferometry measurement of the depth of a sputter crater produced in an optically flat surface by the ion beam under conditions similar to those used for sample analysis. A comparison of the profiles measured for the respective glasses (Figs.1 and 2) show a significant difference in the B_2O_3 distribution with depth. In the case of the PF glasses (Fig.1) the average B_2O_3 concentration is low at the onset of sputtering and increases rapidly to a steady value that corresponds to the bulk composition determined by atomic absorption spectroscopy. The surface inhomogeneity can be attributed to a chemical change caused by atmospheric reaction at the air-solid interface. The B_2O_3 concentration at the MCVD film surfaces (Fig.2) is also low initially, but the profile remains relatively flat and at no point does the concentration reach the calculated value. The B_2O_3 deficiency observed here in comparison with the PF glasses is not clearly understood. It may be that insufficient control of the MCVD processes during the last stage of sample preparation results in significant loss of B_2O_3.

It is clear from this experiment that the composition and uniformity of the MCVD film fabricated for this study does not compare favorably with that found for PF glasses. Secondary ion mass spectroscopy was shown to be sensitive to small changes in the Si-B content across a depth of at least 1 μm and is useful for the study of glass surfaces.

References

1. G.W. Tasker, The Physiochemical Nature of Glass-Polymer Interfaces - Research Proposal, Department of Material Science and Engineering, M.I.T., Cambridge, Mass., May 10, 1978.
2. W.G. French, J.B. MacChesney, P.B. O'Connor and G.W. Tasker, BSTJ 53, 951, (1974).
3. M.J. Vasile, D.L. Malm, Int. J. Of Mass. Spec. and Ion Phys. 21, 145 (1976).

SIMS Study of Metallized Silicon Semiconductors

K.L. Wang
General Electric Company, Corporate Research and Development
Schenectady, NY 12301

H.A. Storms
General Electric Company, Vallecitos Nuclear Center
Pleasanton, CA 94566

1. Introduction

Transition-metal silicides have been used extensively as contact layers in semiconductor applications [1,2]. These applications are becoming more important as electronic circuit complexity and scale of integration increase.

Elevated temperature annealing thin metal films on silicon substrates is generally used to form transition-metal silicides [1,3-5]. More recently, ion-implant-induced formation of silicides has been reported [5-8]. In this study, secondary ion mass spectrometry (SIMS) is used to investigate the formation of silicides resulting from P^+ implantation of thin metal films (i.e., Mo, Pd, and Ta) on silicon substrates.

2. Experiments

Polished <111> or <100> oriented silicon substrates were used. Thin films of Mo (1000Å), Pd (1000Å), and Ta (550Å) were evaporated onto silicon substrates. The metallized silicon wafers were implanted with 250-keV $^{31}P^+$ ions at controlled temperatures of -195, 25, and 150°C. At this energy, the peak of the damage distribution was near the interface. Implant flux density was kept below 1.3×10^{13} cm^{-2} s^{-1}; temperature changes during implants were <25°C. Implant fluence within the beam averaged 1×10^{17} P^+ cm^{-2}. Outside the area of the beam, P^+ fluence decreased rapidly with distance to $<1 \times 10^{15}$ cm^{-2}, as determined by SIMS. Regions interior and exterior to the P^+ beam area were depth profiled by SIMS.

A commercial ion microprobe mass analyzer (IMMA) under computer control was used for depth profile analysis. A 13.5-keV $O_2^=$ ion beam was rectangular-spiral rastered over ~5×10^{-5} cm^2 area of the sample surface while up to eight secondary-ion species were sequentially monitored. Secondary ion counts were collected from the central 10% of the raster. Effective O_2^+ current densities ranged between 0.04 and 0.08 mA/cm^2.

3. Results

Typical depth profile plots are given in Fig.1 and 2. Fig.1a shows the presence of Pd silicide thin film on the surface of a silicon substrate. The uniform intensities of Pd^+, Si^+, and $PdSi^+$ profiles in the thin film indicate a constant stoichiometry. However, lack of suitable standards inhibits accurate assessment of this stoichiometry; Pd_2Si, PdSi, or $PdSi_2$ are likely silicides. Diffusion toward the surface and long-range inward diffusion are indicated by the bulge on the left and extended tail on the right of

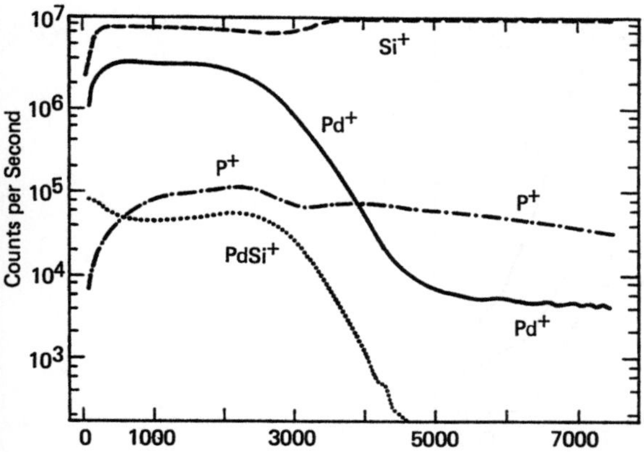

a. With 1×10^{17} P^+ cm^{-2}; depth profiled with 0.041 mA cm^{-2} O_2^+

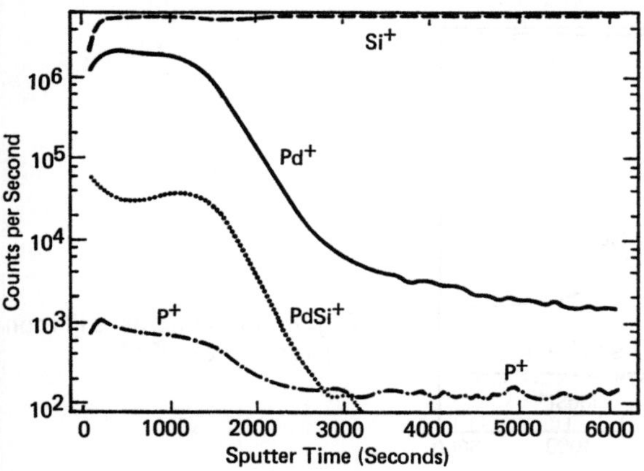

b. With $\sim 2 \times 10^{15}$ P^+ cm^{-2}; depth profiled with 0.051 mA cm^{-2} O_2^+

Fig.1 Pd film on silicon implanted at -195°C

the flat P^+ maximum. Little difference in stoichiometry or silicide thickness was observed for implants at 150 and -195°C.

Figure 1b shows comparable profiles to those in Fig.1a on a region of the same wafer outside the beam area, where P^+ implant fluence is $<2 \times 10^{15}$ cm^{-2}. Intensity ratios Si^+:Pd^+ and $PdSi^+$:Pd^+ are substantially equal to ratios observed for 1×10^{17} cm^{-2} implant fluence. Silicide formation is apparent. However, film thickness was only 80% the higher implant fluence value.

Molybdenum silicide formation resulting from 1×10^{17} cm^{-2} P^+ implantation at 25°C is shown in Fig.2a. Likely silicides are MoSi and $MoSi_2$. The P^+ profile structure indicates two directional diffusion. Film thickness of silicide formed at -195°C appears 60% greater than silicide formed at 25°C. Relative Mo^+, Si^+, and $MoSi^+$ intensities are roughly comparable.

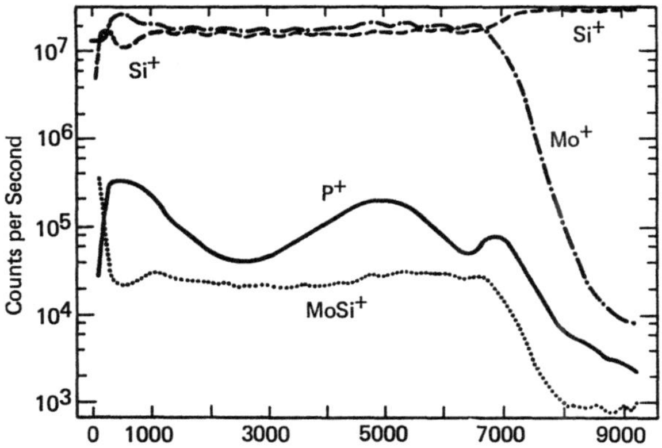

a. With 1×10^{17} P^+ cm^{-2}; depth profiled with 0.074 mA cm^{-2} O_2^+

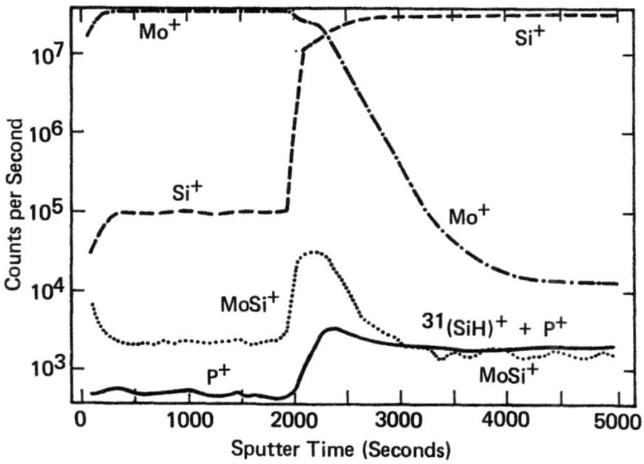

b. With ~2×10^{14} P^+ cm^{-2}; depth profiled with 0.074 mA cm^{-2} O_2^+

Fig.2 Mo film on silicon implanted at 25°C

At low P^+ fluence, possible formation of Mo silicide (Fig.2b) is indicated only in the immediate vicinity of the interface. Surface film thickness, in terms of O_2^+ exposure, is one-third that observed for high-implant fluence (Fig.2a). The major contributor to the mass 31 signal in the silicon substrate is $^{31}(SiH^+)$.

Formation of silicides resulting from P^+ implantation of Ta thin films (550Å) on silicon substrates was also studied. Variability of Si^+ and $TaSi^+$ intensities indicates complex silicide formation, perhaps two different stoichiometries. Long-range inward diffusion of P was observed. Reduced implant temperature appears not to affect silicide formation. At low P^+ fluence (e.g. $<1 \times 10^{15}$ cm^{-2}) possible silicide formation is observed only in the immediate vicinity of the interface.

References

1. J.M. Poate, et al., Thin Films - Interdiffusion and Reaction (Wiley, New York, 1978).
2. B.L. Crowder and S. Zirinsky, IEEE J. Solid State Circuits SC-14, 291 (1979).
3. J.M. Poate and T.C. Tisone, Appl. Phys. Lett. 24, 391 (1974).
4. H. Muta and D. Shinoda, J. Appl. Phys. 43, 1913 (1972).
5. D.H. Lee, et al., Phys. Statis Solidi (a) 15, 645 (1973).
6. W.F. van der Weg, et al., Application of Ion Beams to Metals, edited by S.T. Picraux, et al. (Plenum Press, New York, 1974), p. 209.
7. B.Y. Tsaur, et al., Appl. Phys. Lett. 34, 168 (1979).
8. K.L. Wang, et al., J. Vac. Sci. Technol. 6, 130 (1979).

IV. Static SIMS

Molecular Secondary Ion Emission

A. Benninghoven
Physikalisches Institut der Universität Münster
4400 Münster, Germany F.R.

1. Introduction

Since the early days of SIMS [1-4] molecular ion emission has been a well-known phenomenon. For example, secondary ions of the general composition $C_mH_n^{\pm}$ were observed in many spectra. These ions disappeared after removal of some monolayers during the sputtering process and were generally attributed to "contaminations" on the surface. There exist, however, some more or less systematic early investigations of molecular ion emission as, e.g., of anion complex emission from inorganic salts [3].

For a long time the systematic investigation of molecular ion emission was inhibited by the fact that during ion bombardment molecular surface structures are destroyed. This surface damaging was the reason for developing static SIMS [4], which enables the investigation of single monolayers on a solid by employing extremely low primary ion dose densities. Many molecular surface structures have since been studied by this technique.

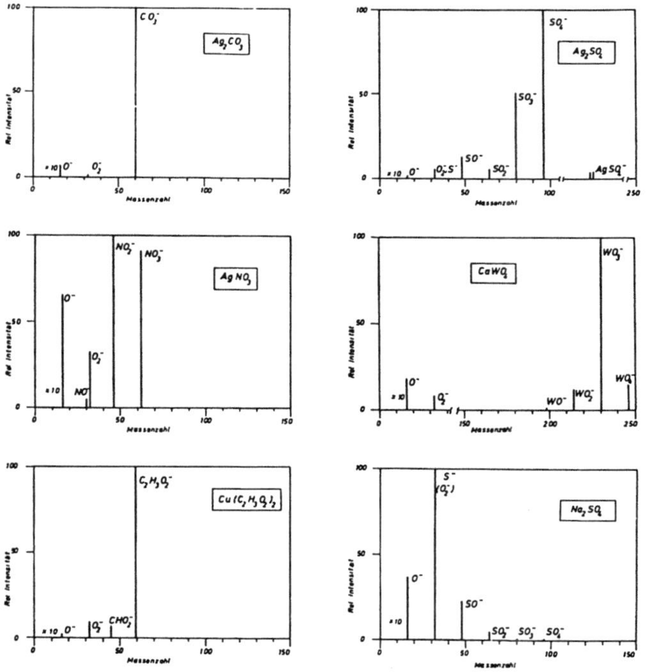

Fig.1 Negative molecul
emission from some inor
and organic salts [3]

Molecular ions are of increasing importance now in SIMS. There are two main reasons for this development: First, molecular ions may provide important insight into the fundamental processes of secondary ion formation and radiation damaging, especially for organic compounds. Second, there is a wide field of application such as in surface reaction studies [4], organic mass spectrometry [5], etc. In this contribution molecular secondary ion emission from clean solid surfaces, from solid surfaces covered by adsorbed atoms or small molecules, and from large organic molecules on solid surfaces will be treated.

2. Molecular Ion Emission from Clean Solid Surfaces

If different types of atoms A,B,C...are incorporated in a solid lattice, the principle secondary ions of the general composition $A_mB_nC_o..^{\pm}$ may be observed. The investigation of simple systems like metals and metal oxides results in two important empirical facts:

1. In general, the formation probability for a molecular cluster decreases with increasing number of incorporated atoms. For metal-oxygen clusters $Me_mO_n^{\pm}$, however, this holds only if n exceeds a certain value, which, in general, is different for positively and negatively charged clusters and, in addition, depends on m [3,4].
2. The observed secondary ions tend to have that charge which is dominating in the corresponding surface cluster before sputtering: Metal ions Me_n will appear preferentially in the positive spectrum. Clusters originating from an oxide and containing metal and oxygen atoms as well will be preferentially emitted as positively charged cluster ions, if there is only a small number of oxygen atoms incorporated, whereas a cluster containing a large number of oxygen atoms will be emitted preferentially as negatively charged ions [4,6].

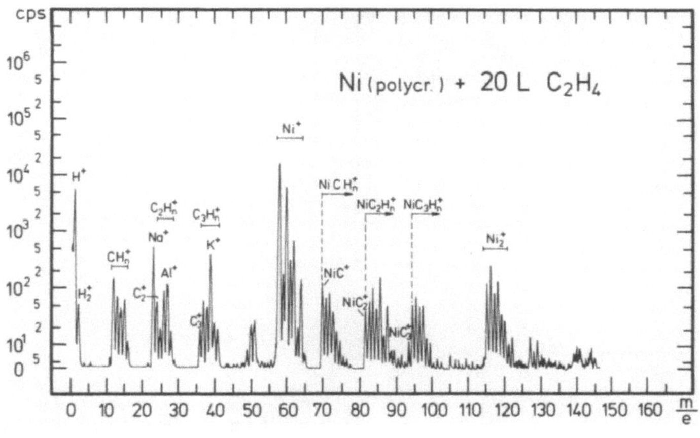

Fig.2 Positive secondary ion spectrum of a Ni-surface after 20 Langmuir C_2H_4-exposure. Primary ions: 3 keV Ar+, $5\cdot10^{-9}$ $A\cdot cm^{-2}$

Cluster ion formation may be explained, in general by two different models:

- Fragment model [3,4]: A complete fragment of the surface lattice is emitted as a secondary ion.

- Recombination model [7,8]: Only single atoms are emitted from the ion bombarded surface. Under certain conditions some of these atomic secondary species may combine in front of the surface, immediately after their emission.

A recent quantitative treatment of the recombination model for a copper single crystal surface results in a certain probability for the formation of metal clusters. Actually, however, there is no experimental decision possible between the two models in the case of metal cluster ions. The relative intensities of the ions $Me_mO_n^{\pm}$ emitted from metal oxides seem to indicate that these ions are produced by direct fragment emission rather than by a recombination process. The same seems to be true for anion complexes like SO_4^-, NO_3^-, etc.

3. Molecular Ion Emission from Adsorption Systems

From metal adsorption systems like Me-O, Me-H, Me-OH, Me-H_2O and Me-CO both complex secondary ions such as MeO^+, MeH^+, $MeOH^+$, MeH_2O^+ and $MeCO^+$ (which characterize the complete adsorption system), and smaller ions such as $O^{\pm}$, $H^{\pm}$, OH^-, H_2O^+, and CO^+ (which represent only the adsorbed species) have been observed. A well-known secondary ion of this type is the ion MeH^+ which is emitted from hydrogen covered metal surfaces [9,10]. For metal-hydrogen systems the ratio MeH^+/Me^+ is proportional to the hydrogen surface coverage, thus eliminating the effect of changing ionization probability for the MeH complex.

As a more complex example Fig.2 presents the positive secondary ion spectrum of a nickel surface after 20 Langmuir ethylene exposure. Molecular ions of the general composition $C_mH_n^+$ and $NiC_mH_n^+$ are emitted from this surface, which is covered by the dissociation products of ethylene.

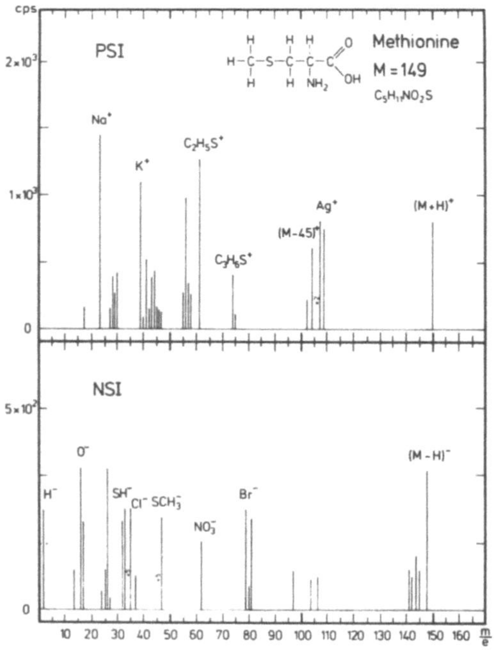

Fig.3 Positive and negative secondary ion emission of methionine on Ag. Primary ions: 2.25 keV Ar^+, $4\cdot10^{-10}$ A on 0.1 cm^2

There is a relatively clear evidence that the emission of adsorbent molecular ions (H_2O^+, OH^-, CO^-, etc.) is the result of a direct fragment emission. To what degree recombination processes between a metal ion and these complexes play an important role in the formation of metal-adsorbate ions cannot yet be determined.

4. Molecular Ion Emission from Large Organic Molecules on Solid Surfaces

Recently we found a strong parent-like secondary ion (Si) emission from a wide variety of organic compounds [11,12]. This organic secondary ion mass spectrometry has attracted much interest. First, investigations in this field originated from the study of secondary ion emission from inorganic and organic anion complexes [3,4]. Following this line we studied systematically the Si-emission of amino acids deposited on metal surfaces from the corresponding aqueous solution. As an example, Fig.3 presents the secondary ion spectrum of leucine on silver. The most important general features of secondary ion emission from metal supported amino acids may be summarized as follows [11,13]:

- Parent-like secondary ions of the general composition $(M+H)^+$, $(M-H)^-$, and $(M-COOH)^+$ are emitted with high yields, which are near 1 in some cases. In addition, cationized parent-like ions appear in the positive spectrum.

- Also, characteristic fragment ions are emitted.

- Similar spectra result for primary ion energies between 500 and 3000 eV and different primary ions as He^+, Ar^+, Ne^+, and Xe^+.

- The acidity of the solution has a strong influence on the intensity ratios $(M+H)^+/(M-H)^-$.

- The same parent-like ions with large differences in the yield, however, are emitted from all investigated metals (Al, Co, Ni, Cu, Ag, Tm, Ta, W, Pt, Au).

- Damage cross sections are in the range of $10^{-14}cm^2$.

Continuing these investigations of amino acids we studied a large number of other compounds (i.e., peptides, purines, vitamins, drugs) with similar results [12]. For all these compounds we observed parent-like ions, mostly as protonated and deprotonated positively or negatively charged particles (Figs. 4-6). In addition, cationized ions were observed in some cases [14].

From all experimental results, the composition as well as the charge of the emitted molecular ions is completely determined by the chemistry of the surface. Only those parent-like molecular secondary ions seem to appear in the spectra which are formed as ions already on the surface. On the other hand, the appearance of these large undestroyed molecular ions can in no way be explained by the existing sputtering models and theories. One has to take into consideration new processes which may be responsible for the formation of such large molecular ions. These may be, e.g.:

1. The formation of a momentary nonequilibrium plasma near the point of impact of the primary ion [15]. From this non-equilibrium plasma the emission of large undissociated molecular particles may become possible.

2. A time and momentum correlated excitation of a limited surface area in

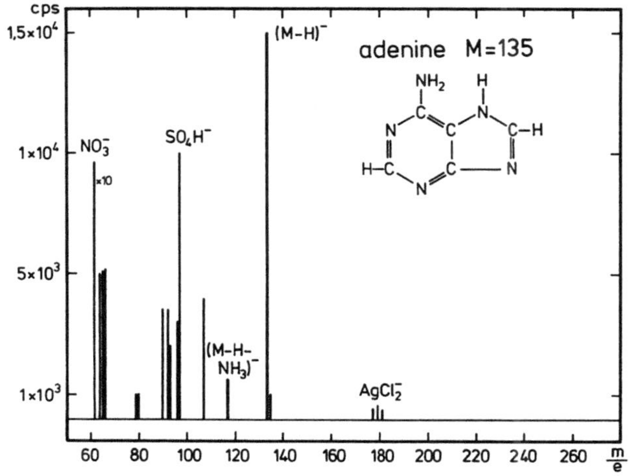

<u>Fig.4</u> Negative secondary ion spectrum of adenine on Ag. Same excitation as in Fig.3

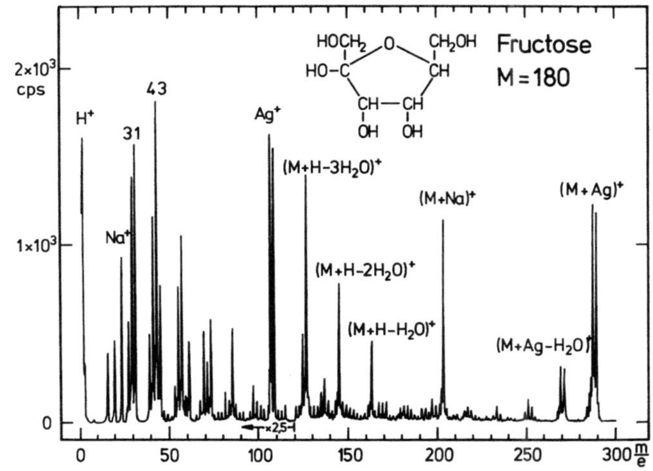

<u>Fig.5</u> Positive secondary ion spectrum of fructose on Ag. Same excitation as in Fig.3

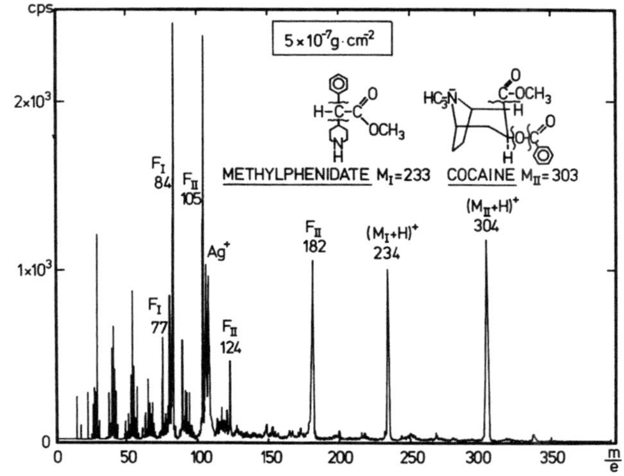

<u>Fig.6</u> Positive secondary ion spectrum of a mixture of methylphenidate and cocaine on Ag. Same excitation as in Fig.3

such a way that a large number of atoms will be knocked on at the same time and in roughly the same direction.

3. Sputtering of a substrate bonding atom, e.g., a metal ion in the case of an ionic bond between the metal surface and the organic molecule.

4. Electronic excitation of a dipole bond on the surface by interaction between this dipole and the excited plasma in the solid [16].

With the experimental material available today it cannot be decided which of these processes may be responsible for the formation of a certain ion. Other processes may be possible. Recombination processes, however, can be excluded.

5. Conclusion

The emission of molecular secondary ions is of increasing interest. One of the reasons for this development is the possibility to study details of the secondary ion formation and emission process. On the other hand, many important applications of molecular secondary ion emission are open in this field of surface reaction studies of organic analytical chemistry. It should be emphasized that also a satisfying application of molecular secondary ion emission in the analytical field requires a better understanding of the ion formation and emission processes itself. Today these processes are far away from being understood, not only for complex organic ions, but also for much smaller particles and even for atomic ions. In all cases, complex processes of momentum transfer to the emitted particles and related electronic interactions have to be considered. For the emission of atomic ions theoretical treatments of both of these aspects exist. A complete understanding of secondary ion emission, however, demands the combination of momentum transfer and electronic interaction aspects. No satisfying approach in this direction has been made so far.

References

1. R.E. Honig, J. Appl. Phys., 29 549 (1958).
2. J.A. McHugh and J.C. Sheffield, J. Appl. Phys., 35 512 (1964).
3. A. Benninghoven, Z. Naturforsch, 24a 859 (1969).
4. A. Benninghoven, Surface Sci., 35 427 (1973).
5. A. Benninghoven, Proceedings of the 9th Materials Research Symposium, NBS Washington, NBS Special Publication 519 627 (1979).
6. C. Plog, L. Wiedmann and A. Benninghoven, Surface Sci., 67 565 (1977).
7. H. Oechsner and W. Gerhard, Surface Sci., 44 480 (1974).
8. N. Winograd, D.E. Harrison and B. Garrison, Surface Sci., 78 467 (1978).
9. M. Someno and M. Kobayashi, Proc. 7th Intern. Vacuum Congr. and 3rd Intern. Conf. on Solid Surfaces, Vienna 1977.
10. A. Benninghoven, K.H. Müller and M. Schemmer, Surface Sci., 78 565 (1978).
11. A. Benninghoven and W. Sichtermann, Appl. Phys. 11 35 (1976).
12. A. Benninghoven and W. Sichtermann, Anal. Chem. 50 1180 (1978).
13. R.J. Colton, J.S. Murday, J.R. Wyatt and J.J. DeCorpo, Surface Sci., 84 235 (1979).
14. H. Grade and R. Cooks, J. Am. Chem. Soc., 100 5615 (1978).
15. C.A. Andersen and J.R. Hinthorne, Anal. Chem. 45 1421 (1973).
16. F.R. Krueger, Z. Naturforsch, 32a 1084 (1977).

Analytical Applications of SIMS

David M. Hercules
Department of Chemistry
University of Pittsburgh
Pittsburgh, PA 15260

The present paper describes the use of SIMS for obtaining analytical data on a number of diverse systems. Of particular importance is a comparison of SIMS with other surface analytical techniques each applied to the same sample set. The emphasis of our work is the importance of multitechnique surface analysis; the present paper focuses on how SIMS complements techniques such as ESCA, AES, ISS and EXAFS. Studies are reported on systems as diverse as boron oxides, organic polymers, trace metal analysis, supported metal catalysts, alloys and biomaterials.

Static SIMS has been used specifically in conjunction with other surface methods such as ESCA and ISS in the analysis of polymeric systems. Our first reported work showed an intercomparison of the three techniques on polymer systems in the study of surface sensitivity of analytical results. We recently have reported the analytical capabilities of static SIMS analysis to the Poly (alkyl methacrylates). After determination of routine static SIMS analysis conditions, systematic studies of such methodological parameters as positive vs. negative SIMS results, charge neutralization phenomena and sputtering effects were undertaken. We found that for these polymers, positive SIMS shows molecular ion "clusters" in the mass spectrum, due to the long chain hydrocarbon backbone and effects that could be linked to the ester side chain. This was not seen in the negative SIMS, although other workers using microprobe conditions on polymers had reported negative SIMS clustering. We attribute this to the analysis conditions of static SIMS, and instrumental factors, combined with the well known phenomenon of low negative ion yield for noble gas ion bombardment used in static conditions. Charge neutralization phenomena were minimal. Sputtering effects were studied because selective sputtering has been reported for this type of organic polymer system under dynamic SIMS conditions. This occurs at the ester bond, as reported by ESCA. Our studies used ISS and ESCA to examine effects of static conditions under typical analysis times. We confirmed that these conditions maintain sample integrity more than long enough to accumulate representative data.

Variation in length and isomeric structure of the alkyl ester group were shown to affect the SIMS spectrum in a number of analytically useful ways. Increase in the length of the alkyl side chain is reflected by a shift in the base peak of the spectrum to higher m/e (amu) value. Previous SIMS analysis of polymers had been limited to teflon and the fluorohydrocarbon series. These exhibit a very small degree of branching, therefore the trend observed has been a decrease in relative intensities for increasing m/e (amu) of molecular ion clusters in the spectrum. Thus our work presents a significant advance for polymers, introducing a fingerprinting technique. Isomeric differences within the side chain, and functionality beyond alkyl groups are reflected by intensity ratio differences with a particular molecular ion cluster. Effects of the structure on the molecular ion spectrum are also important in elucidating the ionization mechanism in static SIMS. Most theoretical mechanisms postulate the atomic ion as the most probable to be ejected. Since this is not true with polymer matrices, various other mechanisms concerning matrix effects must be considered. Organic alkyl polymers show evidence for direct bond breaking through momentum transfer as the primary ionization process, similar to work on other organic systems. Commerical impurities were also monitored with SIMS.

Reactive degradation upon Poly(methyl methacrylate) was studied by static SIMS in conjunction with ESCA and ISS. ESCA has been proven to mild surface effects by observing changes in core level intensity ratios or observing non-polymer hetero-atom signals. Under the mildest of hydrolysis conditions studied, however, ESCA could see no changes. Static SIMS and ISS provide a means to complement ESCA by observing changes in base peak and molecular ion cluster intensities. ISS can show semi-quantitatively changes ESCA could not. Studies on the highly crystalline polymer Poly-(tetramethyl-p-silphenylene siloxane) have been undertaken by ESCA, ISS and static SIMS to determine the folded bond characteristics in the internal stacking within the crystal. Coupled with FTIR, SEM and kinetically controlled reactions the structure and reactivity can be explored by analysis by static SIMS under varying conditions.

SIMS spectra have been recorded of B_2O_3 (boron trioxide) and H_3BO_3 or $B(OH)_3$ (boric acid), whose crystal structures are related, both being composed of BO_3 units. Static SIMS are free of contaminants and very simple. Positive static SIMS showed only B^+ ions for both compounds; negative static SIMS showed differences. The negative spectrum of B_2O_3 had only BO_2^- ions and no O^-, which was the main component of the H_3BO_3 spectrum. No change occurred between static and dynamic conditions for H_3BO_3 but the negative SIMS of B_2O_3 changed drastically becoming very similar to that of H_3BO_3 i.e., O^- and BO^- ions appeared, O^- the most intense.

These results support earlier ESCA and AES studies which concluded that the electronic structure and chemical bonding of B_2O_3 and H_3BO_3 are similar; they are partly covalent and partly ionic. The BO_2^- molecular cluster seen in B_2O_3 negative SIMS shows that B_2O_3 has a covalent nature and the absence of other species indicates a simple breakdown of the polymer-like BO_3 chains. The negative SIMS of H_3BO_3 allows similar conclusions but the appearance of O^-, OH^- and BO^- can be explained by weakening of B-O bonds of H_2BO_3 compared with B_2O_3, allowing the BO_3 units to be easily broken up. This also explains the dynamic SIMS results: no change occurs in H_3BO_3 but the spectrum of B_2O_3 resembles that of H_3BO_3 because of additional energy delivered to the B_2O_3 surface now enables the BO_3 units of B_2O_3 to be broken up.

The use of SIMS in trace analysis is of interest due to the high structural information content and good sensitivity for certain elements of current interest. This has been balanced in the past against experimental difficulties which have produced poor reproducibility. In the present work duplicate samples containing a variety of elements have been analyzed by the four major surface analysis techniques to evaluate the performance of the SIMS method. SIMS was found to be second in sensitivity for transition metals to ISS by at least a factor of 10. ESCA was the next most sensitive while AES was able to only marginally detect these elements on the insulating samples used. ISS and ESCA were the most quantitative techniques, while SIMS and ESCA gave the most chemical information about the trace metal species.

The value of SIMS was clearly established as a sensitive quantitative technique with the ability to provide structural information not available by other methods. However, there are distinct problems with the SIMS method on these types of immobilized chelates, which arise from the physical structure of the surface and the nature of the SIMS phenomena. These problems prevent SIMS from achieving the sensitivity and quantitation expected of the technique based on other work.

Nickel and cobalt supported on γ-Al_2O_3 catalysts have been studied by a variety of spectroscopic techniques in order to ascertain their surface characteristics. It was found that the catalytically active metal species on the surface was dependent upon metal content and calcination temperature as a result of metal-support interactions between nickel (or cobalt) with the alumina support. This interaction is a manifestation of metal ion migration into tetrahedral or octahedral lattice sites in γ-Al_2O_3 during calcination and is limited to the first few layers of the support. As the metal loading is decreased, the percentage of metal ions diffusing into the support increases. Increasing the calcination temperature produces a similar effect. At high metal content, segregation of the metal oxide occurs on the surface.

Because of the surface nature of the interactant, ESCA, ISS and SIMS have proven valuable as probes of these catalyst. ESCA has shown the ability to differentiate between the interaction species and the bulk metal oxide which subsequently forms at high metal loadings. For the cobalt catalysts a semi-quantitative estimate of the relative distribution of surface species is possible. ESCA reduction studies indicate that the degree of metal support interaction has a pronounced effect on the quantity of reducible metal species on the catalyst. For conditions favoring the diffusion of metal ions into lattice sites of the γ-Al_2O_3 the percentage of reducible metal is decreased. The unique surface sensitivity of ISS has provided information concerning surface composition which is unobtainable with any other presently known technique.

A plot of peak height intensity ratio of the metal to alumina signal versus metal concentration shows changes in surface composition which can be correlated to the formation of different metal species. By this method it is possible to determine on a quantitative basis the metal concentration at which various metal species are being formed. SIMS has provided results complementary to those obtained by ESCA and ISS. The differences in surface composition among the catalysts results in the emission of molecular cluster ions which is dependent upon metal loading and calcination temperature.

Recent results from EXAFS have shown changes in coordination number of the metal species in the catalysts which can be attributed to the degree of metal-support interaction. For conditions which favor the diffusion of metal ions into the tetrahedral sites of the support (i.e. low metal concentration and or high calcination temperature) decreases in coordination number have been obtained.

Mo/γAl_2O_3 and Co-Mo/γAl_2O_3 are industrially used hydrodesulfurization catalysts. Current concern over sulfur emissions has focused attention on these catalysts and the ensuing research has resulted in disagreement as to the interactions occuring between the molybdenum and the alumina support.

Initial experiments using ESCA in combination with hydrogen reduction, show the presence of two distinctly different Mo-Al interactions. It has been found that at low molybdenum content (1-4 wt % MoO_3) the majority of the molybdenum will only be reduced to Mo(V). As the MoO_3 content is increased above 4 wt % the amount of Mo(IV) after reduction also increases. ISS has also been used and it shows two distinct regions. One region occurs below approximately 7 wt % MoO_3 and is believed to be due to the molybdenum ion occupying mainly tetrahedral sites which are shielded from detection by ISS. Above 7 wt %, the added molybdenum prefer an octahedral co-ordination which is completely visible to ISS.

Work is continuing on this catalyst system. The cluster formations present in the SIMS spectra will be analyzed and used to unambiguously identify the chemical species present on the suface and also to define the interactions occuring between the molybdenum and the alumina support.

The purpose of the work with biomaterials is to characterize the surfaces of those materials before and after various treatments. The types of treatments in this study are both physical e.g. sterilization, and biological e.g. dynamic contact with blood fluids. The biomaterials studied by SIMS include Haines alloy, a nonferrous metal alloy used in ball and joint heat valves, and Avcothane, a polyurethane used in aortic balloon pumps.

Auger studies of Haines alloy have given elemental analysis and ESCA studies have indicated the oxidation states of the various metals. SIMS shows how oxidation of the various metals in the alloy changes with depth profiling. Static SIMS gives a better indication of the surface contaminants, i.e. from polishing, than either Auger or ESCA. The contaminants may have a marked effect on bioaccumulation, since organic residues could contribute sites for accumulation of blood proteins or, in some instances, organic contamination might retard bioaccumulation by providing a buffer zone between blood proteins and the active sites of the biomaterial.

The electron beam in Auger would tend to destroy the surface structure of Avcothane so that very little valuable information can be obtained from Auger studies of Avcothane. ESCA studies of Avcothane give elemental analysis and oxidation states of the polyurethane. However, SIMS studies of Avcothane are most valuable because an analysis of the functional groups at the surface of the polymer is obtainable. Our previous SIMS studies show sensitivity for differing polymer functionality illustrated directly in the static SIMS spectrum. Since different functional groups at the surface of a biomaterial will interact differently with blood proteins, knowing what functional groups are present on the surfaces of various biomaterials is an important step to understanding bioaccumulation.

Both bulk and surface properties of Raney nickel alloys and catalysts have been studied by many methods. Our research involves the surface characterization of Raney nickel alloys and catalysts. Our ESCA and Auger studies have shown that the initial 50/50 wt. percent Ni/Al alloy has a heavy oxide layer of alumina with some metallic nickel and aluminum on the surface. X-ray diffraction has shown that the alloy exists as Ni_2Al_3, $NiAl_3$ and some eutertic. However, the distribution and amount of the intermetallics on the surface of the alloy is not known. A SIMS study of the Ni/Al intermetallics may give information to distinguish the intermetallics on the surface of Raney nickel alloys.

The small amounts of aluminum and alumina left behind by the activation of the alloy with 20% sodium hydroxide is believed to play a support role for reactive metallic nickel. ESCA and Auger have shown the catalysts to have a large concentration of metallic nickel on the surface with small amounts of aluminum and alumina. However, ESCA and Auger are unable to shown how the aluminum and alumina are linked to the skeletal nickel. A SIMS study of the activated catalyst may give indication of how the aluminum and alumina are bound to the active nickel catalysts on the surface and show whether the aluminum and alumina act as a supportive link. SIMS should also show how the Raney nickel catalysts are poisoned by looking at what surface components poisons are bound.

To summarize, SIMS is applicable to a wide variety of materials and complements other surface sensitive techniques in different ways. The author would like to thank Dr. David Joyner, Dr. Milton Wu, Joseph A. Gardella, Jr., Lawrence V. Phillips, Roland Chin, Derrick Zingg, Susan Graham and Joseph Klein who obtained the data presented in this paper. He would also like to express his appreciation to the sponsors of this research: the National Science Foundation, the Army Research Office, and the Petroleum Research Fund, administered by the American Chemical Society.

Static SIMS Investigations of Amino Acid Mixtures

S. Tamaki
Osaka Prefectural Industrial Research Institute
Nishi-ku, Osaka, Japan

A. Benninghoven and W. Sichtermann
Physikalisches Institut der Universitaet Muenster
Schlossplatz 7, 4400 Muenster, Germany F.R.

1. Introduction

Secondary ions of the general composition $(M+H)^+$, $(M-H)^-$, and $(M-COOH)^+$ are emitted with high intensities from all amino acids [1-3]. Applying HNO_3-etched Ag as supporting metal, the intensity of this emission is proportional to the surface concentration of the corresponding amino acid M up to a concentration of 10^{15} molecules/cm^2. This surface concentration seems to be equivalent to a complete monolayer. Because of a large damage cross section σ the investigation of this type of amino acid layer is only possible with extremely small primary ion current densities (static SIMS) [3].

Recent investigations carried out in our laboratory with a standard mixture of 17 amino acids M_n of the same molecular concentration resulted in different $(M_n \pm H)^\pm$ intensities for different amino acids. In addition, various relative changes of these intensities occur during ion bombardment. To get more information on this behaviour, we carried out some systematic investigations with mixtures of only two amino acids.

2. Experimental

All investigations were carried out with HNO_3-etched Ag as supporting metal. The amino acids were dissolved in 0.1 n hydrochloric acid and deposited with a glass micropipette on the target. The total surface concentration was 10^{15} amino acid molecules per cm^2 in all experiments. The target was bombarded with 4×10^{-10}A Ar^+, 2.25 keV on 0.1 cm^2. The secondary ions were mass analyzed by a 60° magnetic sector field.

3. Results and Discussion

To get more insight into the mixture effects for amino acids, several experimental values are of interest:

- The intensity ratio of different parent-like secondary ions of the same amino acid for different mixtures and concentrations.
- The intensity behaviour of parent-like ions from the same combination of amino acids in dependence on their relative concentration.
- The intensity of parent-like secondary ions from mixtures of different types of amino acids.

We carried out corresponding experiments with mixtures of two amino acids with the following results:

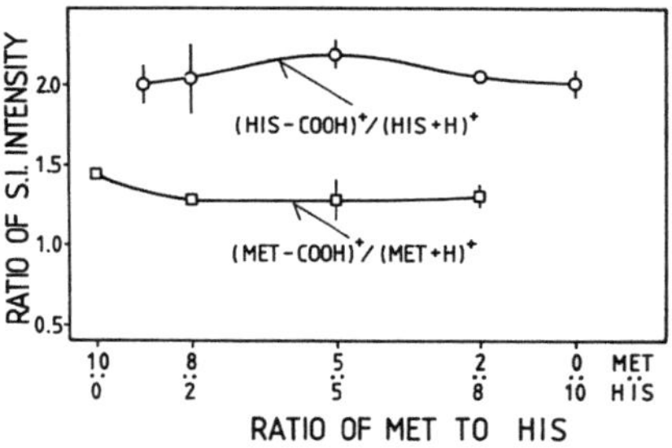

Fig.1 Ratio of secondary ion intensities $(M\text{-}COOH)^+/(M+H)^+$ of various mixtures of methionine (MET) and histidine (HIS)

1. We investigated the intensity ratios $(M\text{-}45)^+/(M+1)^+$ for different concentrations and mixtures. As an example Fig.1 presents the results for a mixture of histidine and methionine. Table 1 summarizes corresponding results for further mixtures. As a general result we found, that the ratio $(M\text{-}COOH)^+/(M+H)^+$ for a certain amino acid M remains constant if one changes the relative concentration of M or the type of the second amino acid.

Table 1 Values of $(M\text{-}COOH)^+/(M+H)^+$ for amino acid mixtures

M	$(M\text{-}COOH)^+/(M+H)^+$	Mixture partner
arginine	0.2	methionine
	0.2	histidine
methionine	1.0	leucine
	1.3	histidine
	1.8	arginine
leucine	2.3	methionine
histidine	2.1	methionine
	2.4	arginine

This general behaviour points out that $(M\text{-}COOH)^+$- and $(M+H)^+$-ions are produced by sputtering of amino acid molecules in the same bonding state on the surface

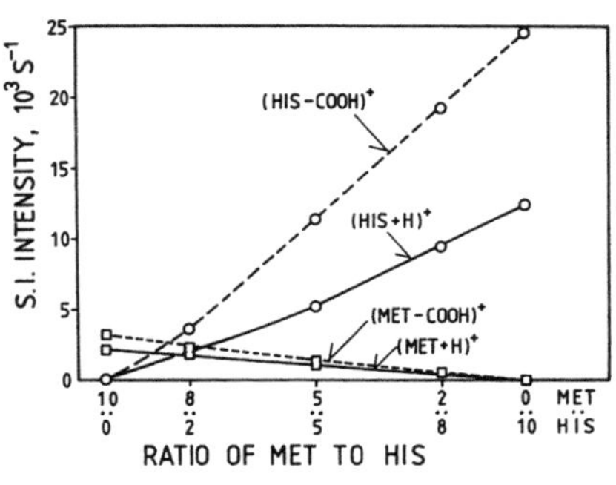

Fig.2 Variation of parent-like ion intensities of methionine and histidine depending on their relative concentration in the mixture

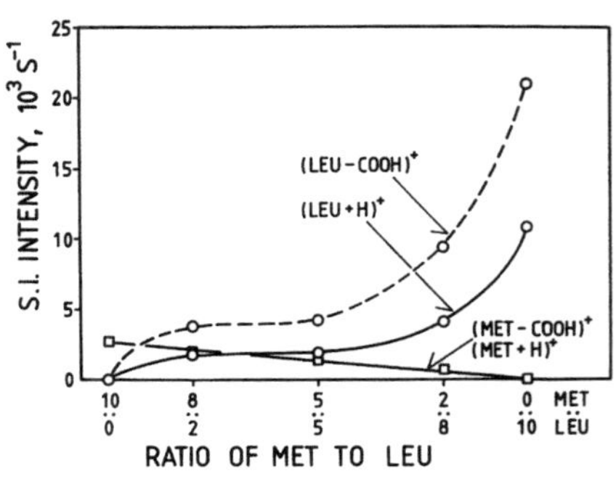

Fig.3 Variation of parent-like ion intensities of methionine and leucine depending on their relative concentration in the mixture

The presence of an additional amino acid on the surface obviously does not influence the nature of the ion emission process but only its probability.

2. For a mixture of two amino acids and various concentrations, one supposes a proportionality of characteristic secondary ion intensity and concentration. Fig.2 shows that this may be true for a certain mixture. On the other hand, for other types of mixtures another behaviour is observed: In the case of leucine plus methionine, a strong increase of leucine characteristic ion emission is observed, if the surface is only covered with relatively small amounts of methionine.

3. For 10 mixtures of different amino acids with methionine we found values for the ratios $(M+H)^+/(MET+H)^+$, which differ by more than one order of magnitude (Table 2). From the results of Fig.3 we expect that this ratio in addition depends on the relative concentrations at least in some cases.

Table 2 Values of $(M+H)^+/(MET+H)^+$ for 1:1 amino acid mixtures

Ala	Glu	Phe	Gly	His	Val	Leu	Tyr	Ser	Thr
0.61	0.35	2.2	0.20	6.9	0.63	0.88	0.16	0.10	~0

4. Conclusion

The investigation of secondary ion emission $(M+H)^+$ and $(M-COOH)^+$ from mixtures of two amino acids on Ag reveals a strong dependence of this emission as well on the type and surface concentration of the corresponding parent compound M as on the added second amino acid. The ratio $(M+H)^+/(M-COOH)^+$ is constant for all combinations and relative concentrations. This indicates the formation of these two types of ions out of the same bonding state of the amino acids on the surface. The absolute probability for the formation of these ions, on the other hand, strongly depends on the presence and relative concentration of the added second amino acid. The degree of this influence varies with the type of amino acids. The experimental results available up to now give no further information on the mechanism of the underlying interactions on the surface during the deposition and drying process on the one hand and the ion formation and emission process on the other.

Acknowledgement

One of the authors (S. Tamaki) expresses his gratitude to the Alexander von Humboldt-Foundation for financial support.

References

1. A. Benninghoven, D. Jaspers and W. Sichtermann, App Phys. 11, 35 (1976).
2. A. Benninghoven and W. Sichtermann, Anal. Chem. 50, 1180 (1978).
3. A. Benninghoven, Proc. 9th Materials Research Symposium, NBS Gaithersburg Md; NBS Special Publication 519 (1979) p. 627.

Static SIMS of Amino Acid Overlayers

R.J. Colton, J.D. Ganjei, J.S. Murday, and J.J. DeCorpo
Chemistry Division
Naval Research Laboratory
Washington, D.C. 20375

The work of BENNINGHOVEN, et al. [1,2], has shown that static secondary ion mass spectrometry (SSIMS) can be used for the identification of organic molecules adsorbed on a surface. In this work we have studied the fragmentation patterns for some simple amino acids - glycine, α and β-alanine, and serine-deposited onto Ag substrates from aqueous solutions of various concentrations. Identification of the mass fragments is aided by use of isotopically labeled compounds and comparison with chemical ionization (CI) mass spectra of amino acids. X-ray photoelectron spectroscopy (XPS) and Auger electron spectroscopy (AES) are combined with SIMS to examine the factors influencing the SIMS fragmentation patterns [3]. For the SIMS work the primary ions are 1 keV Ar^+ incident at 70° to the sample normal with current densities $\sim 10^{-8}$ amps/cm^2. The sample is oriented normal to the axis of an Extranuclear Model 4-162-8 quadrupole mass spectrometer equipped with a modified Bessel box energy prefilter [4].

The general features of the secondary ion (SI) mass spectra include the intense isotopic doublet of Ag^+ at m/z 107 and 109 and a second intense high mass doublet at m/z 182 and 184 for glycine, 196 and 198 for α and β-alanine, and 212 and 214 for serine. These high mass doublets are assigned to the corresponding silver-amino acid complex ion, i.e., $(AgM)^+$ where M is the molecular weight of the parent amino acid. (Similar observations involving transition and alkali metal cation stabilized parents have been made for organics emitted by SIMS [5] and by a laser induced ionization technique [6].) Other mass peaks are present throughout the spectra and their assignments are outlined in Table 1 together with their relative abundances. The only parent-like species are those attached to H or Ag. The fragmentation patterns show the cationized molecular ions, HM^+ and AgM^+, losing H_2, NH_3, H_2O, and/or HCOOH (analogous to CI mass spectra of amino acids [7]). The major fragmentation routes usually result in positively charged amine fragments. These pathways are confirmed with isotopically labeled compounds such as glycine-^{15}N and glycine-1-^{13}C. The majority of ions listed in Table 1 have extremely low intensities due to severe mass discrimination in the quadrupole spectrometer. Comparative SIMS studies on Ag with a recently constructed magnetic [8] and the quadrupole instrument show discrimination by the quadrupole of $\sim$ 8% between the m/z 107 and 109 isotopes of Ag^+. The magnetic instrument gave the isotopic abundances as 51.1 and 48.9 while the quadrupole gave 53.6 and 46.4. The accepted values are 51.35 and 48.65. Therefore, the data in Table 1 are significant despite their seemingly low abundance.

Table 1 Secondary ion mass spectra of amino acids

Amino acid (mol.wt.)	HM^+	$(M-H)^+$	$HM^+ - NH_3$	$HM^+ - H_2O$	$HM^+ - 2H_2O$	$HM^+ - HCOOH$	$HM^+ - HCOOH - H_2O$	Other
Glycine (75)	6	2	6	15	-	1000	-	
α-Alanine (89)	14	9	-	10	-	1000	-	m/z 30 = 110
Serine[b] (105)	220	-	-	45	160	670		m/z 30 = 730
β-Alanine (89)	44	180	12	48	-	250	-	m/z 30 = 1000
	AgM^+	Ag^+	$AgM^+ - NH_3$	$AgM^+ - H_2O$	$AgM^+ - 2H_2O$	$AgM^+ - HCOOH$	$AgM^+ - HCOOH - H_2O$	Other
Glycine (75)	10	400	< 1	< 1	-	5	-	m/z 138, 140 = 6
α-Alanine (89)	4	410	-	1	-	8	-	m/z 124, 126 = 13
Serine[b] (105)	30	2900	-	20	-	60	-	m/z 124, 126 = 50; m/z 150, 152 = 50;
β-Alanine (89)	10	1130	17	-	-	38	-	m/z 123, 125 = 44; m/z 135, 137 = 32

[a] Abundances relative to base peak assigned intensity of 1000.
[b] Data from low resolution spectra.

The SI mass spectrum of serine demonstrates the high degree of fragmentation caused by the ion beam where further decomposition of the initial fragment ions occurs, e.g., $HM^+ - 2H_2O$ and $HM^+ - HCOOH - H_2O$. (This type of fragmentation is also caused in serine by H_2 CI mass spectrometry.) The fragmentation of β-alanine is considerably different from the other amino acids (particularly α-alanine). The SI mass spectrum shows a significant H_2 loss from HM^+ plus the loss of both NH_3 and H_2O but unusually little loss of HCOOH. In β-amino acids the NH_3 group is lost from a terminal position. In α-amino acids the loss of NH_3 is restricted since the charge would be located next to the polarized carbonyl group resulting in an unstable ion. Cationization by Ag (AgM^+) on the other hand favors the loss of NH_3 and HCOOH suggesting a cationization site or mechanism differing from H.

The XPS observations of the C:O:N relative intensities are used to monitor the stoichiometry of the adsorbed species. The ratio of the X-ray excited Ag MNN Auger (∿ 350 eV) intensity to the Ag $3d_{5/2}$ photoelectron (∿ 1100 eV with the AlK_α X-ray source) intensity provides a measure of the extent of the organic overlayer. When the Ag substrates are prepared by acid etching in dilute nitric acid, the relative mass peak intensities of the parent-like and fragment ions mentioned above, with the exception of the HM^+ ion, are independent of the surface concentration as measured by XPS. However, the total ion emission does decrease as the organic films increase in thickness. When the Ag substrate is prepared by brushing the surface, the organic ion intensities are reduced by at least a factor of two. For a given solution concentration, XPS observations show a more extensive organic overlayer on the Ag for the brushed samples when compared to the acid etched samples. Scanning electron micrographs of the etched sample show a highly porous structure. Removing an Ag substrate from solution causes a film of water to be retained. As that film evaporates, its solute is deposited on the surface of the sample. On the etched samples, it is postulated that the solute concentrates into the pores as evaporation proceeds; any large deposits are thereby formed in recesses which are not seen by either the XPS or SIMS. On the brushed sample the large deposits are exposed. Where the deposits occur, the organic is not in contact with the Ag and the SIMS signal decreases.

References

1. A. Benninghoven, D. Jaspers, and W. Sichtermann, Appl. Phys. 11, 35 (1976
2. A. Benninghoven and W. K. Sichtermann, Org. Mass Spectrom. 12, 595 (1977)
3. R. J. Colton, J. S. Murday, J. R. Wyatt, and J. J. DeCorpo, Surf. Sci., 85, 235 (1979).
4. J. R. Wyatt, R. J. Colton, J. D. Ganjei, J. J. DeCorpo, and J. S. Murday, to be published.
5. H. Grade and R. G. Cooks, J. Am. Chem. Soc. 100, 5615 (1978).
6. M. A. Posthmus, P. G. Kistemaker, H. L. C. Meuzelaar, and M. C. Ten Noeven de Brauw, Anal. Chem. 50, 985 (1978).
7. C. W. Tsang and A. G. Harrison, J. Am. Chem. Soc. 98, 1301 (1975).
8. J. R. Wyatt, R. J. Colton, and J. J. DeCorpo, to be published.

Static SIMS Studies of Metal-Covered W(110) Surfaces

S. Prigge and E. Bauer
Physikalisches Institut, TU Clausthal
3392 Clausthal-Z., Germany

1. Introduction and Experimental

It is well-known that ion bombardment of solids leads to the emission of atomic and molecular secondary ions (SI). According to the models proposed to explain the emission of molecules from metal surfaces [1-3], the constituents of the molecules come predominantly from the topmost layer. In order to test these models we have grown metals monolyer by monolayer (ML) and studied the evolution of the atomic and molecular SI yields with increasing thickness, as well as a function of emission angle. The layer/substrate systems Cu/W(110) and Pd/W(110) were chosen because they have been studied in considerable detail with AES, LEED, $\Delta\phi$ and TDS [4,5].

The general set-up of the multi-analysis method system has been described repeatedly before so that only details used for the SI emission studies will be presented here. The primary 1keV Ar^+ ion beam was focussed into a spot of 2mm diameter. A typical ion beam current was $i_0 = 4x10^{-9}$A. In our experiments SI emitted normal to the surface were analyzed with a quadrupole mass spectrometer (1-300amu) and an off-axis channeltron in a counting mode. The adsorbate/substrate systems were characterized by AES and work function change measurements.

2. Results

Fig.1 shows that the yield of all Cu SI depends in a piece-wise linear manner on coverage. Upon completion of each ML a change of slope of the SI yield is seen. An additional slope change occurs at about 1.8ML. All yields reach saturation at about 3ML. No Cu_2^+ and Cu_3^+ emission is found up to the completion of the first and second ML, respectively, within the detection limit of .5 cps/nA which is determined by the background count noise. In the case of Pd (Fig.2) only Pd^+ and Pd_2^+ could be studied because of mass range limitations. Again, the yields increase in a piecewise linear manner and saturate at 2 and 3 ML for Pd^+ and Pd_2^+, respectively. Below 1ML the Pd_2^+ signal is below the detection limit of .5 cps/nA. The (normalized) yields as a function of the polar emission angle θ in the $[1\bar{1}0]$ azimuth are shown in Fig.3. All yields have maxima for emission in the most densely packed direction in this azimuth, the [100] direction (θ = 45°).

3. Discussion

The most striking feature of the results is the absence of dimer and trimer emission below about 1 and 2 monolayers, respectively. Most other aspects such as the slope change at 1.8 monolayers and in part also the relative

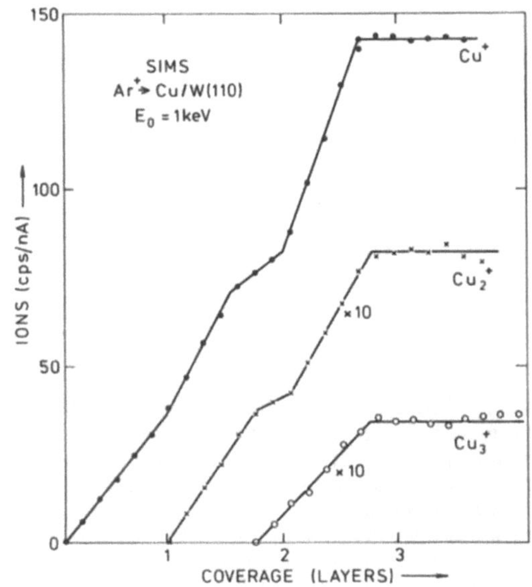

Fig.1 SI yield (Cu^+, Cu_2^+, Cu_3^+) as a function of coverage

slopes for the various monolayers are largely determined by sputtering rates, dipole moments and work function and are discussed elsewhere [6].

Unless the thickness thresholds for molecular SI formation are attributed to specific substrate/layer interaction processes, none of the existing molecular ion formation models can explain them. This is illustrated in Fig.4 for Me_3^+ emission. In the emission model (a) clusters with the largest number of bonds are most likely emitted such as a triangular trimer from the topmost layer. In the association model (b) the emitted monomers which form the molecular ion outside the crystal also originate most likely from the topmost layer. Therefore, neither of the two models can explain the absence of dimer and trimer emission below about 2 and 3 monolayers, respectively.

Apparently, a third model (indicated in Fig.4c) is necessary: each layer contributes only one atom to the molecule. Thus, only atomic ions are

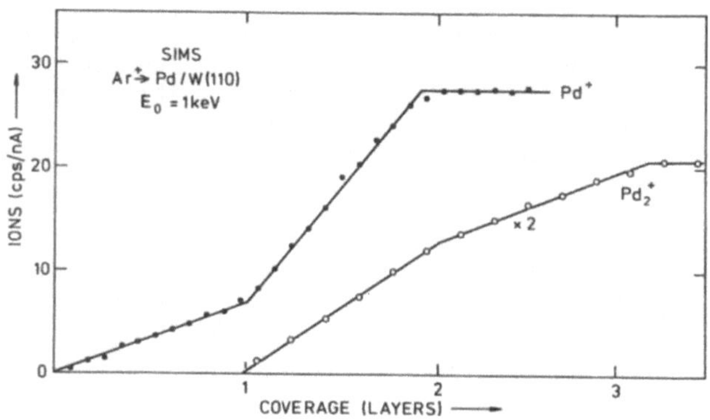

Fig.2 SI yield (Pd^+, Pd_2^+) as a function of coverage

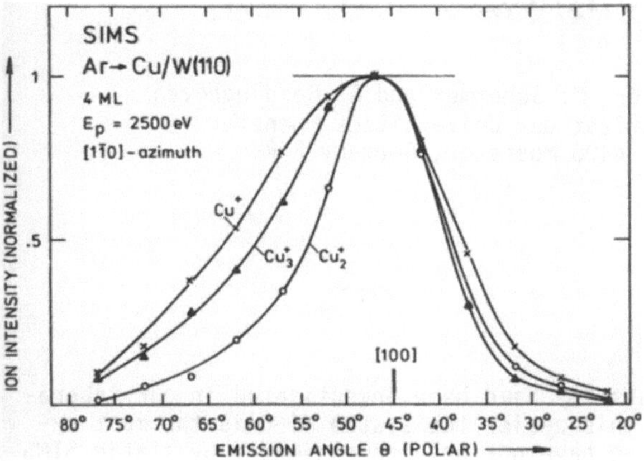

Fig.3 Normalized SI yields as a function of the polar emission angle θ

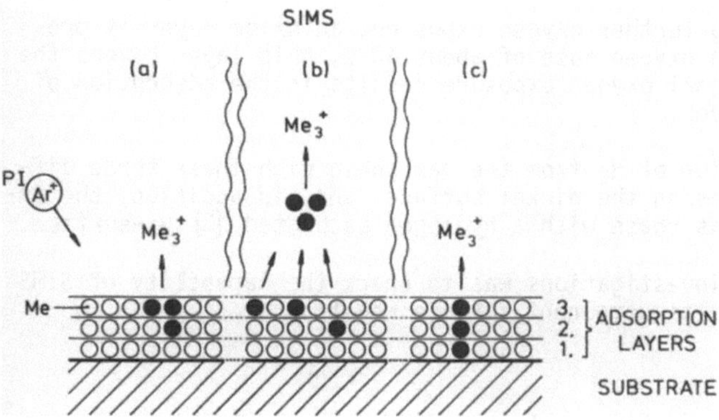

Fig.4 Models for for emission of Me_3^+ molecules (schematic)

emitted below 1ML, monomers and dimers between 1 and 2 ML; the onset of the trimer emission varies somewhat from one deposition sequence to the other (between 1.8 and 2 ML) which is attributed to the incorporation of Cu atoms into the third ML before the second is completed.

The differential emission yields of Fig.3 are also compatible with model c (although they do not prove it because model b predicts similar results [7]): strongly preferred molecule emission in the densely packed direction. In conclusion it should be stated that the proposed model (c) does not invalidate the others but suggests an additional mechanism for molecule emission. Future work with other film/substrate pairs, alloy films and other orientations will have to decide the relative importanceof the various mechanisms.

References

1. G.P. Konnen, A. Tip, A.E. de Vries, Radiation Effects, 21 269 (1974).
2. W. Gerhard and H. Oechsner, Z. Physik B, 22 41 (1975).
3. N. Winograd, B.J. Garrison, D.E. Harrison, Jr. Phys. Rev. Lett., 41 1120 (1978).
4. E. Bauer, H. Poppa, G. Todd and F. Bonczek, J. Appl. Phys., 45 5164 (1974).
5. W. Schlenk and E. Bauer, Surface Science, in print.
6. S. Prigge and E. Bauer, Advances in Mass Spectrometry, in print.
7. T. Fleisch, N. Winograd, W.N. Delgass, Surface Sci., 78 141 (1978).

Investigation of Surface Reactions by SIMS: Nickel-Oxygen-Hydrogen-Interaction

P. Beckmann, K.H. Müller, M. Schemmer and A. Benninghoven
Physikalisches Institut der Universitaet Muenster
Schlossplatz 7, 4400 Muenster, Germany F.R.

1. Introduction

A great number of metal oxygen systems have been investigated in our laboratory during the last decade. In this series the system Ni-O is the most extensively studied system, which we have not only investigated by static SIMS and TDMS, but in addition, by combining SIMS, XPS, and AES. The interaction of O_2 with Nickel at room temperature, can be divided into three different steps. Below an oxygen exposure of 10 L, the surface is covered by a chemisorbed oxygen layer. During further oxygen exposure, an oxide layer is produced on the surface. At an oxygen dose of about 40 L, this layer covers the surface completely. Additional oxygen exposure results in the adsorption of O_2 on top of this oxide layer.

We studied the interaction of H_2 from the gas phase with these three different oxygen bonding states on the nickel surface, and, in addition, the interaction of O_2 from the gas phase with a hydrogen saturated [3] Ni-surface.

The intention of these investigations was to check the capability of SIMS for the investigation of multi-component surface reactions.

2. Results and Discussion

2.1 O_2-Exposure of a Hydrogen Covered Nickel Surface

O_2-exposure of a hydrogen saturated surface, on which only the stronger bound β_2-hydrogen is present at room temperature [4] , results in an undisturbed formation of the well known oxygen binding states [1,2] , described in the introduction. No OH^--emission, which is characteristic for hydroxide groups, and no H_2O TDMS signal were observed. The continuous decrease of the intensity ratio Ni_2H^+/Ni_2^+, indicating a decreasing concentration of metal-hydrogen bonds [5] , shows the complete displacement of hydrogen during oxygen exposure (Fig.1).

2.2 H_2-Exposure of Oxygen Covered Nickel Surfaces

H_2-exposure of the different oxygen binding states...[1,2] results in a formation of hydroxide groups, indicated by OH^-, $NiOH^{\pm}$, NiO_2H^- and Ni_2OH^+ emission. The intensity of these signals depends on the amount of oxygen preexposure as well as on the H_2-dose.

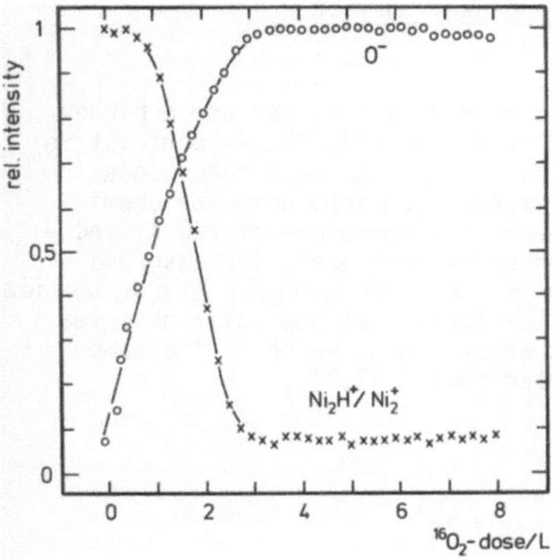

Fig.1 Behaviour of oxygen (O^-) and hydrogen (Ni_2H^+/Ni_2^+) specific SIMS-signals during 8 L O_2-exposure of a 3 L H_2-preexposed Ni-target

2.2.1 H_2-Exposure of a 1 L O_2-Exposed Ni-Surface (Chemisorption I)

H_2-exposure of a nickel surface covered with chemisorbed oxygen results in the formation of OH groups (indicated by a strong OH^--emission), as well as nickel hydrogen bonds (indicated by Ni_2H^+-emission) (Fig.2). If the target is heated at 350 K, after saturation of OH^--emission, an H_2^- as well as an H_2O-flash signal appears, correlated with an intensity decrease of OH^- and Ni_2H^+ emission. Obviously the hydroxide is transformed into desorbing H_2O molecules at this temperature. H_2O complexes are not observed in the SIMS-spectra.

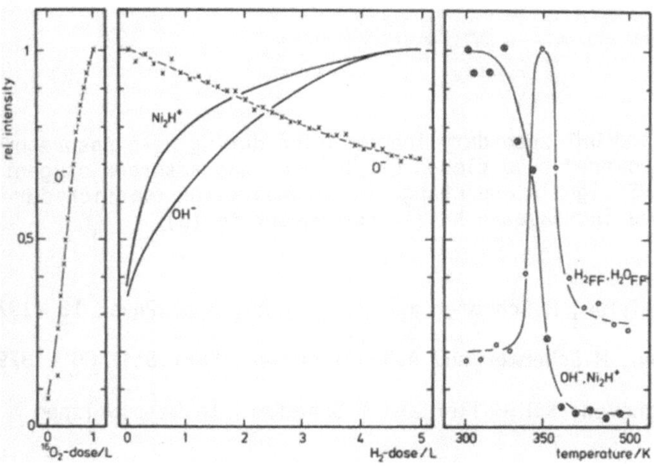

Fig.2 (a) Change of O^--emission during 1 L O_2-exposure of a clean nickel target;
(b) Result of subsequent 5 L H_2-exposure on the emission of characteristic secondary ions;
(c) Thermal behavior of characteristic secondary ion emission and related TDMS signals during temperature increases (5 $K \cdot s^{-1}$) subsequent to (b)

2.2.2 H_2-Exposure of an Oxygen Saturated Nickel Surface (Oxide + Adsorbed Oxygen)

On this surface a strong hydroxide formation occurs too, but now a higher H_2-dose is required for saturation of OH^--emission (Fig.3). In contrast to the chemisorbed oxygen layer, there are no direct metal-hydrogen bonds (no Ni_2H^+ emission). Another significant difference compared with the chemisorption state is the increased H_2O desorption temperature of 700 K, indicating a higher binding energy of the hydroxide complexes. The adsorbed oxygen, which seems not to be attacked by H_2, is still present on a hydroxide covered NiO-layer. This is indicated by the fact, that one still observes thermal desorption of adsorbed oxygen at about 400 K, which is the same temperature as for an only oxygen saturated target [1,2] .

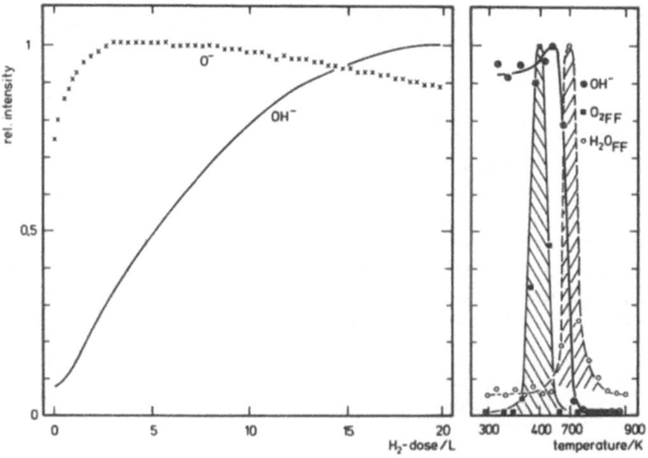

Fig.3 (a) Change in O^- and OH^- secondary ion emission during H_2-exposure of a Ni-surface covered by a closed oxide layer and adsorbed oxygen; (b) O_2 and H_2O TDMS signals and change in secondary ion emission during temperature increase ($5\ K \cdot s^{-1}$) subsequent to (a)

References

1 A.Benninghoven, K.H.Müller, M.Schemmer and P.Beckmann, Appl.Phys. 16 (1978) 367

2 K.H.Müller, P.Beckmann, M.Schemmer and A.Benninghoven, Surf.Sci. 80 (1979) 325

3 P.Beckmann, A.Benninghoven, K.H.Müller and M.Schemmer, to be published in Surf. Sci.

4 see for example:
T.N.Taylor and P.J.Estrup, J.Vac.Sci.Technol. 11 (1974) 244
K.Christmann, O.Schober, G.Ertl and M.Neumann, J.Chem.Phys. 60 (1974) 4528

5 A.Benninghoven, K.H.Müller, C.Plog, M.Schemmer and P.Steffens, Surf.Sci. 63 (1977) 403

Study of Inorganic Salts by Static and Dynamic Secondary Ion Mass Spectrometry (SIMS)

E. De Pauw and J. Marien
Institut de Chimie (B-6)
University of Liège
Liège, B-4000, Belgium

1. Introduction

Up to now, many works in SIMS (see [1] and Refs. 1,2,5,8,9 therein) have been devoted to the study of the various phases (oxidation state of the metal) present in an oxidized metal. We have extended this kind of study to inorganic salts in which the metalloid is in various oxidation states. Therefore, we have investigated sulfate, sulfite, nitrate and nitrite salts using static and dynamic SIMS. Our purpose was to check the usefulness of the molecular information obtained from these systems and especially to see if the oxidation state of the metalloid can be deduced from the spectra. We also looked for possible artifacts induced by the ion beam and tried to follow chemical reactions on the surface of the salts.

2. Experimental

The experiments were performed in a RIBER SIMS apparatus equipped with a large quadrupole and a differentially pumped argon ion gun. The energy of the primary ions was 4.5 keV and current densities ranging from $6x10^{-9}$ A cm^{-2} (static) to $3x10^{-7}$ A cm^{-2} (dynamic) were applied. The salt samples being insulating pellets, it was necessary to flood the surface with 200 eV electrons. In some experiments, oxygen was introduced by a capillary at a rather high pressure directly onto the sample.

3. Results and Discussion

The SIMS spectra of sulfite and sulfate of sodium are qualitatively alike whether obtained in the static or dynamic SIMS mode. The main peaks observed are shown in Fig.1. One notices that there are two rather intense ion clusters at mass 149 $(Na_3SO_3)^+$ and mass 165 $(Na_3SO_4)^+$ which correspond to the cationized parent molecule. As illustrated in Fig.2, the intensity ratio R $(Na_3SO_3^+/Na_3SO_4^+)$ depends on the nature of the solid (sulfite or sulfate, the primary ion dose already seen by the sample, and the partial pressure of oxygen above the solid. For slight ion doses, in vacuum, the R value is 2.7 for sulfite and 1.6 for sulfate. Those values decrease as the primary ion dose increases. Equilibrium values of 1.07 and 0.60 are respectively reached for the sulfite and sulfate samples after submitting them to argon ion doses higher than $2x10^{14}$ ions cm^{-2}. This result may be explained by an oxygen enrichment of the solid due to recoil implantation. Besides, it has been observed that an oxygen-doped sulfate sample releases its oxygen when the ion gun is switched off. The relaxation phenomenon does not occur in the case of an enriched sulfite. It is thus suggested that the implanted oxygen is chemically bound in sulfite while it is rather labile in an oxygen saturated

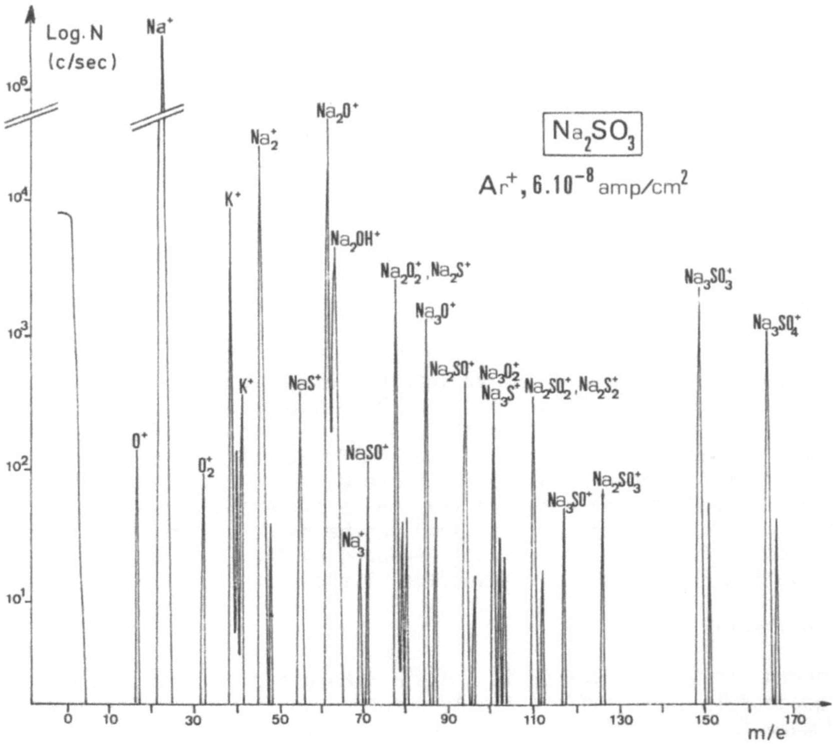

Fig.1 Positive SIMS spectrum of sodium sulfite in ultra-high vacuum (Ar^+, 4.5 keV, $6x10^{-8}$ A cm^{-2})

compound like sulfate. The chemical affinity of the solid towards oxygen could thus play a great role in the relaxation following the recoil implantation.

When static SIMS is performed under an oxygen partial pressure of about 10^{-6} Torr, one observes a strong decrease in R for both solids (see Fig.2) and the reversal in the relative peak heights takes place even in the case of sulfite. Experimental evidence shows that this R decrease is to be attributed to the adsorption of an oxygen monolayer on the solid. In fact, this adsorption on sulfite and sulfate of sodium is an activated adsorption because it takes place only in the presence of the ion beam. The surface of Na_2SO_3 and Na_2SO_4 is insensitive towards oxygen at 10^{-6} Torr and room temperature but it becomes active owing to the ion bombardment. For other oxygenated inorganic salts such as $AgNO_3$, $AgNO_2$, $NaNO_3$, $NaNO_2$, the ion induced adsorption of oxygen does not occur.

Reduction of the salts into metallic sodium has also been observed. This could be tentatively interpreted by a mechanism according to which sulfur dioxide is preferentially sputtered into the gas phase and oxygen is implante[d] into the solid.

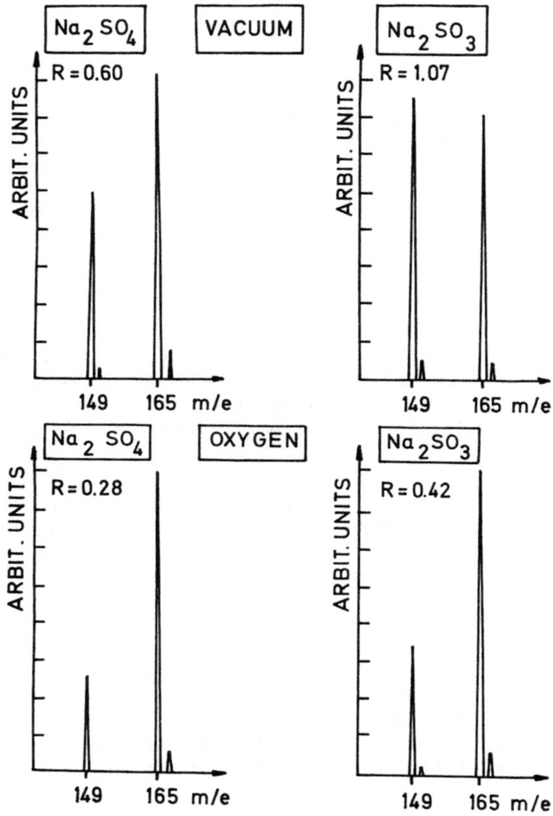

Fig.2 Recording of the equilibrium relative heights of $(Na_3SO_3)^+$ and $(Na_3SO_4)^+$ in vacuum after an exposure of 10^{15} ions cm^{-2} in vacuum (a)(b) and after an exposure of $5x10^{12}$ ions cm^{-2} in $4x10^{-7}$ Torr of oxygen (c)(d). See the definition of R in the text.

In conclusion, we may say that, under well defined experimental conditions, SIMS allows one to distinguish between sulfite and sulfate but ion induced artifacts must be taken into account.

References

1. J. Marien and E. De Pauw, Bull. Soc. Chim. Belg. 88, 115 (1979).

Secondary Ion Mass Spectrometry of Amino Acids by Proton and Alkali Ion Attachment

W. Sichtermann and A. Benninghoven
Physikalisches Institut der Universitaet Muenster
Schlossplatz 7, 4400 Muenster, W. Germany

1. Introduction

For several years the investigation of nonvolatile organic compounds has developed into a new field of secondary ion mass spectrometry (SIMS). In contrast to conventional methods, such as electron impact or chemical ionization, SIMS has the advantage of directly ionizing the substances from the solid state.

Up to now a great number of organic compounds as amino acids, barbiturates, opiates, sugars, etc., have been investigated with static SIMS [1]. In the present work the dependence on concentration and temperature of alkali ion and proton attachment to amino acids has been studied in more detail.

2. Experimental

The samples have been prepared by depositing 10^{13} to 10^{15} molecules of the amino acid and the alkali salt, included in 1 µl of 0.1 n hydrochloric acid, on 1 cm^2 of an etched clean silver foil. The corresponding temperature during annealing of the target has been applied for two minutes in an argon atmospher

The investigations have been carried out in an UHV magnetic sector field mass spectrometer focussing a 2.25 keV argon ion beam on an area of 0.1 cm^2 of the sample. With regard to the large cross section for desorption and damaging of more than 10^{-14} cm^2 [2] , an extremely low primary ion current density of $4x10^{-9}$ Acm^{-2} has been applied for all experiments with the exception of the disintegration measurements.

3. Results and Discussion

Common characteristic of all amino acids (basic formula $R{\cdot}CH(NH_2){\cdot}COOH$) is the emission of $(M\text{-}COOH)^+$ fragments, $(M\text{-}H)^-$ and $(M+H)^+$ quasi molecular ions [2] , and, by reaction with the supporting material, cationized molecules $(M+Ag)^+$. In the presence of alkali (C) halides we observed additional cationized molecules as $(M+C)^+$, $(M+2C\text{-}H)^+$, $(M+C+Ag\text{-}H)^+$ etc..

As shown in Fig.1 for leucine-LiCl mixtures, the intensities of $(M+Li)^+$, $(M+2Li\text{-}H)^+$ etc. increase with increasing temperature at the expense of $(M+H)^+$ ions. The thermal treatment obviously promotes the exchange of protons by Li ions.

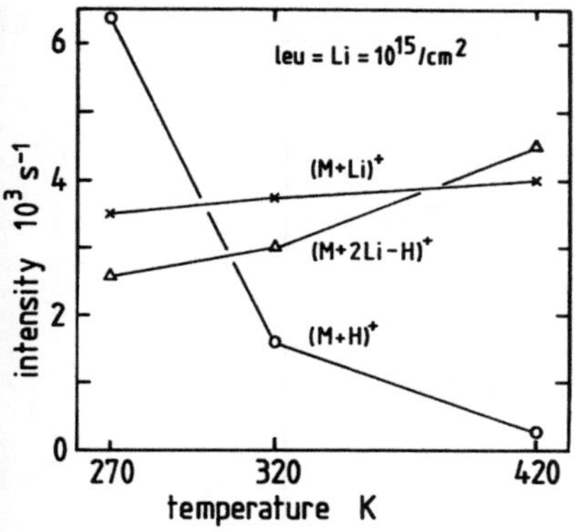

Fig.1 Cationization for a 1:1 leucine-LiCl mixture by annealing for 2 minutes at various temperatures.
Primary ion current density:
$4x10^{-9}$ Acm-2

Similar variations of molecular ion intensities are achieved by increasing the concentration of LiCl in the solution of leucine. Fig.2 shows that a Li excess advances the attachment of Li to the leucine molecules, simultaneously reducing the intensity of protonated parent ions $(M+H)^+$drastically.

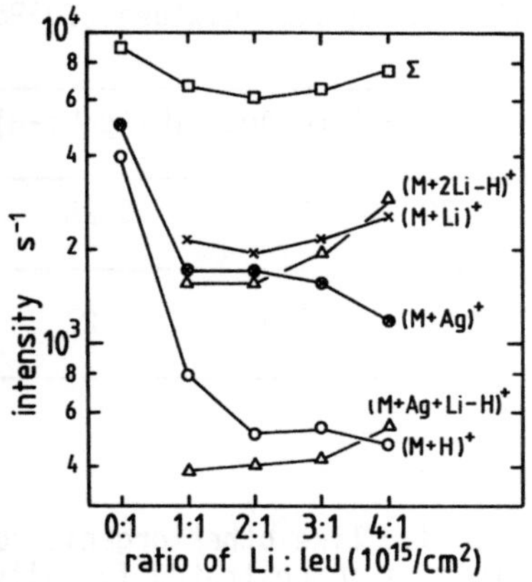

Fig.2 Cationization for leucine-LiCl mixtures of various relative concentrations

Figure 3 shows the decrease of $(M+H)^+$ and some cationized ions with increasing primary ion dose density. From this kind of experimental data, damage cross sections can be calculated. Table 1 presents cross sections and relative intensities for the most important molecule-like positive ions emitted from a 1:1 mixture of LiCl and leucine. It appears that the damage cross sections for all cationized ions are small compared with the protonated parent ion. It may be of some interest for the analytical application that these small damage cross sections enable higher primary ion current densities, which result in a corresponding increase of cationized secondary ion intensities.

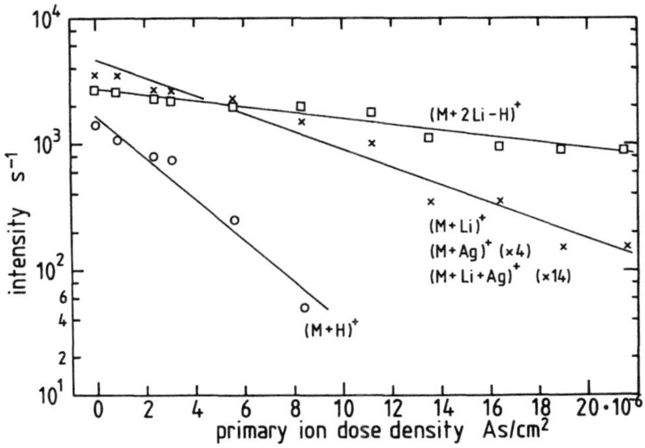

Fig.3 Disintegration and sputtering of a leucine-LiCl mixture (10^{15} cm^{-2} each) on Ag during ion bombardment

Table 1 Initial intensities N_0 and damage cross sections σ for molecule-like secondary ions emitted from a 1:1 leucine-LiCl mixture ($10^{15}cm^{-2}$ each) on Ag

Secondary ion	$(M-COOH)^+$	$(M+H)^+$	$(M+Li)^+$	$(M+2Li-H)^+$	$(M+Ag)^+$	$(M+Ag+Li-H)^+$
N_0 (10^3 s^{-1})	3.98	1.66	4.06	2.70	1.03	0.28
σ (10^{-14} cm^2)	1.16	6.00	2.50	0.86	2.45	2.27

4. Conclusion

Beside protonated molecular ions $(M+H)^+$, amino acids, like other organic compounds[3], produce cationized secondary ions. This is demonstrated for silver supported leucine-LiCl mixtures. The process of cationization is stimulated as well by LiCl excess as by annealing at relatively low temperature. Compared with $(M+H)^+$, the damage cross section σ is largely reduced for all cationized secondary ions as $(M+Li)^+$, $(M+2Li-H)^+$, etc..

References

1 A.Benninghoven and W.K.Sichtermann, Anal.Chem. 50 (1978) 1180
2 A.Benninghoven, D.Jaspers and W.Sichtermann, Appl.Phys. 11 (1976) 35
3 H.Grade and R.G.Cooks, J.Anal.Chem.Soc. 100 (1978) 5615

V. Metallurgy

Application of SIMS to Analysis of Steels

K. Tsunoyama, T. Suzuki, Y. Ohashi and H. Kishidaka
Research Laboratory
Kawasaki Steel Corporation
1, Kawasaki-cho, Chiba 260, Japan

1. Introduction

The mechanical and chemical properties of steels are largely influenced by the concentrations and distributions of impurity elements. A variety of analytical techniques have been applied to analysis of these elements, and it has been proven that secondary ion mass spectrometry (SIMS) is one of the most useful. SIMS has great sensitivity with excellent lateral resolution and can accomplish surface analysis. These capabilities make SIMS a unique technique for solving several metallurgical problems such as corrosion, passivation, toughness and brittleness. The present paper is a brief review of studies performed in our laboratory using an IMMA made by Applied Research Laboratories.

2. Sample Preparation

Since sputtering and ionization are surface sensitive phenomena, it is of primary importance to establish the method of sample preparation. Several kinds of polishing techniques were applied and it was found that mechanical polishing with the fine Al_2O_3 powder was most suitable for iron base alloys [1]. The other abradants, Cr_2O_3 and diamond paste, left large amounts of Cr, Al or Si on the surface of pure irons. These contaminants could not be eliminated by ultrasonic cleaning.

In addition to mechanical polishing, chemical and electropolishing were applied to Fe-0.1%Al alloys. Using the specimen polished with Al_2O_3 powder as reference, the ion image of Al was observed. In the case of chemical polishing the intensity of Al ions was enhanced at grain boundaries, while the intensity was different with grains in the electropolished specimen. The inhomogeneity of the ion intensity is considered to be caused by selective etching along the grain boundaries or preferential oxidation of grains.

On the surface of the sample polished with Al_2O_3, there remained several impurity elements, Na, Ca, K, Mg and Al. But the thickness of the contaminated layer is less than 20Å and they are easily eliminated by preliminary sputtering.

3. Bombarding Ion Species

When we make an ion microprobe analysis, it is necessary to select appropriate primary ion species, accelerate them to proper energy and control their intensity along with beam diamter. In the hollow cathod duoplasmatron ion source a variety of ion species can be generated, but O_2^+, N_2^+, Ar^+, are generally used for metallurgical studies.

We have applied these ions to the analysis of steels and found that O_2^+ is the most appropriate. It enhances the yield of sputtered positive ions [2], erodes the surface of iron quite uniformly [3], and reduces the variation of the relative sensitivity of alloy elements [4]. In addition to these advantages, the ratios of ion intensities obtained for O_2^+ bombardment are independent of the energy, current intensity and beam diameter of the primary ions.

Such stability of ion intensity ratios cannot be obtained for Ar^+ bombardment. As pointed out by many investigators [5], the relative intensities depend not only on the bombarding conditions but also on the oxygen partial pressure in the sample chamber of the spectrometer.

In the case of N_2^+ bombardment, relatively stable ion intensity ratios can be obtained provided the beam diameter is large enough. If the beam diameter is reduced below a certain value, then the ion intensity ratios vary. Fig.1 shows the variation of the intensities of Fe^+ and Cr^+ from stainless steel with the beam diameter of N_2^+.

We therefore adopt O_2^+ as the basic bombarding ion species for metallurgical studies and N_2^+ as the second choice. All of the experiments described below were performed with oxygen primary ions.

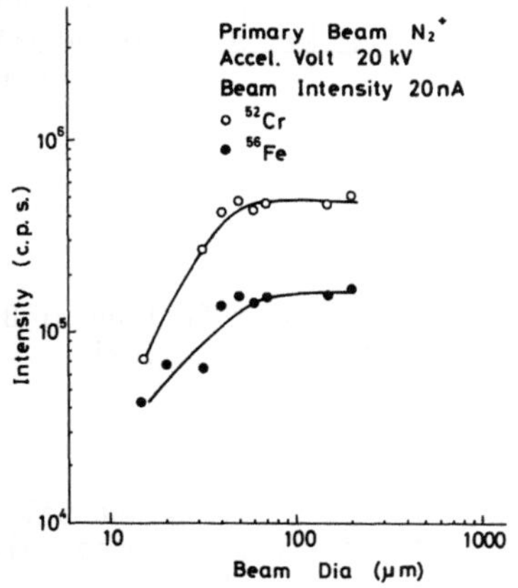

Fig.1 Variation of the intensities of Fe^+ and Cr^+ with the beam diameter of N_2^+

4. Quantitative Analysis

For the quantitative analysis of iron base alloys we have applied the calibration curve approach. This procedure is quite practical if homogeneous and well-characterized standard alloys can be obtained.

After preliminary sputtering, the primary ion beam is focused at the center of the pre-sputtered area and the intensities of secondary ions are measured in vacuum of ~10^{-7} Torr. To reduce the effect of the inhomogeneity of the sample, the diameter of the incident ion beam is defocused to more than 100 μm. A preliminary experiment showed that the relative standard variation of the ion intensity ratio of Mn, Ni, or Cr to Fe, obtained by bombarding 5 points in NBS 466 standard alloy with an O_2^+ ion beam of 100 μm in diameter, was up to 4%. Calibration curves were obtained by using several dilute binary iron alloys made by the Iron and Steel Institute of Japan.

Table 1 shows a typical example of quantitative analysis of low alloy steel. NBS standards 1261, 1262, 1263 were bombarded with O_2^+ at the energy of 20 keV. The diameter of the incident ion beam was 200 µm and the current was 50 nA. Relatively good analytical values are obtained for 42 elements in three samples. The relative errors of 22 elements are within 30%. The poorer results obtained for P, Zr and Mo may be caused by precipitates in ISIJ standards.

Table 1 Quantitative analysis of NBS standards

	1261		1262		1263	
	standard value	analytical value	standard value	analytical value	standard value	analytical value
B	0.0005	0.0005	0.003	0.004	0.0010	0.0015
Al	0.028	0.028	0.08	0.07	0.25	0.18
Si	0.20	0.22	0.39	0.34	0.78	0.54
P	0.017	0.007	0.043	0.014	0.026	0.007
Ti	0.019	0.020	0.09	0.12	0.06	0.05
V	0.011	0.011	0.04	0.03	0.29	0.24
Cr	0.69	0.56	0.29	0.24	1.33	1.05
Mn	0.66	0.65	1.03	1.01	1.49	1.48
Co	0.024	0.033	0.29	0.31	0.05	0.04
Ni	1.99	2.34	0.59	0.59	0.31	0.39
Cu	0.048	0.057	0.49	0.62	0.10	0.15
Zr	0.01	0.03	0.21	0.62	0.048	0.081
Nb	0.02	0.01	0.28	0.26	0.049	0.020
Mo	0.20	0.35	0.070	0.109	0.029	0.070

(wt %)

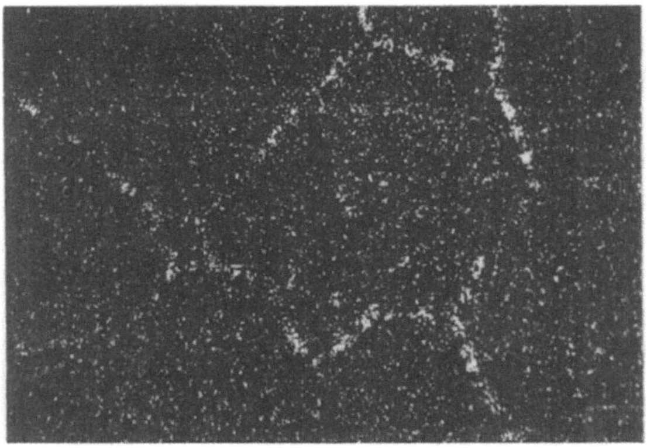

Fig.2 Distribution of B in high strength steel

5. Microanalysis

One of the merits of IMMA is the possibility of high sensitivity spatial analysis of alloy elements. The distribution of light elements, especially H and B, is quite interesting for the metallurgist. Fig.2 shows the secondary ion image of B in a high strength steel. Much instructive information has been obtained for understanding the role of these light elements.

In addition to these visual observations, quantitative spatial analysis is required in several cases. To test the possibility of such an alaysis, we have proceeded with the following basic study.

On the surface of pure iron, a thin film of manganese was deposited by vapor deposition. The specimen was annealed in a vacuum of 10^{-4} Torr at 1200°C for 24 hr. and its cross-section was analyzed by IMMA and an electron probe micro analyzer (EPMA). This ion microprobe analysis was performed with O_2^+ at 20 keV. The beam diameter was 3 µm and the current intensity was 4 nA. Fig.3 shows the variation of the intensity of Mn ions obtained at the intervals of 5 µm. The same locations were analyzed by the electron micro probe with the same beam diameter and the analytical values obtained

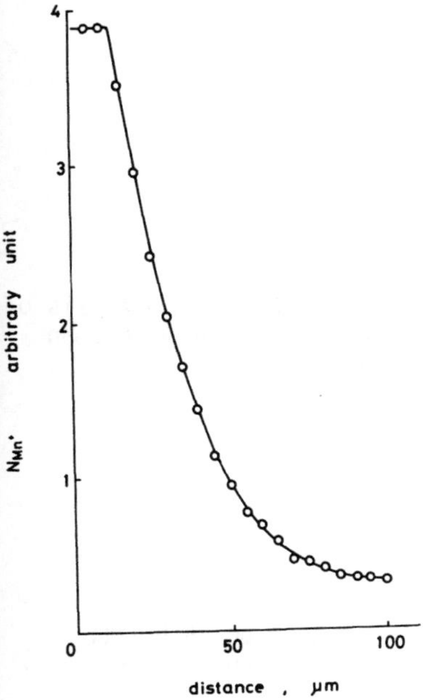

Fig.3 Diffusion of Mn into iron matrix

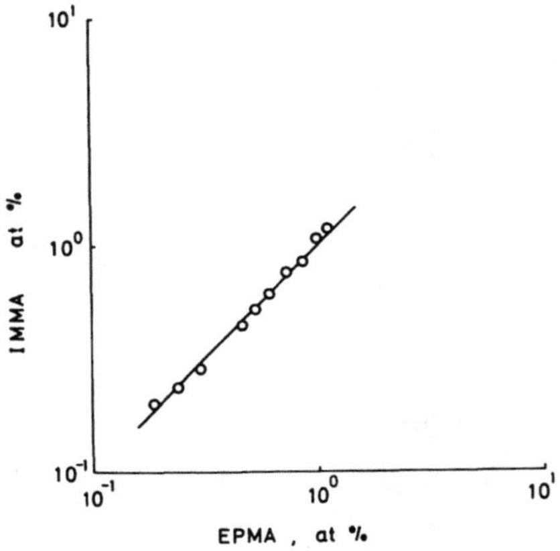

Fig.4 Comparison of the analyses of IMMA with those of EPMA

by both instruments were compared. As can be seen in Fig.4, the analytical results show good correlation. One may therefore say that IMMA can achieve similar accuracy to that of EPMA.

6. In-depth Analysis

During heat treatment of metals, the impurity atoms dissolve in the matrix and tend to segregate to the surface. The segregation of impurity elements affect the adherence to coatings of the oxidization resistance of metals. SIMS has high potential for providing sensitive depth profile information on these problems.

In the case of steels, however, the surface layer is more or less oxidized and the interpretation of the measured depth profile is somewhat complicated. We have recently found that the calibration curve approach can be successfully applied for quantitative estimation of minor element concentration in oxide layers [6].

We have also found that the relative ion intensity of Fe^+ to O^+ is different for oxide film and base steel. This difference makes it possible to distinguish the oxide layer from base metal despite the incidence of oxygen ions. Fig.5 shows the in-depth profile of Fe, Mn, and O obtained for Fe-0.2% Mn alloy annealed at 450°C for 24 hr. in laboratory air. The relative intensity of Fe^+ to O^+ is ~14 in Fe_3O_4 layer and then increases up to ~23 as the oxide layer is eroded away.

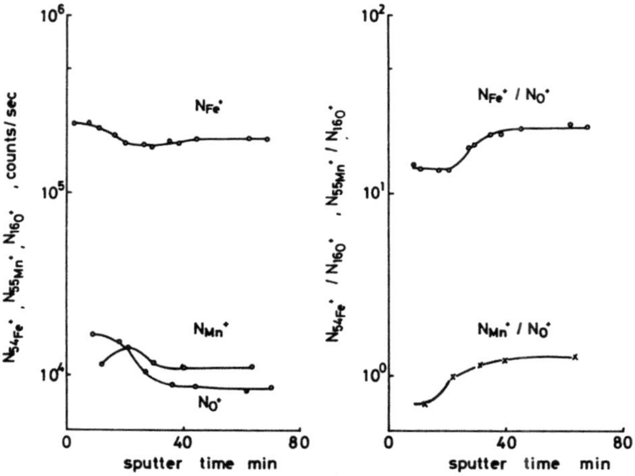

Fig.5 In-depth analysis of Fe_3O_4 layer on Fe-0.2%Mn alloy

References

1. K. Tsuruoka, K. Tsunoyama, Y. Ohashi and T. Suzuki, "Proc. 6th Int. Vacuum Cong., Kyoto, March 1974," Jpn J. Appl. Phys., Suppl. 2, Pt. 1, 391 (1974).
2. C.A. Andersen, "Third Natl. Electron Microscope Conf., Chicágo, Ill., 1968," Int. J. Mass Spectrom. Ion Phys., 2 61 (1968).
3. K. Tsunoyama, Y. Ohashi and T. Suzuki, Jpn J. Appl. Phys., 13, 1683 (1974
4. K. Tsunoyama, Y. Ohashi and T. Suzuki, Anal. Chem., 48 832 (1976).
5. For example, K. Kusao, Y. Yoshioka and F. Konishi, National Tech. Rep., 23 14 (1977).
6. K. Tsunoyama, T. Suzuki and Y. Ohashi, "Second Japan-United States Joint Sem. On SIMS, Takarazuka, Oct. 1978" p. 42

Investigation of Metal Corrosion Mechanisms Using Stable Isotopes with the Ion Microprobe

S.S. Cristy and J.B. Condon

Y-12 Development Division
Oak Ridge Y-12 Plant
Oak Ridge, TN 37830

The ability of the ion microprobe or SIMS instrument to distinguish between isotopes and to analyze thin layers sequentially by sputtering provides unique methods well suited for studying oxidation and corrosion mechanisms. Although many SIMS studies of oxidation have been reported, very few have taken advantage of isotope substitution [1-4]. The ability to directly observe the order and direction of events in a corrosion reaction in relatively thin films should be of importance to metallurgists and material scientists, particularly those who are interested in oxidation or corrosion at relatively low temperatures or under mild conditions.

One method which was used to study the oxidation and fluorination of nickel consisted of depositing a thin layer (∿100 A) of ^{62}Ni enriched nickel on the surface of a nickel substrate, exposing to the reactant, and depth profiling the $^{62}Ni/^{58}Ni$ isotopic ratio [1]. In the oxidation case, the nickel substrate atoms diffused through the oxide layer to react with oxygen at the surface, thus leaving the ^{62}Ni enriched layer beneath the last oxidized nickel atoms from the normal substrate. However, in fluorination, the ^{62}Ni layer was not dislocated, showing that the fluorine diffused through the fluoride to react at the metal-to-fluoride interface.

A second method requires exposure of the reaction surface to an isotopically altered reactant followed by exposure to natural abundance reactant (or vice versa) before SIMS depth profiling. The mechanisms of the reactions of water and oxygen with uranium are being investigated using this method.

Our first experiment involved exposing a polished uranium coupon at 80°C to 610 Pa of $^{16}OH_2$ for four hours followed by exposure to 610 Pa of $^{18}OH_2$ for four hours. The profiles showed that the water adsorbed last produces the oxide nearest the metal. Thus, oxygen (in some form) diffuses to the oxide/metal interface to react. The profiles also indicated a limited exchange between the mobile and immobile oxygen species.

Also of interest is the role of hydrogen in the reaction. OH^- and $OH\cdot$ are suspected diffusing species in water oxidation of uranium, and the formation of uranium hydride as an intermediate in the oxidation or as a side product is a possibility. To test this, uranium was exposed to $^{18}OH_2$ followed by $^{16}OD_2$. The deuterium profile was found to mimic the coprecipitant oxide, ^{16}O. No buildup of deuterium was seen at the oxide/metal interface as would be expected if a hydride of finite lifetime were an important intermediate in the oxidation process. The deuterium concentration in the oxide was ∿0.4% atomic. Thus, hydrogen was carried into the oxide, but only a very little did not return to the surface to be expelled as H_2 gas. The deuterium level in the first oxide layer showed that the mobile species trapped at the end of the exposure is no more than 0.2% atomic. This result affirmed an oxygen exchange rather than merely a mixing of mobile species.

Multiple exposure experiments with sequential exposures to $^{16}OH_2$, $^{18}OH_2$, $^{16}OH_2$, $^{18}OH_2$, and $^{16}OH_2$ have provided additional insight. These exposures showed that the OH species migration utilized interstitial positions with the permanent lattice positions essentially unaffected by their passage. The slope of the $^{16}O^-$ signal between the first and second layers was the same as between the third and fourth, and the slope of the $^{18}O^-$ signal between the second and third layers was the same as between the fourth and the fifth layers. Thus, the slopes or "tail-offs" between layers were due to the final usage of the migrating or exchangeable species and not to interdiffusion in the oxide lattice (no counterflowing or coflowing ions or vacancies on the oxide lattice to balance the inward migrating species).

The multiple exposure experiments also revealed that the rate-limiting step must be at the oxide/metal interface. If the rate-limiting step were at the outer oxide surface, a concentration gradient of migrating species and little backflow of these species would be expected. This would cause the diffusing species to be more enriched in ^{18}O at the end, after passing through two ^{18}O rich layers, than at the beginning. This was not observed. With the rate-limiting step at the oxide/metal interface, the migrating species have time to backflow and to establish an equilibrium between diffusing and exchangeable species. Under this condition, a relatively equal level of ^{18}O in the ^{16}O layers would be expected and was observed.

Multiple exposure experiments showed linear reaction rates after an incubation time associated with formation of the first layer. This incubation time also varied depending upon the crystal orientation of the uranium lattice.

Exposure of uranium to oxygen reveals some similarities and some distinct differences from water exposure. After exposing uranium to 1.3 kPa of $^{16}O_2$ for 48 hours at 80°C followed by 1.3 kPa of $^{18}O_2$ for 48 hours at 80°C, the following was noted. 1) The total quantity of ^{16}O detected in the oxide layer was approximately five times the quantity of ^{18}O. This is consistent with a reaction rate law that relates total oxygen uptake versus time to a fractional power such as the "parabolic" law. 2) Interstitial migration of reactant oxygen is involved. 3) Oxygen concentration into the metal decreased much more gradually than for water exposure. 4) When the sample was held at temperature, the migrating oxygen species continued to migrate through the film and react with the metal. These features favor a migration mechanism whereby oxygen migrates into the metal at a constant average velocity and is absorbed. Similar features and conclusions are evident for the reaction of oxygen with U-6Nb and thorium.

Investigations of the water and oxygen reactions with uranium are continuing with more detailed and quantitative ion microprobe studies utilizing isotopic labeling planned. Computer modeling is expected to aid in interpretation of profiles and in designing of critical experiments.

References

1. S.S. Cristy, D.V. Ferree, T.A. Nolan, and W.H. McCulla, Studies of Oxide and Fluoride Films on Metals Using an Ion Microprobe Mass Analyzer, Y-DA-4815, Union Carbide Corporation-Nuclear Division, Oak Ridge Y-12 Plant, Oak Ridge, TN; 1972.
2. M. Croset and D. Dieumegard, Corrosion Science, 16, 703 (1976).
3. C.A. Evans, Jr., and J.P. Pemsler, Anal. Chem., 42, 1060 (1970).
4. J.A. McHugh, Methods of Surface Analysis, A.W. Czanderna, ed., p. 269 (Elsevier Scientific Publishing Co., New York, 1975).

Investigations of Corrosion Layers on Mild Steel with a Direct Imaging Mass Spectrometer [1]

A. Pebler and G. G. Sweeney
Westinghouse Research & Development Center
Pittsburgh, Pennsylvania 15235

1. Background

The growth of magnetite (Fe_3O_4) on mild (carbon) steel surfaces immersed in high temperature water is of considerable practical importance to the electrical power industry because of the extensive use of mild steel in boilers. In neutral or mildly alkaline oxygen-free water iron spontaneously forms a corrosion-resistant, fine-grained Fe_3O_4 barrier without significant dimensional change. About half the iron consumed remains fixed in the position it occupied before being oxidized by the inward migration of oxygen-bearing anions, thus establishing a tightly-adherent film. The remaining oxidized iron diffuses, as cations, outward to the oxide-water interface where reaction with water produces a loose, coarse-grained Fe_3O_4 that offers no protection and is largely flushed away. When the aqueous environment carries a sufficient concentration of dissolved iron, the outward movement of iron cations through the incipient Fe_3O_4 film is suppressed, so that the film continues to grow almost entirely at the metal/oxide interface. The stresses resulting from the specific volume mismatch between Fe and Fe_3O_4 are relieved by reoccurring failures (breakaway) in the oxide, which render it nonprotective to continued oxidation of iron at rates that may be controlled by the cathodic half-cell reaction at available external surface sites [1]. An acid chloride solution, in particular, can keep iron cations in solution. Such conditions may arise in industrial boilers, when initially small sea water impurities (from condenser inleakage) are concentrated by several orders of magnitude in flow-starved regions near heat transfer surfaces, such as in crevices and underneath loose deposits of corrosion products.

This work is part of an investigation to simulate in the laboratory a unique case of crevice corrosion on mild steel in an acid chloride-containing medium that occurs in certain utility boilers and to find practical solutions to alleviate or arrest the concern [2]. A CAMECA Direct Imaging Mass Analyzer (DIMA) proved to be a valuable tool for the examination of crevice specimens for the following reasons:

- High detection sensitivity for lightweight elements of interest, including H, Li, B, O.
- High lateral resolution in the order of µm.
- Equally sensitive to electropositive and electronegative species, such as OH^-, Cl^-, BO_2^-, PO_2^-.

[1] This work was supported in part by EPRI under contracts RP 699-1 and S-112-1.

2. Experimental

Accelerated corrosion of mild steel was initiated in a crevice, formed between an Inconel 600[2] tube and a surrounding mild steel ring, by hydrolyzable metal chlorides, such as $MgCl_2$ (in sea water), $NiCl_2$, or $CuCl_2$. The corrosion tests were performed in autoclaves at about 300°C either under isothermal conditions, in which case the corrodants were added as fairly concentrated solutions, or under conditions of high heat flux, in which case initially low levels of impurities were concentrated in the crevice area. The progression of linear corrosion was recorded by periodically measuring the deformation of the Inconel tube. Upon termination of the tests, the crevice specimens were cut up and mounted for metallographic and DIMA examination. Ar^+ and O^+ primary beams were applied for the secondary ion mass analysis to generate both positive and negative secondary ions. The use of Ar^+ instead of O^+ enables one to discriminate better between metallic and oxidic phases.

3. Results

A typical set of ion images taken across the crevice area of a corrosion specimen is shown in Fig. 1. The original crevice appears to be completely

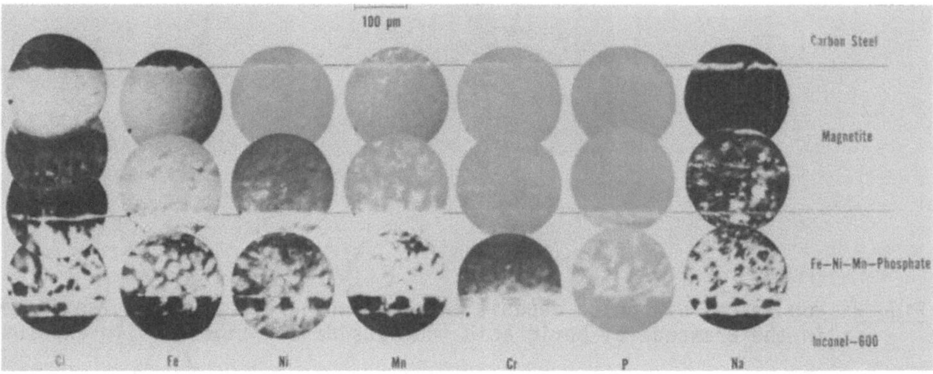

Fig. 1 Series of ion images of corrosion products in a mild steel/ Inconel 600 crevice

packed with corrosion products including a Fe_3O_4 layer on the mild steel and a layer of low soluble heavy metal phosphates between the Fe_3O_4 layer and the Inconel tube. Of particular interest is the accumulation of chloride next to the Fe/Fe_3O_4 interface. Since Na^+ is practically absent from the same area, one concludes that Cl^- is present as acid chloride. This environment maintains a sufficiently high concentration of Fe ions in solution resulting in the observed nonprotected iron corrosion. The presence of a mixed Fe-Ni-Mn phosphate can be concluded by comparing the Fe^+, Ni^+, Mn^+, and P^+ images.

When boric acid was added to the corrosive environment in sufficiently high concentration, the corrosion rate of mild steel was considerably reduced [2]. It was therefore of interest to identify the presence, distribution, and chemical nature of boron in crevice specimens that had

[2] Inconel is a nickel base alloy with 15.7% Cr, 9.5% Fe.

corroded in the presence of boric acid. Through systematic DIMA-aided microchemical and selective X-ray diffraction analyses, we established that an Fe-Mg-Mn-boracite with the gross composition $(Fe_{1-x-y}Mg_xMn_y)_3B_7O_{13}\cdot X$ (where X can be Cl or OH) precipitated in breakaway cracks in the Fe_3O_4 layer near the corroding mild steel. The constituents that form the mixed boracite phase are the corroding mild steel (Fe and Mn) and sea water impurities (Mg and Cl) in addition to boric acid. We think that insoluble borates plug up residual pores and cracks in the magnetite scale and thereby inhibit the mass transfer of corrosive species to the iron-magnetite interface.

A series of ion micrographs of a borate deposit and the surrounding area is shown in Fig. 2. The identity of the borate deposit is amply demonstrated

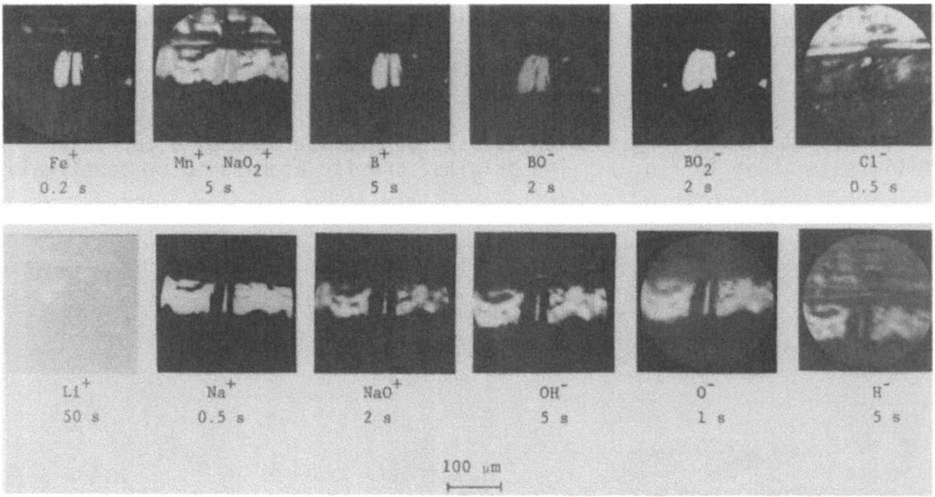

Fig. 2 Ion images of borate deposit and adjacent areas in magnetite formed in the presence of boric acid (numbers denote time of film exposure)

by the Fe^+, B^+, BO^-, and BO_2^- images. Chloride is present in the area between the borate deposit and the mild steel. Sodium hydroxide (NaOH) appears to be accumulated in areas surrounding the borate deposit, as seen by the close similarity of all major ion fractions of NaOH. The presence of NaOH can be associated with the cathodic half-cell reaction of the corrosion process.

4. Conclusions

A direct imaging mass analyzer has contributed to the identification and distribution of corrodants and corrosion products in the crevice corrosion of mild steel in aqueous acid chloride environments. The identification and distribution of borate deposits has led to a plausible explanation for the inhibiting effect of boric acid on acid chloride corrosion of mild steel.

References

1. E. C. Potter and G.M.W. Mann, Brit. Corrosion J., 1, 26 (1965).
2. M. J. Wootten, G. Economy, A. R. Pebler, and W. T. Lindsay, Jr., Materials Performance, 17, 30 (1978).

The Use of SIMS in the Oxidation of Metals

J.S. Sheasby and J.D. Brown
Faculty of Engineering Science
The University of Western Ontario
London, Ontario, Canada N6A 5B9

The study of the modes of growth of oxide scales has hitherto been made difficult by the absence of a radioisotope of oxygen with a convenient half-life. The advent of SIMS has largely superseded this deficiency and now tracer studies can be performed with the convenience and resolutions of a conventional electron-probe analyzer. The concept of the tracer technique applied to oxidation is straightforward. Specimens are oxidized first in ^{16}O and subsequently in ^{18}O. SIMS is then used to locate the ^{18}O to determine the location of new oxide growth. To a first approximation if the metal is the most mobile species in the oxide the new oxide will be sharply concentrated at the oxide-gas interface; whereas, if oxygen is the more mobile, an error function type of decay profile from the oxide-gas interface would be anticipated.

In reality other factors have been observed and deduced to affect scale growth and these include grain boundary diffusion of the slower moving species, grain growth, scale flow and spalling, and the development of porosity. To further elucidate modes of scale growth it is therefore necessary that the SIMS analysis has as high an accuracy and resolution as possible. These processes are illustrated with analyses of CoO formation in which the metal ion is the more mobile; and with Nb_2O_5 formation in which oxygen is the most mobile species.

1. Experimental

Specimens were oxidized in a system designed to minimize oxygen exchange with the furnace walls. To achieve this the specimens were heated by a quad-ellipticallight furnace and the silica reaction tube maintained cool by multiple air jets. Specimens were heated to temperature in vacuum, oxidized for the selected period in ^{16}O which was then pumped out and replaced within 60s by ^{18}O. Oxidation was continued for a period estimated to give either 5 or 10% further scale growth whereupon the specimens were cooled in the ^{18}O atmosphere. Every second specimen was oxidized only in ^{18}O to provide concentration standards. The ^{18}O was 92% enriched and the concentrations reported are fractions of this value.

2. Analysis of $^{18}O/^{16}O$ Gradients Using an ARL IMMA

The oxidespecimens were analyzed using a number of different techniques in an Applied Research Laboratories Ion Micrprobe Mass Analyzer (IMMA). The primary beam was $^{28}N_2^+$ ions at an energy of 21.5 keV, the energy being the sum of the duoplasmatron voltage plus the 1.5 kV negative applied to the specimen for extraction of negative secondary ions. The primary beam current was typically 1-2 nA with the beam diameter of approximately 1-2 μm. $^{16}O^-$ and $^{18}O^-$ secondary ion intensities were measured either by setting the secondary magnet for mass 17 and using the ratio plates to deflect mass 16 or mass 18 ions into the detector or by setting the secondary magnet to the mass 16 or 18 position.

Three different methods of analysis were used to determine the full $^{18}O/^{16}O$ concentration profiles, a surface analysis, depth profiling and step scans. To determine the surface $^{18}O/^{16}O$ ratio, the primary beam was rastered over an area of 50 μm x 50 μm while measuring the ^{16}O and ^{18}O intensities by switching using the ratio plates about 10 times per second. Total analysis time was 2 seconds.

The depth profiling was accomplished again with the primary beam rastering over an area of 40 μm x 40 μm but the secondary ion intensities were measured by changing the magnet current under computer control so that the ^{16}O and then the ^{18}O peak intensities were measured alternately. Each pair of measurements required about 10 seconds. An electronic aperture was used to ensure that data originated from the central 30% of the crater area.

Step scans were used to establish the concentration gradients over larger distances on either taper-sectioned or cross-sectioned specimens. Again using computer control of the instrument, each analysis point was measured using the following sequence. The primary beam was blanked and the stage moved a preset distance in a selected direction. The beam was unblanked and a burn in delay time of 15 seconds allowed to pass before the measurement of $^{16}O^-$ and $^{18}O^-$ ion intensities. Typical counting times were 10 seconds on each peak. After measuring, the beam was again blanked and the process repeated. An important consideration in the step scans is the primary beam diameter which determines the resolution of the analysis. To maintain high secondary ion intensities, a higher current but elliptically shaped primary ion beam was used giving a lateral resolution of approximately 1 μm in the step scan direction.

3. Results and Discussion

Niobium

In the temperature range 740-840°C, Nb_2O_5 on flat areas of specimens is solid and protective and grows approximately at a parabolic rate. Oxygen is the more mobile species in this oxide and it is therefore anticipated that the ^{18}O distribution will be of the erfc type being 100% at the oxide-gas interfact falling to zero within the scale. This distribution was basically that observed except in detail near the oxide-gas interface. In this region the ^{18}O distribution peaked at ~90% and actually fell very close to the interface. Further consideration was given to these observations to determine whether they were an artifact of the analysis, of the oxidation technique or intrinsic to the oxidation reaction. It was noted that the standard specimens, nominally oxidized only in ^{18}O, also showed a fall in ^{18}O concentration very close to the interface. However, if those same specimens were polished to remove the surface layers, remeasurement of the ^{18}O showed no surface dilution by ^{16}O. The observations were therefore not an artifact of the analysis. It is therefore deduced that the ^{18}O depleted surface layer formed in the imperfect vacuum on heating the specimens to temperature. It is of interest that this oxide was not absorbed on subsequent oxidation and it is suggested that this is due to the mode of movement of oxygen in the surface layers of the oxide. In this layer oxygen ions move down crystallographic tunnels before incorporation in the oxide.

At temperatures below 740°C, Nb_2O_5 scales form by a para-linear mechanism in which scale layers grow at a parabolic rate to a critical thickness whereupon they detach and a new layer forms at the metal-oxide interface. This layer in turn reaches the critical thickness and detaches and the process repeats. In this mode of scale growth the newest oxide is near the metal-oxide interface not the oxide-gas interface. The measured ^{18}O distribution across layered scales grown at 600 and 720°C confirmed this model.

Cobalt

CoO scales basically grow in a protective manner on cobalt so that the kinetics are approximately parabolic with time. In detail the scales are duplex, that is, with time the initially solid oxide develops porosity at the metal-oxide interface. While it is well established that cobalt is much more mobile than oxygen in the CoO lattice, there is less certainty of their relative mobility in duplex scales. In such scales oxygen movement has been postulated to occur by grain boundary diffusion, open porosity, and within closed porosity by scale dissociation or by reaction with C (present as an impurity).

To distinguish between these alternatives, a cobalt specimen was oxidized in ^{16}O and ^{18}O to no further than the compact stage. At this point, if only cobalt is mobile, the new oxide will be sharply segregated at the oxide-gas interface, whereas if the scale is in fact porous, new oxide will form near the metal-oxide interface, and if oxygen is mobile in grain boundaries, new oxide will form throughout the scale.

To avoid the formation of a duplex scale the oxide could be grown to no thicker than 15μ and it was therefore necessary to perform the SIMS analysis on a taper-section of the oxide. On this section the ^{18}O was sharply segregated at the oxide-gas interface at essentially 100% concentration. To confirm this distribution analysis was performed in depth profile from the oxide-gas interface. This was done with and without the electronic aperture described in the experimental section, clearly demonstrating the value of this technique when analyzing through a sharp change in concentration. With the window the ^{18}O distribution fell sharply within the oxide confirming that the CoO was non-porous and that oxygen mobility in the scale grain boundaries is low.

Influence of Atomic Concentrations on Ion Emission Yields of Alloys Flooded with Oxygen

C. Roques-Carmes and J.C. Pivin
Laboratoire de Métallurgie Physique, Bât. 413
Université Paris-Sud, 91405 Orsay Cedex, France

G. Slodzian
Laboratoire de Physique du Solide, Bât. 510
Université Paris-Sud, 91405 Orsay Cedex, France

1. Introduction

It has previously been shown that, for binary FeNi, FeCr and NiCr single phase alloys sputtered by argon ions and flooded with oxygen, the ionization probability P^M_{NM} of M atoms varies linearly with C_M or C_N [1]. Evidences for this law were given by the values determined for the following ionization coefficient :

$$\frac{P^M_{NM}}{P^M_M} = \frac{I(M^+)_{NM}}{I(M^+)_M} \cdot \frac{S_M}{S_{NM}} \cdot \frac{1}{(C_M)_{NM}} \qquad (1)$$

It was verified that S_{NM} is always equal to S_M in FeNi, FeCr, NiCr alloys flooded with oxygen. Since P^M_M is a constant and C_M is known, variations of P^M_{NM} could be deduced from measurements of the ionic intensities.

The linear variations of P^M_{NM} with C_M, C_N was written :

$$P^M_{NM} = P^M_M \cdot C_M + P^M_N \cdot C_N \qquad (2)$$

where P^M_M is the ionization probability of M in the oxide film developed on pure metal M, and P^M_N that of M atom isolated in the oxide film developed on a N matrix. This law was further extended to a ternary system FeNiCr [2] .

A more practical ionization coefficient for analysis is k^M_N :

$$k^M_N = \frac{P^M_{NM}}{P^N_{NM}} = \frac{I(M^+)_{NM}}{I(N^+)_{NM}} \cdot \frac{C_N}{C_M} \qquad (3)$$

which does not require measurement of sputtering yields nor intensities of standards M and N . The hyperbolic variation of this coefficient observed in FeCr, FeNi, NiCr alloys could be explained, substituting (2) for P^M_{NM} and P^N_{NM} :

$$k^M_N = \frac{P^M_{NM}}{P^N_{NM}} = \frac{P^M_M \cdot C_M + P^M_N \cdot C_N}{P^N_N \cdot C_N + P^N_M \cdot C_M} = \frac{P^M_M}{P^N_N} \cdot \frac{C_M + P^M_N/P^M_M \cdot (1-C_M)}{(1-C_M) + P^N_M/P^N_N \cdot C_M} \qquad (4)$$

Using this relation, 3 coefficients must be determined to obtain the variation of k^M_N in the whole range of concentrations : P^M_M/P^N_N ,P^M_N/P^M_M and P^N_M/P^N_N . For practical purpose measurements of k^M_N on only 3 standard alloys would be sufficient. IN this paper measurements of these 3 coefficients in other single phase alloys of two 3d transition metals (except Mn) and CuAl, NiAl, FeAl alloys will be presented and compared.

2. Results

Experimental procedure was defined in Ref. [1]. Fig. 1 shows the measured variations of k_{Ni}^{Al} with C_{Al}/C_{Ni} (full circles). The three coefficients P_{Al}^{Al}/P_{Ni}^{Ni}, P_{Ni}^{Al}/P_{Al}^{Al}, P_{Al}^{Ni}/P_{Ni}^{Ni} were fitted to obtain a variation of calculated k_{Ni}^{Al} (open circles and continuous curve) in agreement with measured values. Similar results were obtained on binary alloys VCr, VFe, VCo, VNi, FeTi, FeV, FeCo, NiCo and NiV [4] . An example is given in Fig. 2 for the calculation of k_{Ni}^{V} in the whole range of concentration with measurements of only 4 values.

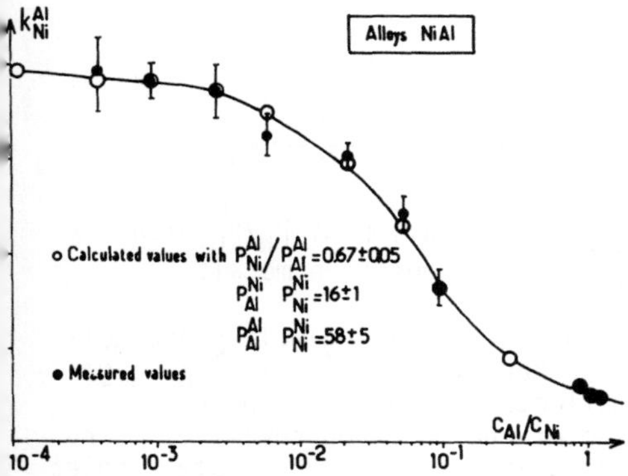

Fig. 1 Variation of k_{Ni}^{Al} with C_{Al}/C_{Ni} in NiAl alloys

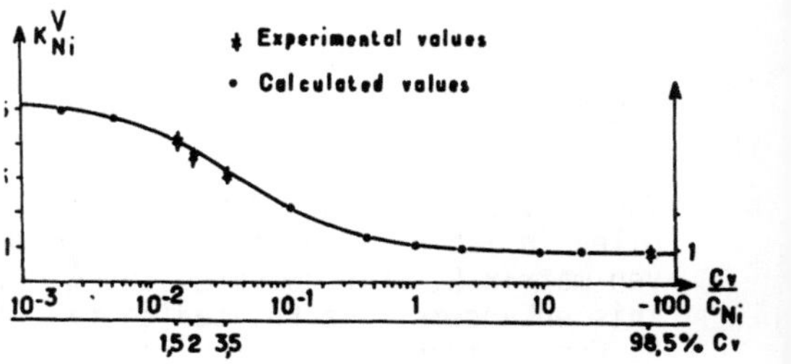

Fig. 2 Variation of k_{Ni}^{V} with C_V/C_{Ni} in NiV alloys

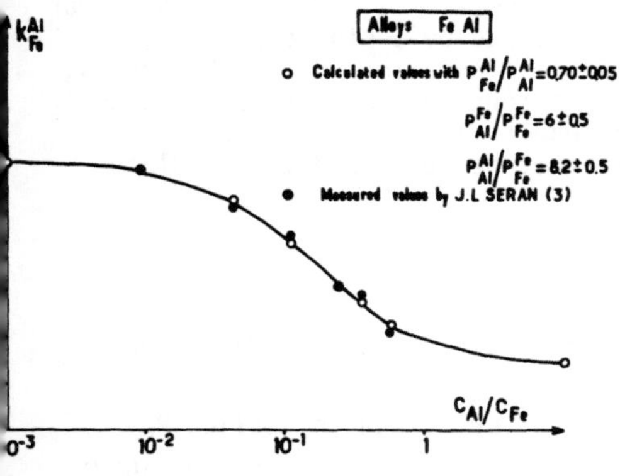

Fig. 3 Variation of k_{Fe}^{Al} with C_{Al}/C_{Fe} in FeAl alloys

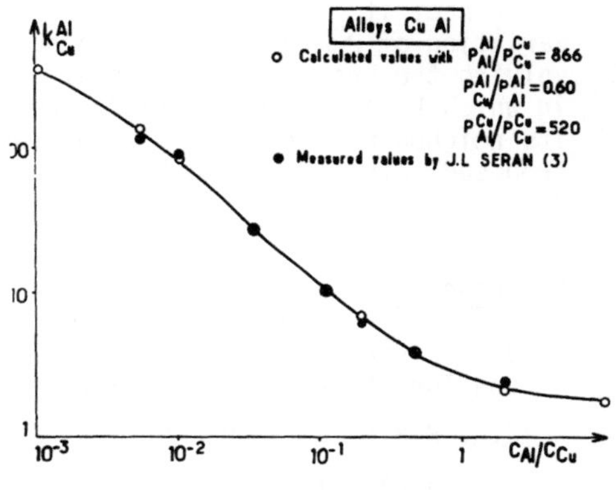

Fig. 4 Variation of k_{Cu}^{Al} with C_{Al}/C_{Cu} in CuAl alloys

Moreover the experimental variations of k_{Fe}^{Al}, k_{Cu}^{Al} obtained by J.L.SERAN [3] are well fitted by such calculations (Fig. 3 and 4). Note the important variation of k_{Cu}^{Al} (more than 100).

3. Discussion and Conclusions

P_N^M/P_M^M	M = Ti	V	Cr	Fe	Co	Ni
N =						
Ti	1			2.0		
V		1	1 9	3.5	15	23
Cr		0.45	1	2.7		20
Fe	0.88	0.49	0 67	1	2.4	6.1
Co		0.25		0.55	1	2.7
Ni		0.25	0.29	0 23	0.48	1
P_M^M/P_{Ni}^{Ni}	20	20	39	8.8	3.0	1

Table 1

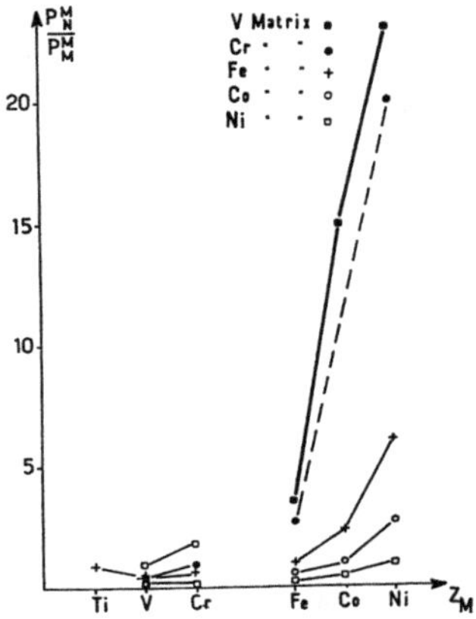

Fig. 5 Variation of P_N^M/P_N^M with Z_M in 3rd transition metals

The values of P_M^M/P_N^N and P_N^M/P_M^M for alloys of 3d transition metals are compiled in Table 1 and Fig. 5. The chemical enhancement of P^M in a given matrix N, when compared to P_M^M, is reflected by the value of P_N^M/P_M^M. This enhancement is related to the filling of the electron shells of M and N. For a given matrix (for instance V) the emission yield of solute species increases with Z_M ; the chemical enhancement of a given solute (Ni for example) decreases with Z_N from V to Ni.

If some of the properties of NO_x oxides formed on N matrix flooded with oxygen (i.e. energy for breaking a M-O bond, length of bonds in the oxide) vary in the same way with Z_N than P_N^M/P_M^M, there is no direct relation between P_M^M, P_N^M and these properties.

In fact it has been outlined in previous papers [1] [2] that the linear variation of P_{NM}^M with C_M, C_N cannot be explained by a static process, involving the breaking of M-O chemical bonds. A dynamic " binary " process, in which P^M is statistically determined by strong interactions of M with its neighbours during collisions (which lead to the sputtering of M) was proposed . In this case, a dynamic model must be developed to calculate P^M.

References

1 J.C. Pivin, C. Roques-Carmes, G. Slodzian , Int. Jour. of Mass Spect. and Ion Phys., 26, 219 (1978).

2 J.C. Pivin, C. Roques-Carmes, G. Slodzian .Submitted to Applied Phys.

3 J.L. Seran, Thesis Orsay 1978, and Jour. Spect. Electron. 2,323 (1977).

4 G. Bellegarde, Thesis Orsay 1978 (Laboratoire de Métallurgie Physique Université Paris-Sud, 91405 Orsay).

Application of SIMS and AES to Environmental Studies of Fatigue Crack Growth in Aluminum Alloys

Anna K. Zurek and H.L. Marcus
Department of Mechanical Engineering/
Materials Science and Engineering Program
The University of Texas at Austin
Austin, Texas 78712

1. Introduction

The purpose of this investigation was to study the effect of hydrogen, deuterium, water vapor and oxygen on the fatigue crack growth of high strength aluminum alloys. Fatigue crack growth tests in a controlled environment were followed by Secondary Ion Mass Spectrometry (SIMS) and Auger Electron Spectrometry (AES).

Influence of humid air on fatigue crack growth in high-strength aluminum alloys is an important practical problem well documented in literature [1-5]. The studies show an increase of crack growth rate of aluminum alloys in water vapor. In many cases it is attributed to hydrogen embrittlement. On the other hand gaseous hydrogen has no effect on fatigue crack growth [4].

Several models of hydrogen embrittlement for various materials have been considered [6-10]. The sweep-in mechanism during fatigue crack growth proposes combination of diffusion and dislocation sweeping of hydrogen into the plastic zone. This results in a hydrogen penetration deeper than $\sqrt{Dt}$, the characteristic depth for diffusion, and a consequent increase in hydrogen concentration that in some manner weakens the material [9,10]. The hydrogen effect is even more pronounced in the presence of various trapping sites such as particle interfaces, which are made more effective by the deformation associated with the propagating crack [9,10,11].

2. Experimental Procedure

The age hardenable aluminum alloys used in this study are 7075-T651 (chemical composition weight percent: 5.6% Zn; 2.5% Mg; 1.6% Cu; 0.3% Cr; and σ_y=505 MPa; K_{IC}=25 MPa$\sqrt{m}$) and 2219-T851 (chem comp.: 6.3% Cu; 0.3% Mn; 0.18% Zn; 0.1% V; 0.06% Ti; and σ_y=350 MPa; K_{IC}=35 MPa$\sqrt{m}$). The compact tension fracture mechanics type specimens were used in the fatigue crack growth experiments [12]. The fatigue crack was first grown in a 1 μPa vacuum at a constant ΔK (ranging from 10 to 21 MPa$\sqrt{m}$) and frequency (6 or 10 Hz). The chamber was then filled with one of the gases hydrogen, water vapor, oxygen-18, deuterated water vapor or deuterium gas. Fatigue crack growth was then continued in the environment, up to a certain crack length and then the specimen was fast fractured. Crack propagation rate was monitored optically. Specimens cut from the fracture surfaces were then transferred to AES/SIMS vacuum system. Hydrogen, deuterium and oxygen 18 concentration profiles normal to the fracture surface were investigated by inert ion sputtering combined with SIMS. Negative SIMS was utilized in case of deuterium due to the higher yield of D^- over D^+ ions. For oxygen AES was used in parallel. In addition deuterium ion implanted standards were prepared using a high energy ion implantation. These were used for calibration purposes of the SIMS system and as a source of measuring induced trapping effects.

3. Results and Discussion

Experimental difficulties were encountered in measuring the atomic hydrogen profile with SIMS in the presence of the $3x10^{-5}$ torr argon sputtering gas. Therefore, oxygen 18 and deuterium as isotope markers in SIMS profiling were used.

Fatigue crack propagation rate for dry oxygen 18 did not significantly increase in comparison to vacuum (1.1 times). The ion sputtering profiles determined by AES and SIMS for 2219 and 7075 aluminum alloys have been published elsewhere [12,13]. A significant enhancement of the thickness of the oxide for the region fatigued in oxygen 18 by a factor of about 2 to that fatigued in vacuum and subsequently exposed to oxygen 18 occurred in both cases. Dry hydrogen or deuterium also show no change in fatigue crack growth rate when compared to vacuum.

Water vapor (H_2O or D_2O) strongly influences the fatigue crack growth rate The ratio of crack growth rate in water vapor to that in vacuum varied from 1.8 to 3.6 depending on the ΔK value. The effect was more pronounced for intermediate ΔK values.

The SIMS profile of deuterium in the fracture surface for the specimen tested in vacuum represents deuterium for which diffusion occurred when this surface was exposed to D_2O atmosphere subsequent to the fatigue crack growth (Fig.1, curve 1). The surface profile for the segment cracked in the D_2O vapor represents diffusion and any additional deuterium transport mechanism occurring during the fatigue cycling process (Fig.1, curve 2). Deeper penetration of deuterium for the portion of specimen where fatigue crack growth was in D_2O vapor to that in vacuum was observed (Fig.1). This is in qualitative agreement with dislocation sweeping mechanism [9,10].

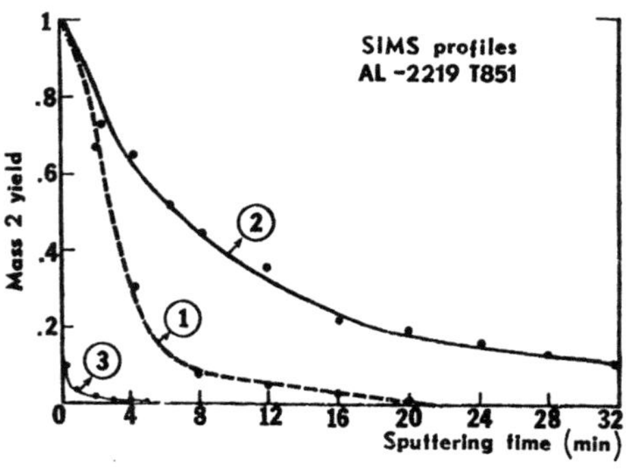

Fig.1 SIMS profiles of D^- in Al-2219 T851 alloy on fractured surfaces and dummy specimen

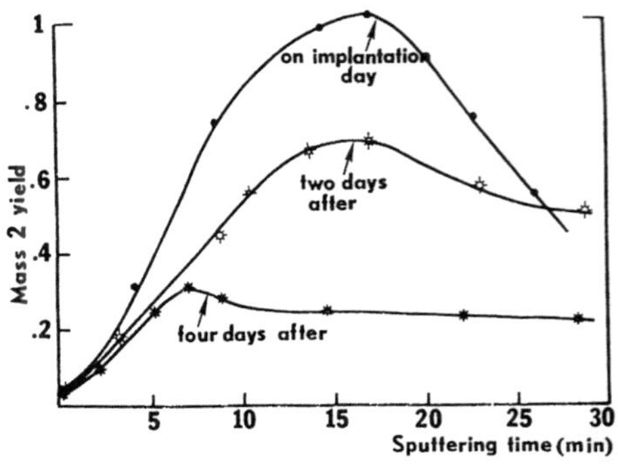

Fig.2 D^- SIMS profiles of deuterium ion implanted Mg-base Mg-Zn alloy

The deuterium background level in the AES/SIMS system was obtained using the dummy specimen. This specimen was exposed to D_2O environment for the same period of time as the others but was not fatigued and only a prior existing oxidized surface was exposed. The deuterium SIMS profile from the dummy specimen is initially only about 10% of the signal obtained from the D_2O and vacuum fractured specimens and disappears within 3 to 4 minutes of sputtering time (Fig.1, curve 3). This shows that deuterium contamination from the AES/SIMS vacuum is very low.

Preliminary attempts at calibrating the SIMS system were done on the Mg-base Mg-Zn alloy ion implanted specimens. To simulate the fatigue fracture surface conditions, the specimens were ion bombarded with neon at energies 130 keV and 150 keV. By such multi-implantation, radiation damage was deep and uniform. Then a known dose of deuterium was implanted into the specimen within the damaged region. The profiles of deuterium determined by SIMS are shown in Fig.2. Fig.2 provides two other profiles taken two and four days later. These experiments will be repeated in the near future using aluminum alloys of our interest. Hydrogen implanted into the single crystal of aluminum has been reported to diffuse out within half an hour [15]. Therefore, presence of trapping sites is required for retarding the diffusion of hydrogen and/or deuterium atoms [11,14]. During the deformation associated with the plastic zone of propagating crack in fatigue fractured specimens large amounts of various types of trapping sites are created. Using radiation damage we hope to simulate the destruction of the material of the character created in fatigue fracture. This could then be used to evaluate the hydrogen present and its transport mechanism.

4. Conclusions

(1) SIMS can be applied to study the influence of water vapor on the fatigue crack growth of aluminum alloys. (2) Dry hydrogen (or deuterium) does not affect fatigue crack growth in 2219 and 7075 aluminum alloys. (3) Oxygen penetrates deeper in surfaces formed during fatigue than vacuum fatigued and oxygen exposed surfaces. It does not play a significant role in fatigue crack growth. (4) Hydrogen (or deuterium) from water vapor influences the fatigue crack growth rate and increases penetration of hydrogen and/or deuterium into the material. (5) Ion implanted specimens can be used for the SIMS calibration and for the trapping analysis.

Acknowledgements

This research was sponsored by the Air Force Office of Scientific Research, NE Contract AFOSR 76-2995. The authors want to thank Dr. R. Walser for discussions, Mr. Joe Cecil for preparing the ion implanted specimens and D. Mahulikar for assistance in experimental procedure. Ion implantation was done at Accelerators Inc., Austin, Texas.

References

1. M. Nageswararao and V. Gerold, Metallurgical Transactions, 7a, 1847 (1976).
2. W.L. Morris, Met. Trans., 8A, 589 (1977).
3. J.S. Enochs and O.F. Devereux, Met. Trans., 6A, 391 (1975).
4. M.O. Speidel, Proc. of an Inter. Conf., Champion, Pa., AIME (1973).
5. J. Albrecht, et al., Scripta Metallurgica V.II, 893 (1977).
6. R.A. Oriani, Technological Aspect, 8, 848 (1972).
7. J. Schijve, Eng. Fracture Mechanics, V.11, 167 (1979).
8. C.D. Beachem, Met. Trans, 3 (1972).
9. J.K. Tien, Proc. of an Inter. Conf., Moram, Wyoming, AIME (1975).
10. J.K. Tien, et al., Scripta Metallurgica, 9, 1097 (1975).

11. C.A. Wert, 1978 Topics in Applied Physics, 29, 305.
12. J.W. Swanson and H.L. Marcus, Met. Trans, 9A, 291 (1977).
13. H.L. Marcus, Environmental Degradation of Eng. Materials Conf., Va. (1977).
14. W.M. Robertson, Met. Trans, 10A, 489 (1979).
15. J.P. Bugeat, et al., Physics Lett., 58A, 127 (1976).

Application of Ion Microprobe to Surface Properties of Cold-Rolled Steel Sheet

T. Shiraiwa, N. Fujino, J. Murayama and N. Usuki
Central Research Laboratories
Sumitomo Metal Industries, Ltd.
1-3 Nishinagasu Hondori,
Amagasaki, Japan

1. Introduction

It has been reported that several elements segregate on the surface of cold-rolled steel sheets during annealing in a hydrogen reducing atmosphere [1,2]. In the present report, two-dimensional distributions, depth-profiles and the chemical state of the segregation were investigated by IMMA (ARL), ESCA (V.G.) and XRF.

Specimens were sampled from annealed commercial cold-rolled sheets and prepared under similar conditions in simulated atmospheres in the laboratory.

2. Experimental Results

In Figs.1 and 2, ion microprobe images of segregated elements of 100Å and 1000Å depth are shown.

IMMA depth-profiles are shown in Figs.3 and 4, comparing different annealing conditions [3].

It was experimentally found that the extent of the segregation is affected by annealing time, temperature, and the partial pressure of oxygen, as exemplified in Figs.5,6 and 7.

The chemical state of the segregated elements was investigated by ESCA, and representative ESCA spectra are shown in Fig.8. It was found that the segregated elements Mn, Si, Cr, Fe, P are all in the oxidic state.

3. Quantitative Analysis

It is well known that secondary ion intensities from oxides are enhanced with respect to the metallic form. In the present work, the analyzed surfaces consit of segregated oxides and metallic iron. The enhancement coefficient of oxides of several elements were experimentally determined [4] using N_2^+ as primary beam. The observed secondary ion intensities were corrected using these enhancement coefficients in order to obtain quantitative surface compositions. These quantitative IMMA results were compared with those obtained by ESCA, using a simplified ESCA quantification method. This comparison yields a fair agreement, as shown in Fig.9.

4. Conclusion

The cause of segregation is due to selected oxidation of the elements at

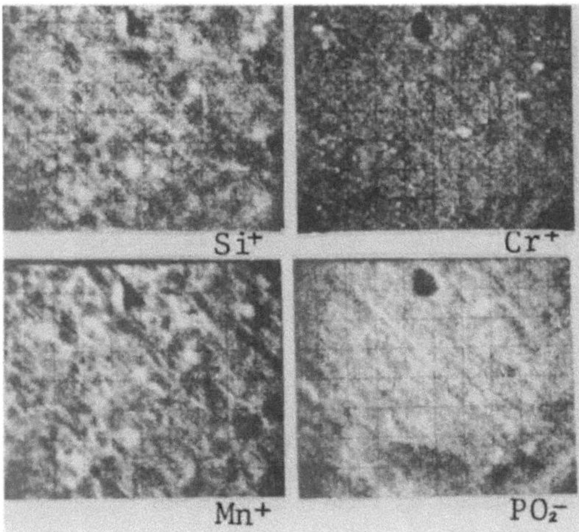

Fig.1 Secondary ion images at 100 Å depth

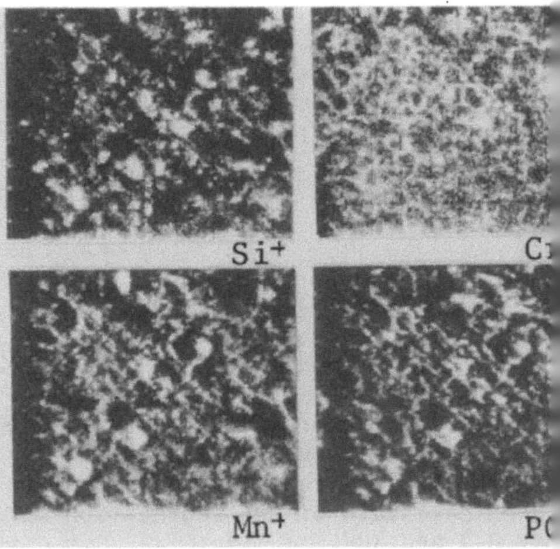

Fig.2 Secondary ion images at 1000 Å depth

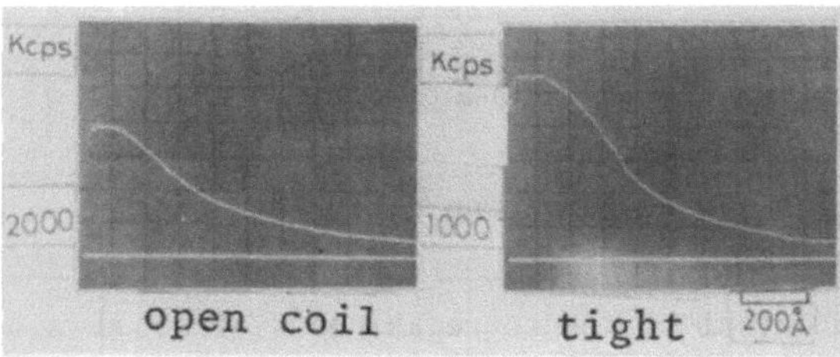

Fig.3 Depth profiles of Mn^+

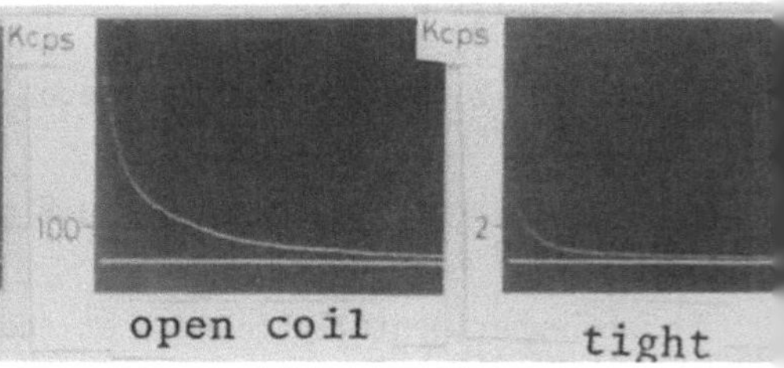

Fig.4 Depth profiles of PO_2^-

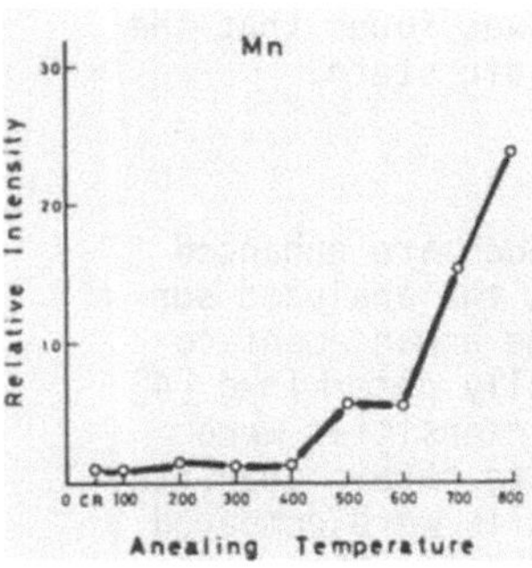

Fig.5 Intensity of Mn^+ vs annealing temperature

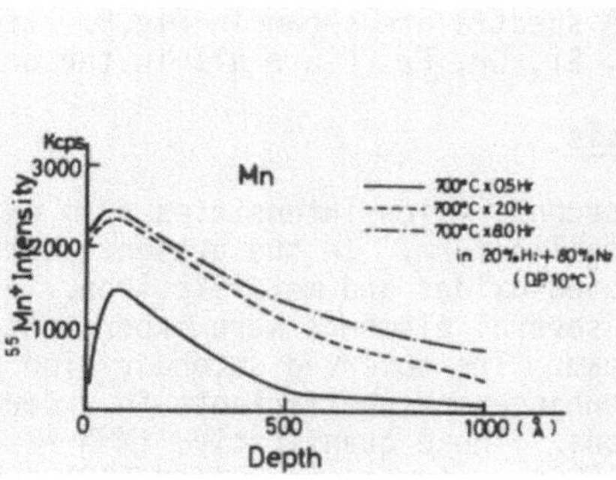

Fig.6 Depth profiles of Mn^+ for various conditions

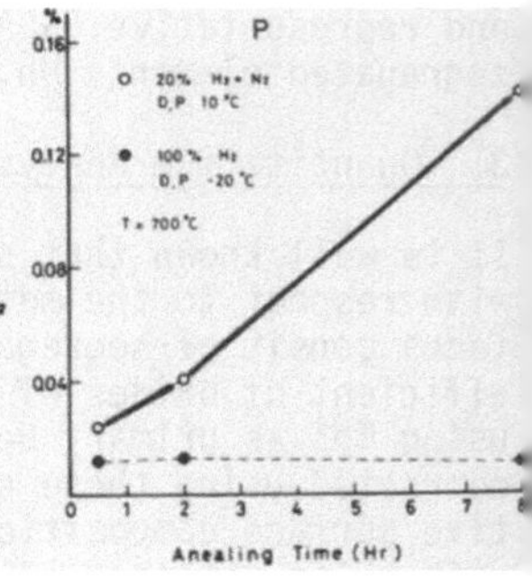

Fig.7 Relationship between segregation of P and atmosphere (F.X)

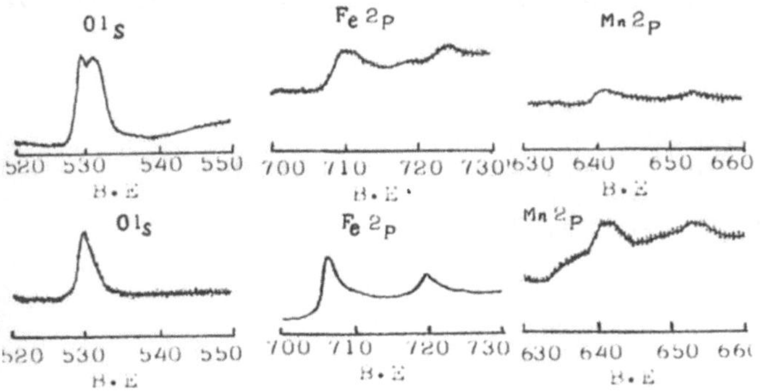

Fig.8 ESCA spectra of O1s, Fe2p, Mn2p, P2p on the cold-rolled steel at the surface and about 500 Å depth

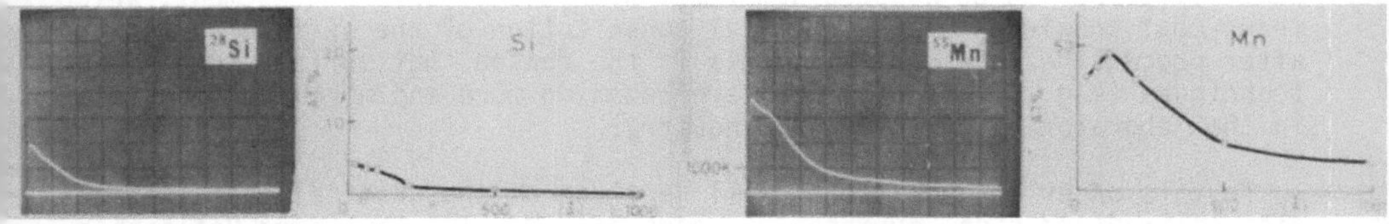

Fig.9 Quantitative results of Si and Mn by IMMA on the cold-rolled steel surface after annealing (700°C x 2 hr. in Ar)

the surface during annealing. The annealing gas is usually 10% H -90% N_2^+ containing H_2O (D.P. -30°C), which is believed to be a reducing atmosphere with respect to iron. However, the segregated elements are oxidized by H_2O because their oxidation potentials are higher than H_2 and Fe.

Cold-rolled steel sheets are usually used after surface treatment such as painting, galvanizing or enamel coating. The quality of bond of the surface treatment to the steel is affected by the surface reactivity of steel which is severely influenced by selective-oxidation of the surface. Relationships between the surface reactivity and oxide layers have been investigated.

References

1. J.P. Servais, et al., CRM Rept. No. 44, 29 (1975).
2. T. Shiraiwa, et al., Tetsu-to-Hagané, 63 S 871 (1977).
3. N. Fujino, et al., Tetsu-to-Hagané, 64 A 167 (1978).
4. T. Shiraiwa, et al., Rept. 2nd Japan-U.S. Joint Seminar on SIMS. M. Someno and D.B. Wittry, Ed., Japan Soc. Promotion of Sci., Tokyo, 1978.

Some Applications of SIMS and Other Surface Sensitive Techniques for the Chemical Characterization of Industrial Steel Surfaces

E. Janssen
Research and Development Laboratories
Department of Product Technology, Surface Technology
Hoogovens IJmuiden BV
1970 CA IJmuiden, The Netherlands

1. Introduction

It has been known for some years now that the surface properties (e.g., corrosion and oxidation behavior, coating adherence, etc.) of sheet material depend largely on the chemical constitution of the surface during and after the manufacture. This so-called "chemical reactivity" of the surface can even cause undesirable surface properties, which decreases the quality of the sheet product. In order to get a good and reproducible product the "chemical reactivity" has to be adapted sometimes by special surface treatment, which means that knowledge of the chemical constitution of the surface during and after processing is essential. This is the reason that surface sensitive techniques (e.g., SIMS, AES, XPS) are becoming more and more indispensable in the laboratories of the metal industry.

In this paper three examples will be given of the chemical characterization of iron and steel surfaces by means of SIMS, IRS (Infrared Reflection Spectroscopy), GDOS (Glow-Discharge Optical Spectroscopy), DIMA and XPS.

2. Results and Discussion

2.1 Initial Oxidation of Fe-Ti Alloys (SIMS, IRS)

Short time annealing of steel can lead to a drastic change in surface composition. In most cases, the less noble elements are selectively oxidized to a pure oxide or a mixed oxide with Fe [1]. As oxygen can diffuse into the steel matrix internal oxidation can occur. In the case of Fe-Ti alloys, oxidation for 2 seconds at 1273 K in Ar/O_2 = 20 leads to the formation of a Ti-rich oxide layer as can be seen from the SIMS depth profile, Fig.1. The IRS-spectra, Fig.2, of these oxidized surfaces show three strong absorption bands, which can be assigned to a lattice vibration of TiO_2 (820 cm^{-1}, [2], a Ti-O stretch (590 cm^{-1}) and two Ti-O bending vibrations (460, 370 cm^{-1}) as found in ilmenite type of titanates [3]. The Fe-O vibrations interfere with the 590 and 370 cm^{-1} bands of Ti-O [4]. From these results one can conclude that during initial oxidation Ti is selectively oxidized to TiO_2 and probably Ti_2O_3.

2.2 Sea-water Corrosion Layers on Construction Steels (SIMS, XPS)

Additions of Al, Ti, Cr and Mo seem to be beneficial with respect to the seawater corrosion resistance of steel [5]. This effect has been attributed to the formation of a compact adherent magnetite layer between the outer porous corrosion layer and the steel matrix. In this magnetite layer alloying

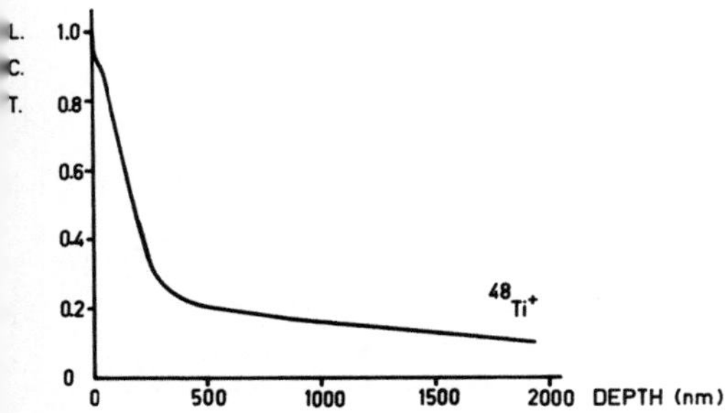

Fig.1 SIMS depth profile of an initially oxidized Fe-2.73% Ti alloy

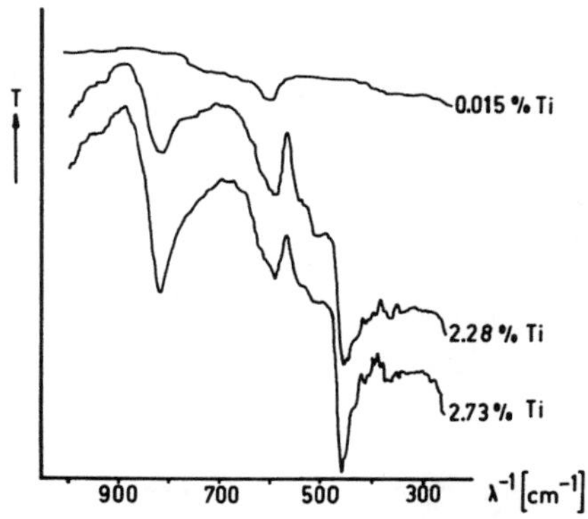

Fig.2 IR reflection spectra of initially oxidized Fe-Ti alloy

elements seem to be incorporated [6]. An Al, Mo, Ti, Cr containing steel has been subjected to sea-water corrosion at 368 K for about 800 h.

XPS powder spectra of the outer porous layer revealed the following elements and state of chemical binding: Fe^{3+}, OH^-, O^{2-} (= FeOOH), Na^+, Mg^+, Al^{3+}, Si^{4+} and Cl^-. IR and XRD spectra showed the presence of Fe_3O_4, α- and γ-FeOOH. The SIMS depth profile of the adherent magnetite layer, Fig.3, shows that Ti and Cr are enriched within the layer, while Al is depleted with respect to the steel matrix, although the course of the Al-profile suggests some concentration level inside the magnetite layer. From these results one can conclude that incorporation of the alloying elements, especially Ti and Cr, occurs and that this may influence the transport properties of the adherent layer, leading to a lower corrosion rate.

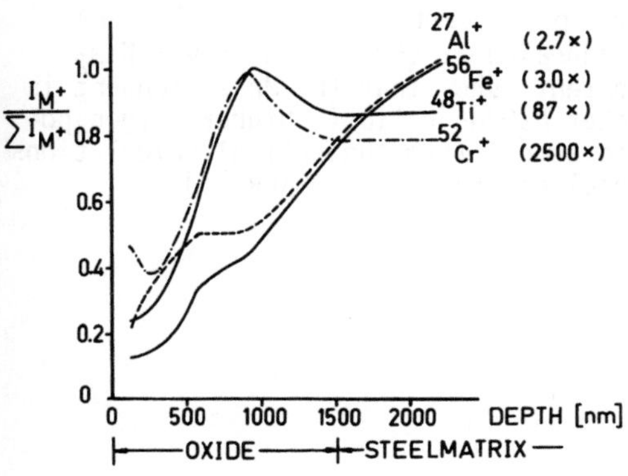

Fig.3 SIMS depth profile of an adherent magnetite corrosion layer on an Al, Cr, Ti, Mo alloyed steel

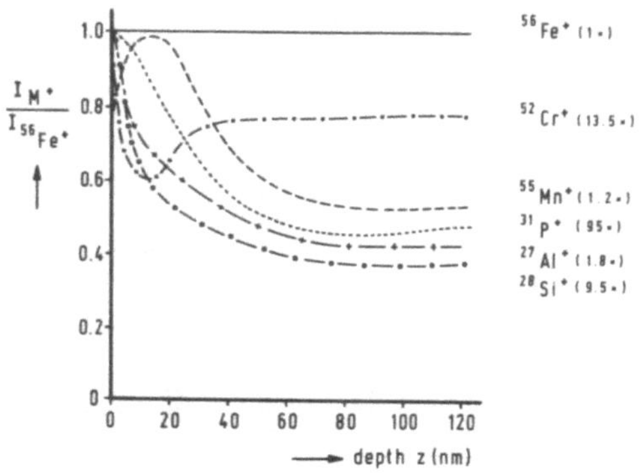

Fig.4 SIMS depth profile of a temper rolled steel sheet

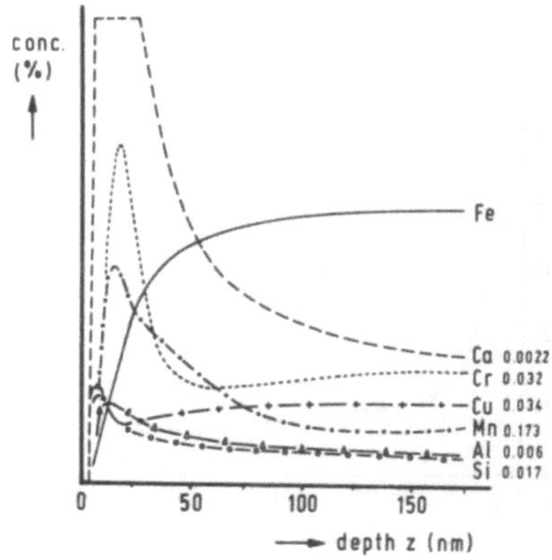

Fig.5 GDOS depth profile of a temper rolled steel sheet

2.3 Surface Enrichments in Batch-Annealed and Temper-Rolled Steel Sheet (SIMS, IMMA, GDOS)

Batch annealing of steel sheet leads to surface enrichment of minor elements due to GIBBS equilibrium segregation and selective or internal oxidation [7]. Also, carbon contamination of the surface can occur caused by graphitization, soot formation and smut formation [8,9]. These reactions lead to a decrease in surface quality of the steel product, which manifests itself in bad adherence of metallic coatings and low corrosion resistance of phosphatized and painted material [10,11]. SIMS and GDOS depth profiles, Fig.4 and 5, show the enrichment of Mn, P, Si, Al, Cr and Cu at the outer surface up to about 50 nm in depth. DIMA images show that Mn, P and O are enriched due to internal oxidation along the grain boundaries, while Cr shows a more even distribution at the surface (selective oxidation). XPS and GDOS measurements show that the surface is enriched by amorphous carbon (+ graphite). SIMS depth profiles, Fig.6, of steel sheet annealed at 953 K in N_2-6 vol % H_2 atmospheres with differences in H_2O content, show that the Mn enrichment is lower but to a greater depth for a higher H_2O/H_2 ratio (higher oxygen potential). These results are in good agreement with theoretical calculations [12] based on an internal oxidation model according to WAGNER [1].

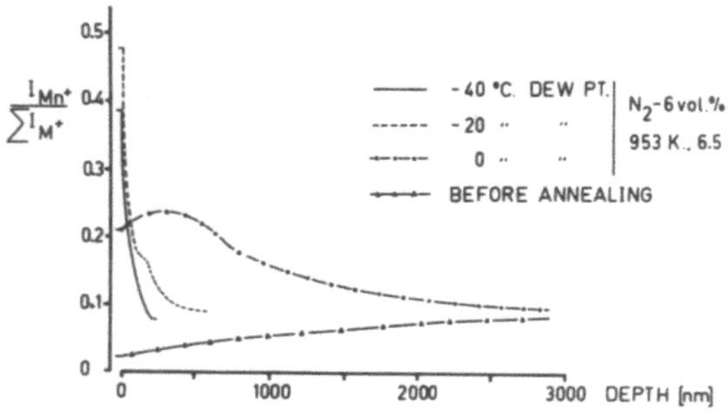

Fig.6 SIMS Mn depth profiles of steel sheet annealed in N_2-6 vol % H_2 atmospheres with different H_2O content

References

1. Z. Wagner, Electrochem., 63, 772 (1959).
2. Phillipi, et al., Phys. Rev., B3, 6, 2068 (1971).
3. Last, Phys. Rev., 105, 6, 1740 (1957).
4. Waldron, Phys. Rev., 99, 1727 (1955).
5. Blekkenhorst, COST progress report 2A - Hoogovens (1976).
6. Tamada, TMS Fall Meeting, Detroit, October 1971.
7. Servais, et al., CRM report, 44, 29 (1975).
8. Inokuti, Trans. Iron and Steel Inst. of Japan, 15, 6, 314 (1975).
9. Huiskamp, Hoogovens research report 46112,(1978).
10. Slane, et al., Met. Trans A, 9A, 1839 (1978).
11. Wojtkowiak, et al., Anti-corrosion, 26, 1, 9 (1979).
12. Huiskamp, private communication.

VI. Instrumentation

Analytical Requirements of SIMS and the Instrumental Implications

H. Liebl
Max-Planck-Institut für Plasmaphysik
8046 Garching bei München, Germany
Association EURATOM-IPP

1. Introduction

Other surface characterization techniques such as electron probe microanalysis or Auger electron spectroscopy have reached a degree of perfection where only minor improvements are yet to be expected. This is not the case with SIMS. One of the reasons is the fact that, contrary to the above methods, where there is only one possible primary species, namely electrons, in SIMS the primary species can be chosen at will, in principle from among the 90 elements, not to mention compound ions. Since they are chemically different, they will yield different SIMS spectra from the same sample.

The high degree of complexity of SIMS, both in instrumentation and methodology, is the price for the benefits of SIMS. Besides the coverage of all elements including H, SIMS has a great potential extending beyond the capabilities of the above mentioned methods:

- Extreme trace sensitivity due to the lack of an inherent background;
- Microcharacterization down to the order of 10 nm both in-depth and laterally due to the short range of ions in solids;
- Isotopic analysis; and
- Compound analysis.

Instrumental developments are still in progress, which can help SIMS to advance towards the fundamental limits, especially in microcharacterization, which is of particular importance in materials science, solid state electronics, geochemistry and biochemistry. These developments include ion sources and ion optical systems.

2. Ion Sources

The ion probe current which can be focused into a small spot depends on the brightness of the primary beam and the aberrations of the ion optical system. The beam brightness in turn is proportional to the current density at the source. For a planar geometry this is space charge limited according to Child's law:

$$J = \frac{5.45 \times 10^{-8}}{\sqrt{M}} \frac{V^{3/2}}{d^2} \text{ [A/cm}^2\text{]}, \qquad (1)$$

where M is the mass number, V the extraction voltage [V] and d the acceleration distance [cm] . This is valid for plasma sources as well as for thermal surface ionization sources. The space charge limitation is overcome by convex emitting surfaces with a very small radius of curvature r. Thereby the effective accelerating distance is of the order of r. This is the case in field ionization sources, which can deliver positive ions from gases such as Ar, O_2, N_2, and can also be done with thermal surface ionization sources [1] (Fig.1), which can deliver positive alkali ions or negative halide ions. The upper limit is then imposed by voltage breakdown. Recent advances in applying field ionization sources to the formation of ion microprobes have been reported [2,3] . Even higher brightness values have been achieved with liquid metal sources [4,5,6]which can deliver positive ions from alkalis or certain other metals.

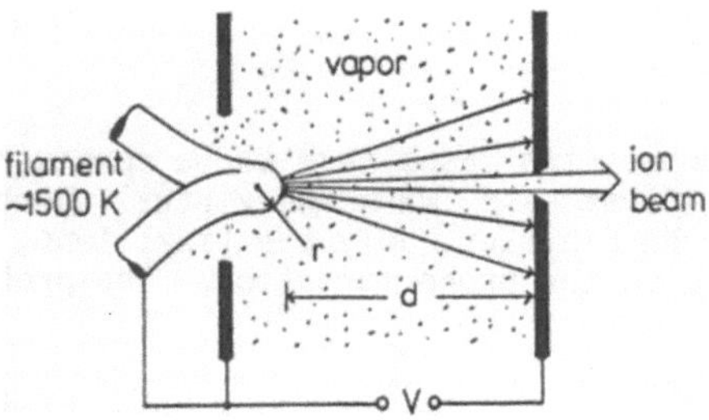

Fig.1 Thermal surface ionization source for positive alkali ions (work function > ionization energy) or negative halide ions (work function < electron affinity). With 0.5 mm $> r \ll d$, $V > 5$ kV, the emission current density is not limited by space charge but by vapor density. With $r > 10$ µm field ionization is avoided [1]

3. Primary Beam Mass Separation

Primary beam mass separation is necessary in the following cases:

- In surface research with a noble gas primary beam: Due to the chemical enhancement the number of secondary ions produced by reactive gas ions present as beam impurities may reach or even surpass the number of secondary ions due to the noble gas ions.

- In trace analysis: Impurity ions stemming from the inner source walls, particularly from the cathode, are implanted and are subsequently resputtered in the same way as sample constituents.

- When a desired primary ion species is obtained as a fragment of a compound supply gas; or when a discharge in the source is stabilized by feeding it with a mixture of gases, one of them being the origin of the desired ions.

Since surface ionization is a very discriminatory process, ion beams from such sources are very pure, making mass separation unnecessary for most applications. This may not apply generally to liquid metal sources. Besides polymer ions impurity ions have been reported [4] .

When mass separation is required, it is generally accompanied by energy dispersion, fractional energy spreads $\Delta E/E$ ranging from 10^{-4} to 10^{-3}. For a microbeam this is undesirable because it transforms a beam which is circular in cross section prior to mass separation into one which is elliptical in cross section with diminished current density. This problem can be overcome by an ion optical trick (Fig.2). For submicron probes this method of mass separation without energy dispersion becomes indispensable.

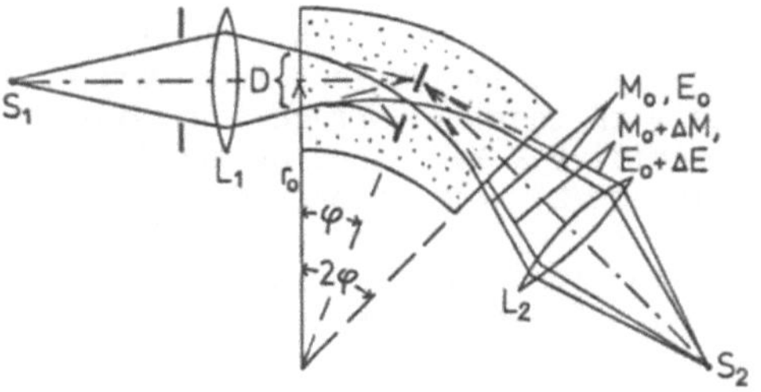

Fig.2 Achromatic and poly-isotopic mass separation. Source S_1 is imaged by lens L_1 into center of magnetic field; virtual image coincides with virtual center of dispersion; lens L_2 forms achromatic image S_2 of S_1. By the same means, the separation of isotopes of one element can be cancelled[7]

4. Sub-micron Ion Probes

For the formation of sub-micron ion probes, two cases have to be considered regarding the focusing system. In the first case, when the virtual source diameter is large compared with the probe diameter, only the final lens, the objective lens, plays the decisive role in the probe formation. The probe diameter d is given by

$$d = (d_g^2 + d_s^2 + d_c^2)^{1/2}, \qquad (2)$$

where d_g is the Gaussian image diameter, $d_s = (C_s/2)\,\alpha_i^3$ is the aperture aberration, and $d_c = C_c(\Delta E/E)\;{}_i$ is the chromatic aberration. Fig.3 shows what can be achieved with typical ion gun brightness values and energy spreads using the lens shown in Fig.5. This lens has very small aberration coefficients C_s and C_c, mainly because of its small dimensions.

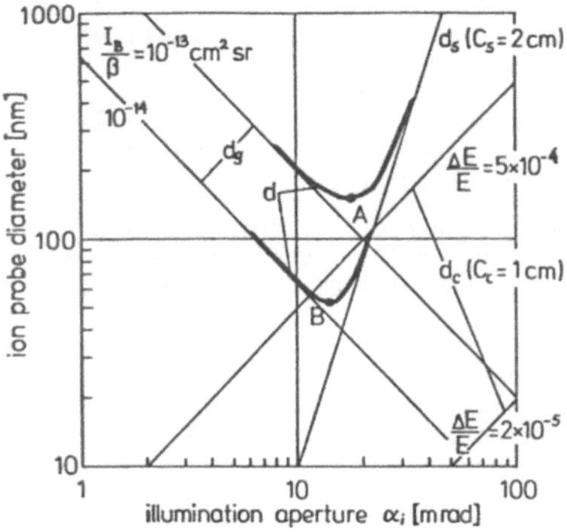

Fig.3 Ion probe diameters according to (2) for 15 keV beam from duoplasmatron or thermal surface ionization source focused by lens of Fig.5. I_B = probe current [A], β = beam brightness [$Acm^{-2} sr^{-1}$]. Point A: I = 10 pA with β = 100 $Acm^{-2}sr^{-1}$, $\Delta E/E = 5x10^{-4}$ (duoplasmatron). Point B: I_B = 10 pA with β = 1000 $Acm^{-2}sr^{-1}$, $\Delta E/E = 2x10^{-5}$ (source Fig.1)

The second case is when the virtual source diameter is negligibly small compared with the probe diameter, which is then determined solely by the aberrations of the focusing system. This applies to field ionization and liquid metal sources. Preferably the optical system consists of a lens close to the source (index e) and the objective lens close to the sample (index i) [3]. The probe diameter d is given by

$$d = (d_s^2 + d_c^2)^{1/2}, \text{ where} \qquad (3)$$

$$d_s = \frac{1}{2}\,[(MC_{se})^2 + (\frac{C_{si}}{M^3})^2]^{1/2}\,\alpha_e^3 = \frac{1}{2}\,\bar{C}_s\alpha_e^3 \qquad (4)$$

$$d_c = [(MC_{ce})^2 + (\frac{C_{ci}}{M})^2]^{1/2}\,\frac{\Delta E}{E}\,\alpha_e = \bar{C}_c\,\frac{\Delta E}{E}\,\alpha_e. \qquad (5)$$

M is the imaging ratio from source to sample. Fig.4 shows what can be achieved, assuming values $dI/d\Omega$ and energy spreads ΔE typical of field ionization and liquid metal sources and an optimally designed source lens [10] .

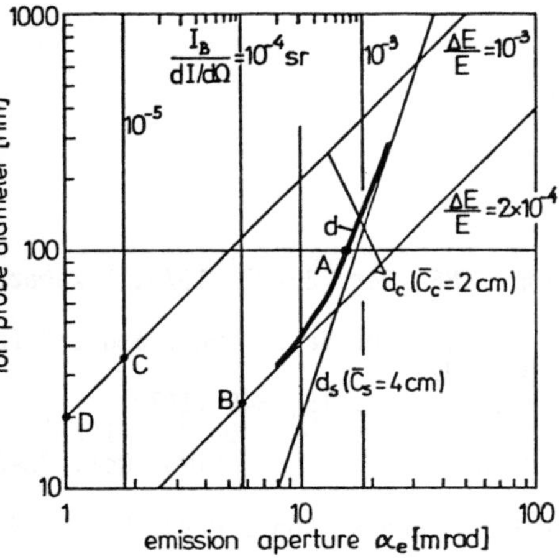

Fig.4 Ion probe diameters according to (3) for 15 keV beam from field ionization or liquid metal source, focused by a source lens and lens of Fig.5.
Points A and B: I_B = 350 pA and 50 pA with $dI/d\Omega = 5\times10^{-7}$ A/sr, $\Delta E/E = 2\times10^{-4}$ (field ionization source).
Points C and D: I_B = 500 pA and 160 pA with $dI/d\Omega = 5\times10^{-5}$ A/sr, $\Delta E/E = 10^{-3}$ (liquid metal source)

5. Secondary Ion Mass Analysis

With the tendency to use mass spectrometers with ever higher mass resolution, optimal matching of the secondary beam emittance to mass spectrometer acceptance becomes ever more important in order not to suffer severe transmission losses and thereby lose sensitivity. This is true with sector field mass spectrometers as well as with quadrupole mass spectrometers.

The lens shown in Fig.5 not only focuses the primary beam, it also acts as an emission lens for the secondary ions [8]. The emittance of a secondary ion beam containing all ions up to the initial energy E_i is proportional to $r(E_i/E_b)^{1/2}$, where r is the radius of the emitting area and E_b is the beam energy after acceleration. At best, for microspots the emitting area is reduced to an apparent emitting area of diameter $\sigma \approx E_i/F$, where F is the field strength on the surface. Thus, with the high field strength possible with this lens, the lowest possible emittance can be achieved. Combined with dynamic emittance matching [9,7], still higher transmission than calculated previously [9] can be realized.

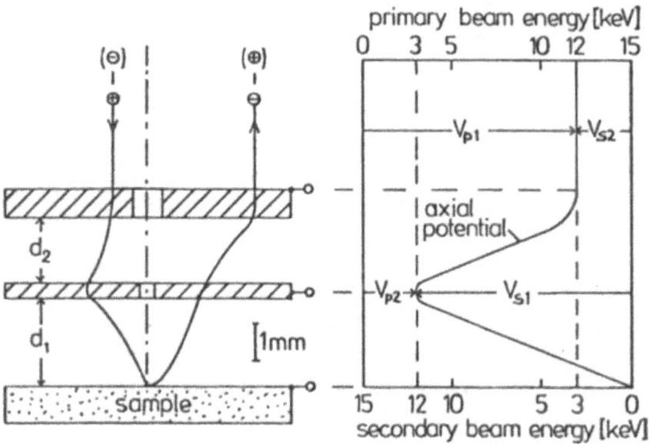

Fig.5 Combined objective and emission lens. With $d_1/d_2 = 4/3$ and $V_{p1}/V_{p2} = V_{s1}/V_{s2} = 4$ the primary beam is focused onto the sample ($f \simeq 3$ mm) and the secondaries of opposite polarity leave as parallel beam [8]

References

1. H. Liebl, Rep. 2nd Japan-US Joint Sem. SIMS, Oct. 23-27, 1978, Takarazuka, Japan, p. 205
2. R. Schwarzer and K.H. Gaukler, Proc. 7th Intern. Vac. Congr. and 3rd Inter Conf. Solid Surfaces (Vienna 1977), p. 2547
3. J. Orloff and L.W. Swanson, J. Vac. Sc. Technol. 15, 845 (1978)
4. R. Clampitt and D.K. Jefferies, Nucl. Instrum. Meth. 149, 739 (1978)
5. V.E. Krohn and G.R. Ringo, Rep. 2nd Japan-US Joint Sem. SIMS, Oct. 23-27, 1978, Takarazuka, Japan, p. 227
6. R.L. Seliger, J.W. Ward, V. Wang and R.L. Kubena, Appl. Phys. Lett. 34, 31 (1979)
7. H. Liebl, "Low Energy Ion Beams 1977," Inst. Phys. Conf. Ser. No. 38 (1978 p. 266
8. H. Liebl, Optik 53, 69 (1979)
9. H. Liebl, Adv. Mass Spectrom. 7A, 751 (1978)
10. H. Liebl, Unpublished

Some Considerations on Secondary Ion Optics

G. Slodzian
Université de Paris-Sud
Centre d'Orsay-Bât. 510
91 405 Orsay, France

1. Introduction

Collecting data suitable for theoretical understanding of the emission phenomena processes and collecting secondary ions with high efficiency to achieve "sensitivity" requirements should concern everyone working in the field of SIMS. Both are of utmost importance because they determine the proper way of using secondary emission as well as the ultimate analytical limitations. Both place the collection system at the center of our concern. Then comes the connection between the collection system and the filter system. This connection is actually realized by a "transfer optical system," which brings about the point that the description will be given with the help of concepts currently used in optics.

The choice of the "illumination" modes of the sample surface, i.e., the primary beam, will appear at the last step, not because the question is of lower importance, but because its place is at the top of the whole construction rather than at its foundations.

Let us recall once more that many of the difficulties we have to cope with arise from the "destructive nature" of the sputtering process. In the field of microanalysis by secondary ion emission, the definition of "useful yield" was introduced nearly 20 years ago in order to evaluate the consequences on the ultimate possibilities. It was dictated by common sense considerations.

2. Collection System

Particles are emitted in all directions and with kinetic energies ranging from zero to several hundred electron-volts. Since the mass spectrometer (filter system) can only operate with a beam of limited geometrical extent, some device is necessary to shape the beam of secondary ions leaving the surface so as to fit the beam size accepted by the spectrometer. Let us denote by dS_0 an area surrounding point A_0 on the surface and characterize by $d\Omega_0$ a solid angle around the direction defined by the angle of emergence α_0 and the azimuthal angle ϕ_0. A particle is characterized by A_0, α_0, ϕ_0, and its initial kinetic energy Φ_0 (in eV) within an interval $d\Phi_0$.

A comprehensive study of the emission processes requires the knowledge of both angular distribution and energy spectrum* and consequently the collection system would have to make possible the selection of ions emitted in various

*In fact, this requirement bears on both ionized and neutral particles.

(α_0, ϕ_0) directions without introducing energy discriminations. Whereas, an analytical utilization will require a collection system capable of gathering secondary ions emitted within a large solid angle in order to reach high useful yields, it should besides leave open the possibility of localizing the emissive area dS_0. We will examine the latter case more extensively. It is clear that the collection process will leave out of the accepted beam those particles which are emitted with both large α_0 angles and high Φ_0 energies, producing that way energy and angular discriminations. These discriminations can be easily described in the case of a sample immersed in a uniform electrostatic field closed by an electrode parallel to the sample and provided with a round hole to let the ions escape from the acceleration space. Such a combination of accelerating and focusing fields is known as a "cathode lens" or "immersion lens." Its properties can be summarized as follows:

- ions leaving A_0 produce a virtual image point spoiled with mixed chromatic and aperture aberrations.
- ions defined by (α_0, ϕ_0, Φ_0) are focused, whatever the position of A_0 on surface (provided it stays in the vicinity of the optical axis), in a virtual "image point." The locus of these image points forms a virtual "illumination pupil" named "cross-over" which "illuminates" the virtual image of the surface previously defined. "Emissions diagrams" corresponding to ions of different types with different energies are to be seen superimposed in the plane of the cross-over.

To follow the transformations that the beam of the collected ions will suffer later on, it is enough to keep track of the positions of the successive images of the illumination pupil and of the surface.

3. Transfer Optical System

In order to fit the geometrical requirements of the spectrometer, defined by the dimensions of the entrance and aperture stops, it is convenient to design a transfer optical device which produces real images of the pupil and the surface situated at the entrance and the aperture diaphragms, respectively. The transfer optical system makes possible further adjustments of the magnification of both images. However, because of the Lagrange-Helmholtz relation, the magnification of the pupil and that of the surface cannot vary independently: any demagnification of the pupil image at the entrance stop level results in an increase of the magnification of the surface image at the aperture stop level.

The size of the entrance and aperture diaphragms depends on the mass resolving power required from the mass spectrometer. Once this parameter has been fixed, the transfer optical system gives some flexibility in regards to the area being analyzed and the solid angles of collection of secondary ions of various energies (through the size of the area carved out of the initial pupil by the entrance stop). Here appears another constraint due to aberrations on each image point of the surface. These aberrations depend upon the solid angle inside which the particles are collected and on the energy pass band. However, a precise picture of the aberration spot cannot be drawn without taking into account both angular and energy initial distributions. In practice, the distribution of intensities in the aberration spot is controlled on one hand by the dimensions of the entrance stop together with the magnification of the pupil at that level and on the other hand by the energy pass band of the spectrometer.

Several cases can be considered. For example, if there is a need to collect ions emitted by a small area dS_0 with the highest efficiency, the transfer optical system will be adjusted so as to demagnify the initial pupil image, i.e., to increase the solid angle of collection. In so doing, the surface will be magnified and, since the solid angle of collection has been increased, the aberration spot will also be increased. An "optimum" is reached when the image of dS_0, including aberrations, fills up the aperture stop. If aberrations are greater than the linear dimension of dS_0, the sample must be illuminated by a fine primary beam (ion probe) which area equals dS_0 in order to preserve the possibility of local analysis.

Another case frequently encountered concerns distribution maps of various elements over extended areas (100 µm in diameter for instance) and with good spatial resolution (about 0.5 or 1 µm). Besides, good statistics (±2% fluctuations) is often required in order to detect small variations of concentration of minor elements (zoning). When the concentrations are not too low (roughly speaking, higher than 0.1%), there is, in general, no need for striving after the highest yield possible, direct imaging, as in an emission microscope, is then well suited. It offers the advantage of giving 10^4 information points at the same time. In such a case, the sample is "illuminated" with a broad uniform primary beam.

In order to compare the two modes, ion microscopy and ion probe, it is necessary to know how the useful yield behaves as the solid angles of collection (or any related parameter) are increased. Let Φ_{om} denote an initial energy such as the emitted ions are entirely collected when $\Phi_0 < \Phi_{om}$, Φ_{om} plays the role of the required parameter. By varying the magnification of the image of the pupil and the size of the entrance stop--as it can be done with the help of the transfer optical system--a relative scale of useful yields has been plotted against Φ_{om} for various elements.

4. Filter System

There is a great variety of possible filter systems, the one mentioned here is suitable for filtering ion images. It is barely necessary to recall that high mass resolving powers are required to filter out polyatomic ions. However, high mass resolving powers are obtained at the expense of the signal. The transfer optics allow the same useful yield to be maintained as the resolving power is increased but this is obtained at the expense of the area being analyzed: on small areas, the resolving power may be increased without any loss of signal.

The filter system is composed of electrostatic and magnetic prisms coupled together by a transfer lens. The whole combination is achromatic for energy variations to first order approximation. Among other features, it allows one ot obtain ion images of the surface at high mass resolving powers and to observe an enlarged projection of a portion of the corresponding mass spectrum. (This latter operation mode enables us to nearly double the resolving power.) The observation device is a channel plate.

5. Conclusion

From the measurements of useful yields, a realistic evaluation of the present possibilities can be made concerning ion microscopy and ion probe working modes as well as an estimate directed towards the future use of very small ion probes. These evaluations have been made while taking into account the

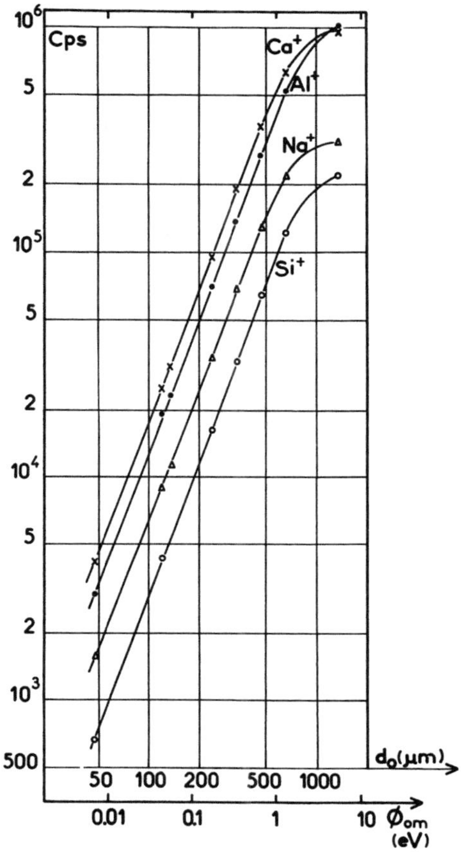

Fig.1

coupling between the mass spectrometer and the collection system by means of a transfer optical device. However, it is worthwhile to mention that improvements are still possible by reducing the aberrations of the filter system.

To end with, it should be emphasized that analysis by a secondary ion emission cannot be reduced to a problem of optics. Sputtering and ionization processes and the physical parameters controlling them, on the one hand, spurious effects like charge effects on insulators, migrations, induced diffusion, polyatomic ions...., on the other hand, have to be considered very carefully.

<u>Table 1</u> The numbers in the column on the right side of the table correspond to signals arriving on the detector under conditions differing as indicated on the left part of the table. Φ stands for the probe diameter, δ_a for the spatial resolution of the microscope and ΔZ for the thickness of the layer being sputtered. These estimates are based on the measurements reported on Fig.1 (see Ref. 3) which were obtained with a cathod lens acting as a collecting system. When the spectrometer is adjusted to give the highest mass resolving power, which can be achieved without reducing the signals here indicated for a 0.5 μm probe size, the probe would have to be static; however, a synchronous scanning of the probe and the ion image can then be considered.

		ATOMIC CONCENTRATION	
ION PROBE	$\Phi \simeq 0.5\,\mu m$, $\Delta Z \simeq 1600$ Å	10^{-6}	
ION MICROSCOPE	$\delta_a \simeq 0.5\,\mu m$, $\Delta Z = 1600$ Å	Na	20 ppm
		Al	35 ppm
		Si	50 ppm
		Ca	25 ppm
ION PROBE	$\Phi \simeq 200$ Å, $\Delta Z \simeq 200$ Å	0.5 %	

NUMBER OF DETECTED IONS	
Na	620
Al	180
Si	40
Ca	450

<u>References</u> (For more detailed considerations, see the following papers)

1. N.B.S. Special Publication (1975) No. 427, edited by K.F.J. Heinrich and D.E. Newbury
2. G. Slodzian and A. Figuerus, 8th Int. Conf. on X-rays and Microanalysis, Boston (1977) (In press)
3. Applied Charged Particle Optics, Ed. by A. Septier, Academic Press, New York (To be published in 1980)

A Compact Cs-Evaporator for High Sensitivity SIMS

T. Okutani, K. Shono and R. Shimizu
Department of Applied Physics
Osaka University
Suita, Osaka 565, Japan

1. Introduction

In recent years Cs^+ primary ions have been used for high sensitivity negative secondary ion mass spectrometry (SIMS) by several workers [1,2,3,], and a Cs ion source assembly for this purpose is now commercially available. To make this ion source interchangeable with other ion sources, e.g., Ar ions, however, one has to employ a specific configuration equipped with isolation valve [3], etc. In this respect, a compact Cs evaporator which can be mounted in SIMS instruments having a duoplasmatro-type primary ion source would be of practical use as well. One can perform high sensitivity SIMS by means of this Cs evaporator which floods the sample surface with a Cs beam while the sample is bombarded with Ar ions as reported by BERNHEIM and SLODZIAN [4]. This paper describes the development of a compact Cs-ion evaporator for high sensitivity negative SIMS which is compatible with those SIMS instruments equipped with a duoplasmatron ion source.

2. Construction of Cs Ion Evaporator

First of all, we aimed at constructing a Cs evaporator which satisfies the following requirements:

(1) Compact size so as to be mounted in a specimen chamber of those SIMS instruments mentioned above.
(2) Deposition rate is high enough to be comparable with sputtering rate by Ar ion bombardment.
(3) Evaporation is stable and the rate is changeable.
(4) The evaporator can be used in UHV without deteriorating vacuum.
(5) Inexpensive.

For this we adopted a compound material, CsI, instead of pure Cs as an evaporating material since the compound is more stable and much easier to handle. The melting point of CsI is 621°C which is much higher than that of pure Cs (28.7°C) and this makes control of the evaporation rate much easier. To heat up a furnace containing CsI we slightly modified the electron bombarding system which has been used for making Cu-Ni alloy films in UHV with considerable success [5]. Namely, as seen in Fig.1a, we positioned a heating coil behind the furnace instead of in front of it in order to avoid additional heating by radiation from the heating coil since this radiation turned out to make the control of the evaporation rate much more difficult. Fig.1b shows a photograph of the evaporator. The furnace is 5x5x5 mm^3 in size and CsI was pressed into the furnace made of stainless steel sheet with an exit aperture 3 mm in diameter. Tungsten wire 0.15 mm in diameter was used as a

(a) 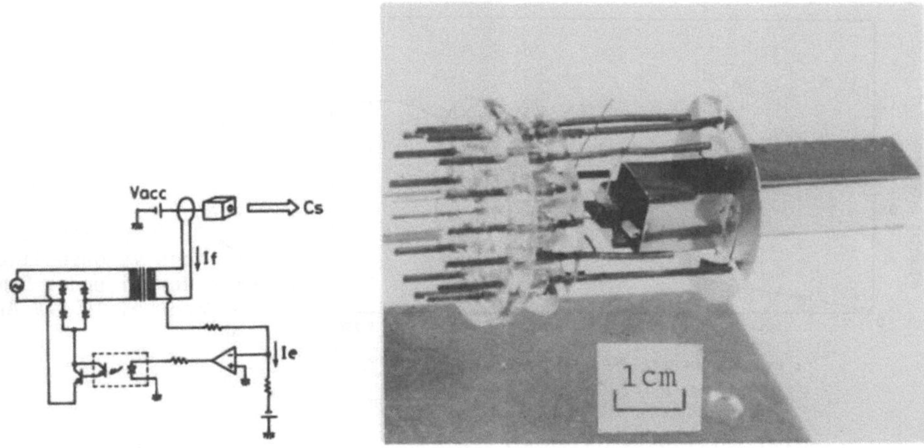 (b)

Fig.1 (a) Schematic diagram of the evaporator and control circuit and (b) photograph of the evaporator

heating coil and electrons emitted from the coil bombard the bottom of the furnace. The evaporation rate was controlled by the emission current of electrons from the W-coil.

3. Results

Figure 2 shows how the evaporation rate (measured with a quartz oscillator after calibration) varies with the current of bombarding electrons. This system allowed us to evaporate Cs (and I) at a constant rate with a variation of several percent and to change the deposition rate from 0.1 to 1 Å/sec. with ease by changing the emission current.

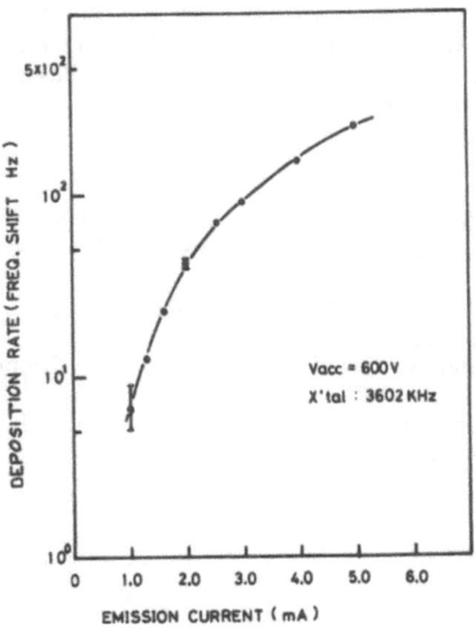

Fig.2 Variation of deposition rate of CsI with change of emission current of electrons bombarding the surface

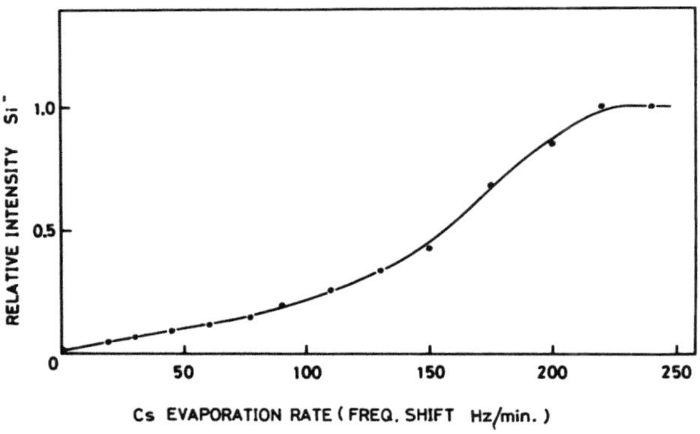

Fig.3 Intensity of Si^- ions vs deposition rate for Si under bombardment of 10 keV Ar ions

Figure 3 shows how the intensity of Si^- ions increases and then reaches a plateau as the Cs evaporation rate increases when 10 keV Ar ions bombard a Si-surface while Cs atoms (and I atoms) are deposited on the sample surface. This result shows that an appropriate Cs deposition rate provides an increase in the intensity of Si^- ions by three orders of magnitude. Consequently, since this system does not at all limit the use of oxygen-enhanced positive ion emission which is widely used in SIMS studies, this technique allows us to perform high sensitivity SIMS of both negative and positive ions in the same apparatus without replacing the ion source assembly. It is also worthy to note that the evaporating system has turned out to be a compact Cs^+ ion source since the system provides a microampere of Cs^+ ions. This allows us to use these Cs^+ ions from the evaporator as primary ions. This new capability of the present Cs evaporator will be published shortly [6].

Acknowledgement

The authors are greatly indebted to Dr. H. Storms, Vallecitos Nuclear Center, G.E. Company, for stimulating discussion and encouragement in developing the evaporator. Thanks are also due to Prof. B.L. Henke, University of Hawaii, for supplying the CsI-sample and helpful advice about Cs evaporation.

References

1. H.A. Storms, J.D. Stein, and K.F. Brown, Anal. Chem. 49, 2023 (1977).
2. P. Williams, R.K. Lewis, C.A. Evans, Jr., and P.R. Hanley, Anal. Chem. 49, 1399 (1977).
3. C.W. Magee and C.P. Wu, Nuclear Instr. and Methods 149, 529 (1978).
4. M. Bernheim and G. Slodzian, J. Phys. 38, L-325 (1977).
5. K. Goto, T. Ishikawa, T. Koshikawa, and R. Shimizu, J. Vac. Sci. Technol. 15, 1695 (1978).
6. T. Okutani and R. Shimizu, to be presented at 40th meeting of Soc. Japan Appl. Phys. (Hokkaido, September 30 - October 2, 1979).

Comparison of Laser Ionization and Secondary Ion Mass Spectrometry for Organic Materials Analysis

L.V. Phillips
Inficon Leybold-Heraeus, Inc.
6500 Fly Road
East Syracuse, NY 13057

This paper compares Laser Ionization Mass Spectrometry (LAMMA 500) with secondary ion mass spectrometry (SIMS) for the analysis of organic compounds. The LAMMA 500 laser microprobe and SIMS 100 mass spectrometer were both developed by Leybold-Heraeus, Cologne, West Germany and are the current state-of-the-art in their respective areas; thus, these comparisons apply primarily to the performance levels achieved by these two unique instruments.

The LAMMA 500 combines a high spatial resolution laser microprobe with a high mass resolution time-of-flight mass spectrometer. Samples are mounted without sample preparation inside the low vacuum (10^{-5} torr) of the mass spectrometer. They can then be observed with any of the techniques of light microscopy and vaporized/ionized with the pulsed laser. The analysis has a maximum spatial resolution of less than 0.5 microns, mass resolution of above 850, mass range of 0-1000 Amu, and sub-ppm sensitivity in the analysis region. The mass analysis can be performed within milli-seconds under these conditions, and both positive and negative ions have approximately equal yield.

In SIMS the sample preparation, though simple, is crucial to the analysis. The sample is coated on an area of approximately 1 cm^2 with a solvent, placed in the vacuum system (10^{-8} torr), bombarded with a low kinetic energy ion beam, and the mass fragments analyzed with a quadrupole mass spectrometer. The analysis takes some tens of minutes, has a mass resolution of 2M Amu for any mass m, a mass range beyond 1000 Amu, and maximum sensitivity of 10-ppm on the wide area analyzed. Negative ion yields are much lower than those of positive ions, magnetic defocusing of stray electrons is necessary for high sensitivity analyses, and charge compensation is needed.

The spectral characteristics of the LAMMA analysis are determined by the laser power density. At high densities, the analysis is highly sensitive mainly for atomic species, since fragmentation is extensive. At low power densities the analysis is structurally sensitive and primarily for molecular species. $(M+H)^+$ and $(M-H)^-$ are predominant along with major fragments, and clusters resulting from geometrical arrangements of molecules are seen. Spectra are simple and directly dependent on structure for low laser power density conditions. At higher densities fingerprinting is possible as with electron bombardment spectra. Structural information on large complexes can be easily obtained, as can fingerprint and trace analysis of polymers.

The SIMS method is less sensitive for fingerprinting organic or polymeric species, but contains more structurally significant fragments than the LAMMA. M^+, $(M+H)^+$ and $(M-H)^-$ are also observed but the occurrence is not consistent. SIMS, however, is a surface sensitive technique, whereas the LAMMA is so only under special conditions. This allows SIMS to perform organic structural analysis on thin surfaces or mono-layers. In summary, the LAMMA method offers unique advantages over traditional techniques for organic compound analysis. The SIMS method offers certain unique possibilities for organic compounds on surfaces or for non-solid organic materials where sample preparation for LAMMA is difficult.

Application to Semiconductor Characterization of a Mass-Filtered, Microfocussed Ion Gun and High-Transmission Quadrupole

A. Diebold
Riber, Rueil Malmaison, France

C.E. Badgett
Instruments SA, Inc.
173 Essex Avenue, Metuchen, N.J.

Abstract

SIMS data, obtained by using a newly developed, microfocussed ion source in combination with a wide dynamic range quadrupole mass analyzer, will be presented, with particular emphasis on profiling of silicon and gallium arsenide semiconductors. The duoplasmatron ion source, having three focussing stages, a 90° magnetic mass selector and 15 keV maximum energy may be focussed to a 5 µm beam diameter, permitting high resolution secondary ion imaging. Rapid profiling will be demonstrated, as well as lower current, static SIMS. Because a high transmission/sensitivity analyzer is needed when using micron-range primary beams, a quadrupole having 15.6 mm diameter rods, driven at 2MHz, was chosen, Comparisons with Auger microprobe analyses of like samples will also be presented.

Transmission of Quadrupole Mass Spectrometers for SIMS Studies

R.-L. Inglebert and J.-F. Hennequin
C.N.R.S., Laboratoire P.M.T.M.
Université Paris-Nord
93430 Villetaneuse, France

Quadrupole mass filters are frequently used in secondary ion mass spectrometry. Unfortunately, quadrupole spectrometers appear less sensitive than magnetic spectrometers. The absolute value of their sensitivity is difficult to theoretically determine because optical properties of a quadrupole are rather intricate [1], and only rough evaluations are published [2]. Calculation of relative sensitivities by means of acceptance areas nevertheless allowed the selection of the optimal design for a given application [3].

As explained in [2], when N_0 atoms of an element M are sputtered from a sample, N ions of the species M^+ are produced, but only p of them are collected at the detector. The *transmission efficiency* of the spectrometer, defined by: $\eta = p/N$, appears essentially as an instrumental parameter; it depends however on the angular distribution and on the initial energy spectrum of the emitted ions [4]. Provided this energy spectrum $f(\varepsilon)$ is independent of the emission direction, the *overall transmission* $\mathcal{C}$ may be defined as the transmission efficiency for each initial energy ε, and $\eta = \int_0^\infty f(\varepsilon)\mathcal{C}(\varepsilon)d\varepsilon$.

The purpose of this paper is to calculate the overall transmission of some quadrupole mass spectrometers from previously published experimental results about the transmission of a narrow ion beam through a quadrupole filter [5].

1. Development of Calculations

The transmission of a quadrupole mass filter was experimentally studied by means of an alkali ion source (Na^+, Rb^+, Cs^+) movable under vacuum in translation and rotation, and giving a narrow ion beam of well-defined energy (between 20 and 100 eV). The experimental acceptance areas for a given resolution and a given transmission were directly plotted [5] and a good agreement was found with DAWSON's theory [1]. The time spent by the ions through the fringing field appears to be a prominent parameter. Thus, the number n of RF cycles experienced by an ion, while it covers a distance equal to the radius r_0 of the quadrupolar field, is used as a measurement of the ion velocity, which appears to be the true ion characteristic. Optimal quadrupole accetances are obtained for n values near 1 RF cycle/r_0 [1].

These experimental results allow us [6] to write the transmission $T(\theta,\phi)$ of a narrow ion beam issued from (or convergent towards) a given on-axis point, according to the empirical formula:

$$T(\theta,\phi) = T_x(\theta)\cos^2\phi + T_y(\theta)\sin^2\phi$$

where θ and ϕ define the direction of the ion velocity when it enters the quadrupolar field (θ : angle with the z-axis; ϕ : azimuthal angle from the x-direction; $T_x(\theta)$ and $T_y(\theta)$ are respectively the transmission laws in the x-plane and in the y-plane, well approached by a constant part at the maximal value 100% followed by a linear decreasing until zero. If, for the sake of simplicity, the angular distribution of secondary ions versus the emission angle α follows a cosine law [2,4], the overall transmission of an ion beam through a spectrometer, for a given n value and the chosen resolution R, is obtained by integration in α and ϕ over all the directions of the half-space in front of the sample:

$$\mathcal{E} = \frac{1}{\pi}\int\int T(\theta,\phi)\ \sin\alpha\ \cos\alpha\ d\alpha\ d\phi$$

where ϕ is a function of determined by the optical system set between the sample and the quadrupole (Fig.1). We call d the distance from the source (a point on the z-axis) to the quadrupole entrance, and a the radius of the entrance aperture.

2. Quadrupole plus Drift Space

When the sample and the axis of the quadrupolar field are both at the same potential, the ion trajectories are straight lines before the ions enter the quadrupole and $\theta = \alpha$. The overall transmission (Fig.2) is limited by the quadrupole acceptance at short distance, and by the entrance aperture at long distance (this effect is more evident with lower values of a, but the finite size of the emitting area would then need to be taken into account). A virtual source would give a much better transmission.

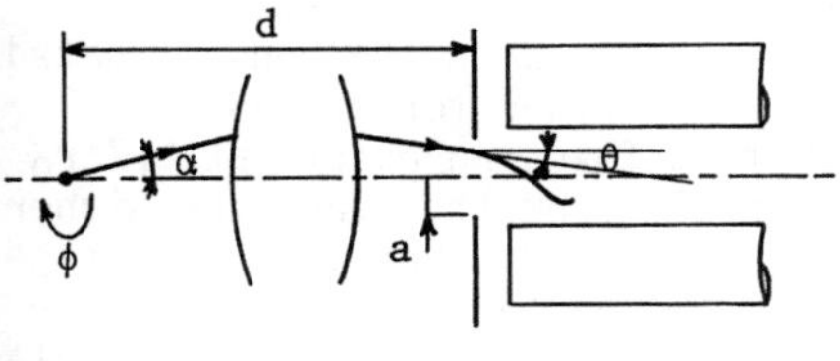

Fig.1 Schema of a spectrometer

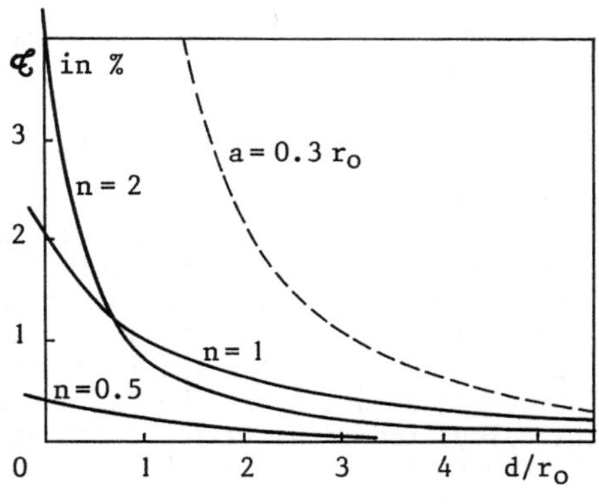

Fig.2 The overall transmission versus the distance source-to-entrance for R = 55 (--- : drift space only)

3. Quadrupole plus Uniform Electric Field

If the normal to the sample surface lies on the z-axis, and if a potential drop is applied between the sample and the quadrupole, the emitted secondary ions are accelerated (or retarded) by a uniform electric field. The ion trajectories are initially parabolic curves, but the hole at the entry into the quadrupole acts as a divergent (or convergent) lens [4]. The potential drop is selected to be such that an ion with the initial energy ε, corresponding to ν RF cycles/r_0, will have the final velocity corresponding to n = 0.5, 1 or 2 RF cycles/r_0. The calculated overall transmission, plotted versus the initial velocity measured by $1/\nu$, shows high values for accelerating fields (Fig.3; $\nu \gtrsim n$: left part of the curves) which unfortunately may be used with very slow ions only. A major advantage of magnetic spectrometers is their ability to accept ions accelerated to several keV, thus allowing high overall transmissions. To attain such values, a very big quadrupole should be used, but its RF power would be quite prohibitive. The transmission is much lower for retarding field (Fig.3; $\nu \lesssim n$: right part of the curves); the hump in the curves, more apparent with larger a or shorter d values, is due to optimal focusing by the entrance hole.

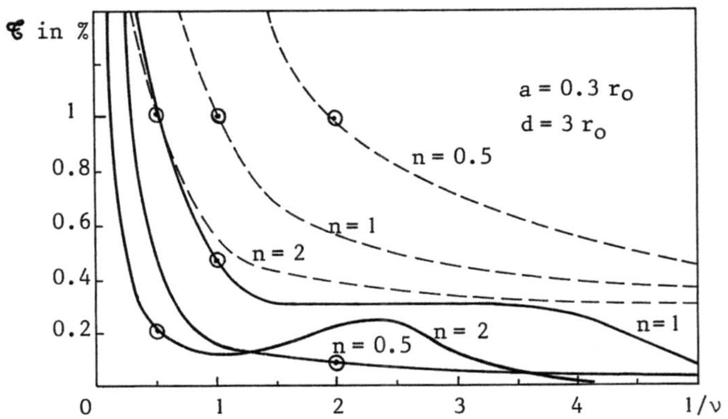

Fig.3 The overall transmission versus the initial velocity measured by $1/\nu$ in r_0/RF cycle units for R = 55 (---: electric field only; o : drift space)

4. A Quadrupole Spectrometer with Focusing

An electrostatic lens assumed to have no aberrations, focuses the secondary ions into the quadrupole. Two concentric spherical grids around the emitting point source would make it possible to adjust the ion velocity at n = 0.5, 1 or 2 RD cycles/r_0 with no change in its direction. With the lens situated at d/2, an overall transmission of about 10% is obtained for R = 55, n = 2 and d = 12 r_0, when the focal length is such that fovusing occurs at 0.7 r_0 inside the quadrupole. The further the quadrupole from the lens, the higher the calculated transmission. It must be kept in mind however that the image size must remain smaller than the entrance cross-over (about 0.15 r_0 in the example [5]) and that lens aberrations should be taken into account, as relatively large angles are involved.

Work is in progress to include some other configurations of quadrupole spectrometers used in SIMS studies, and to introduce the energy spectrum of secondary ions into the calculation of the transmission efficiency [6].

References

1. P.H. Dawson, Quadrupole Mass Spectrometry and its Applications (Elsevier Scientific Publishing Company, Amsterdam 1976).
2. G. Blaise, In Material Characterization Using Ion Beam, ed. by J.-P. Thomas and A. Cachard (Plenum Press, New York 1978) p. 143.
3. P.H. Dawson, Int. J. Mass Spectrom. Ion Phys. 17 447 (1975).

4. G. Slodzian, Surface Sci., 48 161 (1975); see also in Secondary Ion Mass Spectrometry, ed. by K.F.J. Heinrich and D.E. Newbury (Special Publ. 427, National Bureau of Standards, Washington D.C. 1975) p. 33.
5. J.-F. Hennequin and R.-L. Inglebert, Int. J. Mass Spectrom. Ion Phys., 26 131 (1978); Revue Phys. Appl., 14 275 (1979).
6. R.-L. Inglebert and J.-F. Hennequin, Revue Phys. Appl. (in preparation).

SIMS Apparatus to Study Ion Impact Desorption[1]

Robert Bastasz
Physical Research Division
Sandia Laboratories
Livermore, CA 94550

1. Introduction

The desorption of surface impurities on metals by ion impact is a topic of growing interest in fusion energy research [1]. Plasma contamination caused by impurity release from the walls and fixtures in a tokamak can prevent attainment of the temperatures needed to sustain fusion reactions. Consequently, materials are being sought that have minimal impurity release characteristics in a plasma environment. Since ion impact desorption (IID) is a dominant mechanism for impurity release from solids exposed to plasmas, it is important to evaluate IID processes that occur on materials considered for use inside tokamaks. Secondary ion mass spectrometry (SIMS) is well-suited for studying these IID processes; the probing ion beam can be used to simulate the particle flux materials experience from plasmas.

2. Description of Apparatus

A SIMS instrument has been developed to measure the ion impact desorption cross sections of surface impurities on metals. The instrument consists of a low-energy ion source coupled to an analysis chamber containing a quadrupole mass analyzer. The ion source is separated from the analysis chamber by a pumping stage to allow measurements to be carried out under ultra-high vacuum conditions. The instrument is diagrammed in Fig.1.

The low-energy ion source (Colutron) produces a mass analyzed, neutral-free primary ion beam of He^+ or Ar^+ in an energy range 0.5-3 keV [2,3]. This energy range was selected because particles emanating from the plasma in a tokamak have similar energies. Ions are created by a filament discharge, accelerated to the desired energy, and then focused through a velocity (Wien) filter, which functions as a mass selector for the primary ion beam. The filtered ion beam traverses an intermediate pumping chamber where it is deflected by 2° to remove collimated neutral particles. The purified ion beam is again focused by an einzel lens and finally rastered across the sample in the analysis chamber.

Secondary ions emitted from the sample surface are energy selected by a cylindrical mirror analyzer, which also functions as a beam stop for high-energy reflected primaries, and then enter a quadrupole for mass analysis. Detection of the transmitted secondary ions is accomplished with an electron multiplier that is operated in a saturated mode for single ion counting. A

[1]This work was supported by the U.S. Department of Energy.

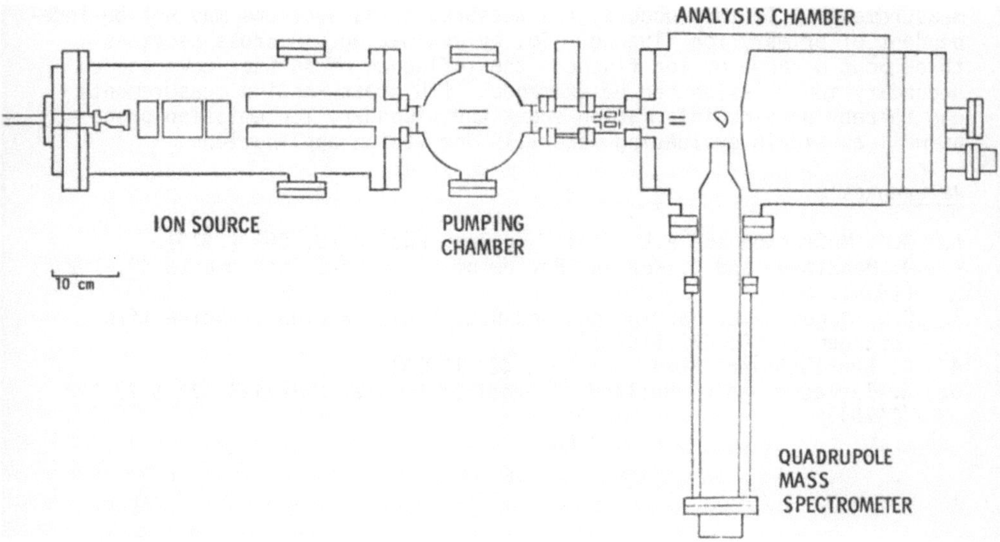

Fig.1 Component arrangement of SIMS for IID measurements

computer is used to control the quadrupole and to process secondary ion signals from a counter connected to the amplified output of the electron multiplier. Gating electronics limit detection so that only secondary ions emitted from the center of the rastered area on the sample are counted. This arrangement permits multiplexing of secondary ion signals with good rejection of spurious signals arising from extrinsic sources.

3. Method of Measurement

Desorption cross sections are measured on surfaces that are covered by monolayer quantities of an impurity. After first cleaning a sample surface as thoroughly as possible by ion sputtering, an impurity layer is prepared by either exposing the sample to a few Langmuirs of a gaseous impurity or heating the sample to promote segregation of a bulk impurity to its surface. Following this preparation, additional impurities are prevented from adsorbing or segregating at the surface. The secondary ion signal from the impurity species is then monitored as a function of primary ion fluence while keeping the energy and impact angle of the ion beam constant. As the primary ion beam removes the impurity layer, an exponential decrease in the secondary ion intensity is generally observed [4].

With first-order kinetics, the desorption cross-section is proportional to the rate of decrease in the secondary ion signal, with a scale factor equal to the product of the incident ion current density and the impurity surface concentration [4,5].This relationship is used to determine IID cross sections from the behavior of the secondary ion signal.

A variation, with impurity surface coverage, in the scattering cross section of the sample material, the surface binding energy of the impurity, or the ion/neutral ratio of the emitted particles can affect IID cross section measurements. If this occurs, the measured cross sections may not be independent of primary ion fluence. So, by evaluating IID cross sections throughout a range of ion fluences the influence of surface coverage on secondary ion emission can be examined. IID cross section measurements can thereby provide information about the secondary ion emission process as well as aid in evaluating materials for fusion applications.

References

1. G.M. McCracken and P.E. Stott, Nuclear Fusion 19, 889 (1979).
2. M. Menzinger and L. Wahlin, Review of Scientific Instruments 40, 102 (1969).
3. C.W. Magee, W.L. Harrington, and R.E. Honig, Review of Scientific Instruments 49, 477 (1978).
4. A. Benninghoven, Surf. Sci. 35, 427 (1973).
5. E. Taglauer and W. Heiland, Journal of Nuclear Materials 76 & 77 328 (1978).

Digital Mass Control for an Ion Microprobe Mass Analyzer

D. B. Wittry and Felix Guo
Departments of Materials Science
and Electrical Engineering
University of Southern California
Los Angeles, California 90007

With a double-focussing mass spectrometer as employed in Applied Research Laboratories Ion Microprobe Mass Analyzer (IMMA) and other SIMS instruments various atomic mass numbers are scanned by varying the magnetic field of the mass spectrometer. The relationship between the mass number M and the magnetic field B is given by

$$M = CB^2 \tag{1}$$

where C is a constant. Also for constant size of the final slit, the mass resolution is given by:

$$M/\Delta M = R \tag{2}$$

where R is typically 300-500. The magnetic field is varied by changing the current to the coils driving the magnet and the field is controlled in the usual case by comparing the output of a Hall Sensor with a reference voltage which is varied according to mass number. This system has two complications which are apparent from the foregoing equation, namely: 1) a non-linear function must be used to program the current drive to the magnet for various mass numbers, and, 2) the control system must provide for greater resolution at low mass numbers.

In order to overcome these problems, we have developed a digital mass control system based on two D/A converters (Analog Devices 7541), a high precision analog multiplier divider (Analog Devices 434B) and a chopper stabilized differential op amp (Datel AM-490-2). The circuit shown in the lower half of Fig.1, operates in the following way: the voltage V_H from a Hall detector sensing the magnetic field is amplified by A_1, and is supplied to the Y and Z inputs of the multiplier divider. The multiplier divider has the following transfer function:

$$e_o = \frac{10}{V_{ref}} \frac{V_Y V_Z}{V_X} \tag{3}$$

V_X in the present case is obtained from a D/A converter with 8 bits used to determine the "coarse" mass setting. The output e_0 is compared with a voltage derived from a second D/A converter with 8 bits used to determine the "fine" mass setting. The difference signal is amplified by A_2 whose output controls the power amplifier supplying the magnet coils. Negative feedback in the system is obtained if increasing output of the Hall detector causes a change in the output of A_2 of the correct polarity to cause the magnetic field to decrease. Thus, the magnetic field will increase as long as e_o is less than e_1, and will settle at value such that

$$(e_o - e_1) = V_o/G \approx 0 \tag{4}$$

where V_o is the output of A_2 and G is the open loop gain of A_2 ($\sim 5 \times 10^8$).

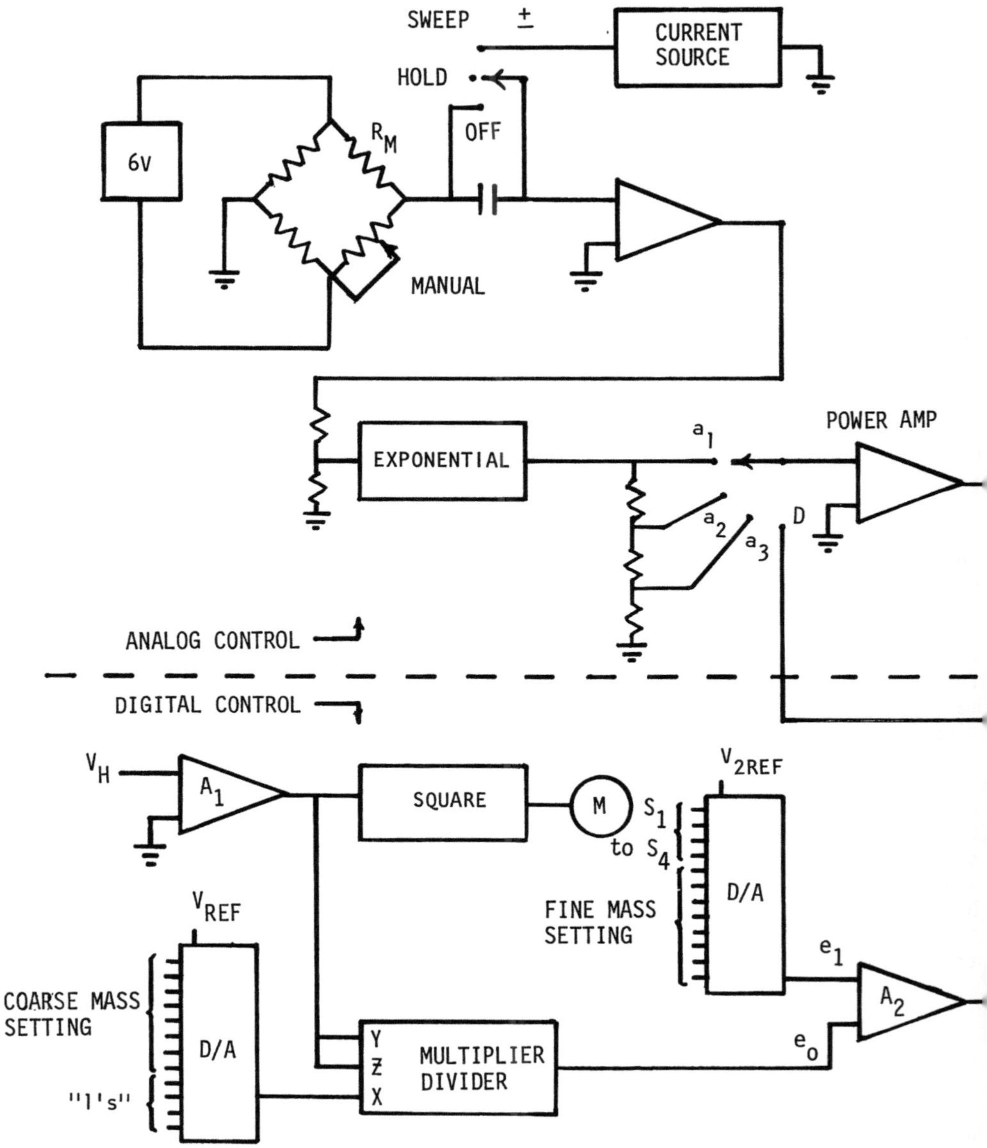

Fig.1 Schematic of a digital control for the IMMA

The upper part of the figure shows the existing analog control system based on a magneto-resistor R_M. The lower part shows the new digital control system based on the output of a Hall detector V_H

Although we require only 8 bits for determining the coarse and fine mass settings (for a range of 1-255 AMU) we have used 12 bit D/A converters in order to provide higher resolution and stability (this is particularly important in setting the "coarse" mass). In practice V_2 ref. is set so that there is a desired number of steps across each mass peak (this depends on the value of $M/\Delta M$ for the instrument and also on the setting of switches of S-1 through S-4 which determine the zero of the "fine" mass setting.

The present system provides simpler programming of mass number and of increments in mass numbers because the output of the Hall detector is squared before comparing with the digital setting of the "fine" mass number *and* because there is a range change for each atomic mass unit from 1 to 255. The latter feature compensates for the constant $\Delta M/M$ of the spectrometer so that the "fine" mass increments will be approximately the same relative to the peak half-width for all mass numbers.

The present system retains the analog mode of setting the masses and also the exponential scan mode as described previously [1](this is indicated in the top part of Fig. 1). The digital control, shown in the lower part of Fig. 1, is based on binary counters which are connected to the D/A converters. In a manual mode, the binary counters are preset by momentary connection to an oscillator. An LED display of the coarse and fine mass settings is provided by binary to BCD decoders, and BCD to 7 segment decoder-drivers. In a computer controlled mode, the binary counters are preset by the computer and the coarse and fine mass settings are still displayed on the LED read out. A digital meter M detecting the square of the Hall voltage V_H provides an indication of the mass number to which the spectrometer has been set. This is redundant but is useful as a diagnostic tool because of the hysteresis of the magnet.

It should be noted that deviations from ideal performance of the multiplier divider as given by(3) does not cause serious errors, but may necessitate a slight adjustment of the starting and ending points of the fine control compared to the values predicted theoretically. Programming of the coarse and fine settings is very simply accomplished when the mass spectrum contains peaks at the desired masses. For peaks intermediate to those present in the spectrum, simple correction factors can be used to take account of deviations from ideal behavior.

Acknowledgements

This work has been supported by NSF under grant no. CHE-77-10133 and by AFOSR under grant no. 77-3419. The United States Government is authorized to reproduce and distribute reprints for Governmental purposes not withstanding any copyright notation hereon.

References

1. D. B. Wittry, Proc. 13th Annual Conference of the Microbeam Analysis Society, Ann Arbor , Michigan, June 19-24, 1978.

Computer-Controlled Peak-Top Search Procedure

D.S. Simons, B.F. Schunicht, and J.A. McHugh
General Electric Company
Knolls Atomic Power Laboratory
Schenectady, NY 12301

Abstract

Computerized acquisition of mass spectral data in the pulse-counting mode requires that the mass position be represented by a voltage that can be controlled by the computer. In a magnetic instrument operating at a fixed accelerating voltage, this is most easily accomplished by using a Hall probe to sense the magnetic field. A calibration function can then be established between analyzed mass and Hall voltage. However, this function may change slightly with time as a result of thermal effects or magnetic hysteresis. It is therefore advantageous to have an automatic procedure controlled by the computer that efficiently seeks the top of a mass peak from a starting point determined by the calibration function. A three-stage procedure has been developed for use on an ARL IMMA, and will be explained in detail in this paper.

A Computer-Based Instrument Control and Data Acquisition System for a Quadrupole Secondary Ion Mass Spectrometry Instrument

B.F. Phillips and N.E. Lares
Perkin-Elmer Corporation, Surface Sciences Division
6509 Flying Cloud Drive, Eden Prairie, MN 55344

The development of Quadrupole spectrometer based SIMS (QSIMS) has been relatively slow compared to the large benefits it offers in the surface analysis field due to several related factors. First, the quadrupole spectrometer has been most carefully characterized in the field of organic analysis and fears of spectrometer and detector contamination were present at the start of the development of QSIMS. Second, electrostatic energy selectors for low energy secondary ions had to be developed through much trial and error. Third, sample surface charge effects had to be better understood in order to allow insulating specimens to be properly analyzed. Fourth, good primary ion sources free of beam contamination problems had to be devised. Finally, data systems that could properly take advantage of the wealth of data available in a SIMS spectrum had to be developed.

The MACS (Multiple-Technique Analytical Computer System) developed by Physical Electronics/Perkin-Elmer for use with Auger, ESCA and UPS spectrometers has now been integrated with the PHI Model 2500 SIMS and 04-303 Differentially Pumped Ion Gun to provide a versatile and adaptable data acquisition system. The computer system is based upon a DEC 11-04 with Model RX01 dual floppy disc drives using a H-P Model 2649C interactive graphics terminal for parameter input, adjustment and display with a Tektronix Model 4632 hard copy unit for output. This gives access to a medium power laboratory computer for use separate from system control as well as a fully integrated instrument control with data acquisition, manipulation and readout capability.

The three common modes through which data are collected are called SURVEY, PROFILE and MULTIPLEX. In the SURVEY mode, preset mass scans of 0-100 or 0-200 AMU can be selected or almost any other range can be covered as long as it will not produce a data file over 2000 points long. As Fig. 1 shows, not only is a linear signal readout available, but by either choosing the mode before the data is collected or redisplaying it after collection, a logarithmic display with a variable number of cycles present is possible. Also, the parameters used to produce the data can be displayed directly on the data file so that they can be reproduced in the future when equivalent samples must be analyzed. The parameters include: IV - Beam Voltage, II - Beam current in microamps, CND - Ion gun condenser lens setting, XR and YR - Beam raster in X and Y direction in millimeters, GP - Raster gating of signal as a percent of the X and Y raster distances, and PE and RE - Pass and Retard energy of electrostatic energy filter. Also shown is the number of points (S3) used in smoothing the data, the data file name and the fact that it is a positive SIMS survey.

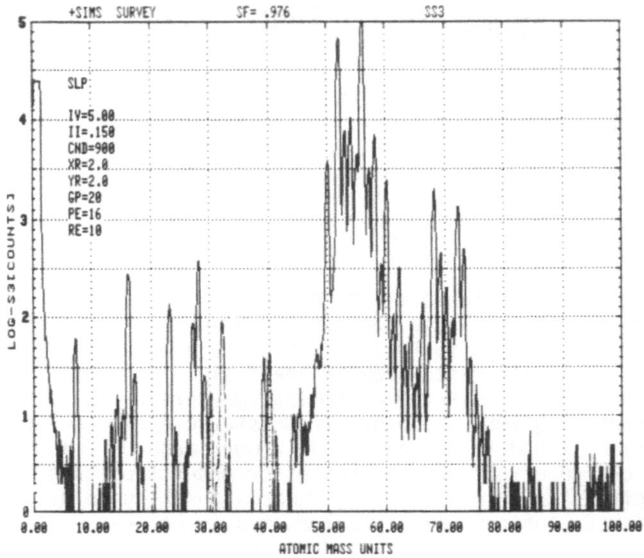

Fig. 1 Stainless Steel Survey Spectrum (Log Scale)

The second data acquisition mode is called PROFILE and is illustrated in Fig. 2. In the profile mode normally one would pick a particular mass number and the spectrometer would be swept from -1/2 AMU to +1/2 AMU for several

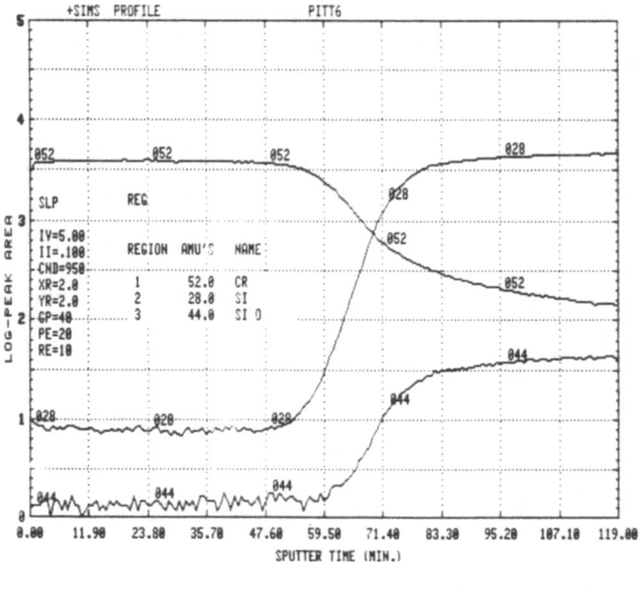

Fig. 2 Cr on Si Depth Profile (Log Scale)

cycles with the total time set at the operator's discretion while the true area integral is calculated. This is then stored and during particular cycles numbers (sputter times) preselected by the operator, the whole peak itself also is stored. The stored peaks, as shown in Fig. 3, can be examined to see if specimen charging, low mass tails from adjacent elements or any other causes are affecting the data. Since a number of spaces are available in the program set up for region names, molecular species (such as SiO in the example shown) can be named and followed in depth without worries of mislabeling at a later date.

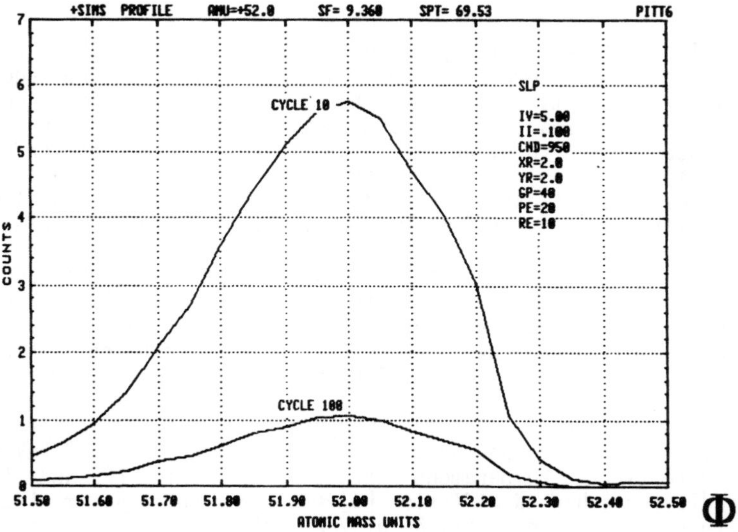

Fig. 3 Region Display Cr on Si - Cr Peak

The last mode of data collection used in the MULTIPLEX routine in which up to 20 individual masses or short range mass spectra can be taken in a repetitive sweep signal averaging mode in order to collect spectra representative of the average composition with depth of a surface being sputtered. This mode is particularly useful in performing bulk analysis when a surface is presputtered for a specific time (depth) and then data collected over a specific time (depth) for comparison to other bulk standards.

A myriad of data display, smoothing, expansion, labeling and readout options are used with the graphics system and contribute greatly to its flexibility. These programs have been designed to be user-oriented with a number of prompting commands coupled with carefully determined parameters to make it easy for the new operator to obtain good data as well as provide the flexibility in instrument operation that the experienced operator demands.

In conclusion, we feel that we have designed an adaptable SIMS data collection system that provides data which can easily be used in qualitative or quantitative calculations. This paper presents our first version of data collection without the refinements of combined AES, ESCA, SIMS profiles or a number of other interrelated modes of operation which will greatly increase the utility of QSIMS.

How to Make the Most of the SIMS Method by Means of the Scanning Ion Microscope A-DIDA

J.L. Maul
ATOMIKA Technische Physik GmbH
München, FR Germany

Secondary Ion Mass Spectrometry has demonstrated extremely high trace sensitivity (ppm to ppb) combined with high depth resolution (nm) and rather universal applicability with respect to sample type (conductor, semiconductor, insulator), sample area (macro- and microanalysis), sample composition (atomic masses from H to U) and depth of interest (monolayer surface analysis and depth profiling). In fact, these benefits are achievable also in somewhat tedious applications, provided the range of experimental parameters available with a particular SIMS instrument is appropriate. A clear distinction has to be made between the limits of the SIMS method and the limitations of the SIMS equipment actually used - for instance, when adding limited SIMS capability to other highly sophisticated surface analysis equipment. In this paper it will be shown on a few examples what capability and flexibility of a SIMS instrument is appropriate according to the literature and how the Scanning Ion Microscope A-DIDA meets these requirements.

As a first example, primary ion energy and current/current density is discussed. The primary ion energy has to be as low as 1 or 2 keV to keep knock-on effects [1], bombardment induced atomic mixing [2] and, in special cases, radiation enhanced diffusion [3] small enough to assure the promised high depth resolution in profiling of very thin layers (100 nm and less). In interface studies interdiffusion can often clearly be distinguished from artifacts only by measuring interdiffused profiles at different primary ion energies down to a few keV and extrapolating to zero ion energy [4]. For static SIMS primary ion energies of less than 500 eV are required to avoid destruction of weakly bound states in adsorption studies [5]. On the other hand, a microspot ion beam is necessary for the analysis of tiny sample areas and for the mapping of the surface composition. Such micrometer diameter beam spots can be obtained only at primary ion energies of 10 keV and more because of the space charge expansion and spherical aberrations of ion optics [6] when maintaining at the same time a primary ion current and current density high enough for maximum trace sensitivity and fast in-depth profiling of µm-thick layers. The Scanning Ion Microscope A-DIDA therefore provides mass analyzed ion beam of argon and oxygen ions in an energy range between 0.1 and 15 keV, a current range from less than 10^{-9}A to several µA and a current density range from less than 10^{-8}A/cm^2 to several 10mA/cm^2. The upper limit of the primary ion energy has been set to 15 keV because depth profile distortion due to knock-on effects make higher primary ion energies undesirable [1].

As the second example, the sputtered area and the detected area of the samples should be discussed. Large sputtered and detected areas of several mm^2 are desirable with respect to trace sensitivity in static SIMS. It is also obvious that depth consumption is increased and hence depth resolution in profiling of very thin layers is reduced accordingly when the detected area is reduced. For ion image mapping,constant transmission has to be assured across the whole detected area, whereas for high dynamic range in depth profiling a transmission decreasing with increasing distance from the center of the sputtered area is advantageous to support edge effect suppression by electronic gating. The Scanning Ion Microscope A-DIDA therefore provides primary ion beam spots between a few μm and several mm in diameter. Transmission of the secondary ion mass spectrometer is constant across the target area of several mm^2, which can be switched to a strongly position dependent one for depth profiling applications [7].

The third example deals with trace sensitivity obtained in practice. The trace sensitivity limits are determined in everyday applications much more often by mass interferences than by transmission of the mass spectrometer. Those interferences may be residual gas mass peaks, molecular and multiply-charged ions, and very intense mass peaks in the neighborhood of the small mass peak of interest. The Scanning Ion Microscope A-DIDA therefore provides an UHV-system differentially pumped with turbomolecular ion gettering and titanium sublimation pump with a base pressure in the $5x10^{-11}$ Torr range and an operational pressure in the 10^{-9} Torr range during analysis. Fig.1 shows a depth profile of a 30 keV ^{31}P implant in Si_3N_4, demonstrating the low level of the $^{30}Si^1H^+$ mass peak [8] interferring with the ^{31}P mass peak. The Scanning Ion Microscope A-DIDA furthermore provides accel-decel type secondary ion optics for increased transmission with an energy window variable in width and center energy [7] to allow the suppression of interferring molecular and multiply-charged ions. The energy band pass can be chosen as small as 0.5 eV, resulting in a mass resolution of more than 2000 [9] and an adjacent mass peak sensitivity of more than 6 orders of magnitude.

Only a few features of the Scanning Ion Microscope A-DIDA can be discussed here in detail, othersonly mentioned: like charge density equilibration on insulating samples to provide the promised universal applicability, one-switch change from positive to negative secondary ion detection without

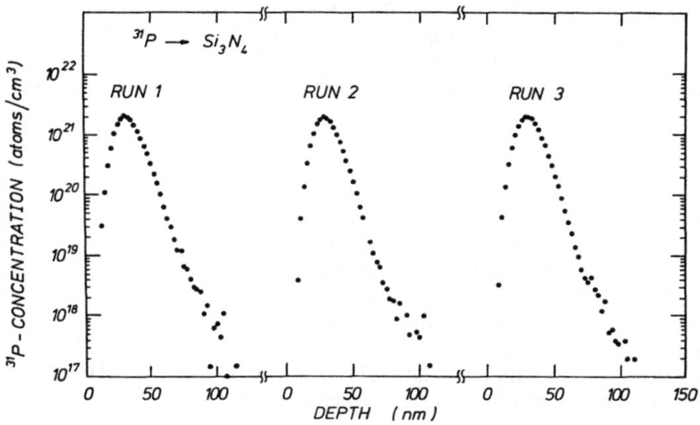

Fig.1 30 keV ^{31}P implantation profiles in Si_3N_4 showing reproducibility of depth profiling in insulating material [8]

Fig.2 The Scanning Ion Microscope A-DIDA

deflecting the primary ion beam to a different target area. The ^{31}P profiles in Fig.1 demonstrate excellent reproducibility in spite of the insulating nature of the Si_3N_4 layer.

Figure 2 shows a standard Scanning Ion Microscope A-DIDA offering all the above-mentioned features. It can be combined with complementary surface analysis techniques - without sacrificing SIMS capability and special versions are available to allow investigation of powder samples or of highly radioactive samples.

Scarcely any SIMS equipment is used at present for just one application. On the contrary, a large variety of different samples and analysis problems have to be handled. The few examples given above show that the Scanning Ion Microscope A-DIDA provides the capability and flexibility necessary to make the most of the SIMS method, covering both microprobe as well as surface analysis.

References

1. F. Schulz, K. Wittmaack, J. Maul, Rad. Effects, 18 211 (1973).
2. H.H. Andersen, Appl. Phys. 18 131 (1979).
3. P. Bank, K. Wittmaack, Rad. Effects Lett.
4. Ch. Kuhl, H. Druminski and K. Wittmaack, Thin Films, 37 317 (1976).
5. K. Wittmaack, ECOSS 2, Surf. Sci., in press.
6. K. Wittmaack, Ion Beam Surface Layer Analysis, Vol. 2 (1976).
7. K. Wittmaack, Proceedings of the 8th International Conference on X-Ray Optics and Microanalysis, Science Press, Princeton (1978).
8. F. Schulz, unpublished results.
9. K. Wittmaack, Proc. 7th Int. Conf. Solid Surface, 2573 (Vienna 1977).

Sputtered Neutral Mass Spectrometry Using a Microwave Plasma

T. Ishitani and H. Tamura
Central Research Laboratory,
Hitachi Ltd.
Kokubunji, Tokyo, JAPAN

Sputtered neutral mass spectrometry (SNMS) has been shown to be a promising new technique for elemental analysis of solid surfaces [1-5]. In SNMS the sputtered neutrals have to be post-ionized after emission, while in secondary ion mass spectrometry (SIMS) the secondary ions are immediately available for mass analysis. As a result, elemental sensitivity in SNMS is predominantly determined by the experimental post-ionization arrangement. Therefore, the sensitivity is largely independent of matrix effect, which has a strong influence in SIMS. The present report describes the preliminary experimental results of SNMS using a microwave plasma.

The apparatus used in the present work is shown schematically in Fig.1. The plasma for post-ionization is excited by the superposition of a 2.45 GHz microwave generated by a magnetron on a DC magnetic field B of up to 1.2 B_c, where B_c is the value of B under the electron cyclotron resonance condition, i.e. 8.75 x 10^{-2} T. The ion source in Fig.1 is arranged for another purpose and was not used in the present. Two kinds of plasma chamber are employed. One is a coaxial chamber which fundamentally consists of a coaxial waveguide. The sample is mounted on the inner conductor end plane of the coaxial line. The other chamber is a corner E-bend of a rectangular waveguide. The sample holder is put in at the corner of the waveguide. Sample exchange is easier in the latter chamber than the former.

In the coaxial chamber a variable DC sputtering voltage V_s of up to +1 kV with respect to the sample is applied to the outer conductor of the plasma wall. The sample current density can be varied from 2 to 20 mA/cm^2. The post-ionized neutrals are extracted from the plasma through a small aperture. Then these ions are deflected by a 90° electrostatic sector and analyzed by a double focusing mass spectrometer. It is noteworthy that only post-ionized neutrals are detected in this arrangement.

The electron density N_e and electron temperature T_e of the plasma were measured with a cylindrical single probe. At Ar pressure P = ~ 10^{-2} Pa, the value of N_e and T_e are (1 - 4) x 10^{11} cm^{-3} and (1 - 2) x 10^5 K respectively. The degree of ionization is estimated to be about 1 %. These N_e values are an order of magnitude larger than those in the HF (27.12 MHz) plasma used by OECHSNER and STUMPE [4]. When P is several 10^{-2} to several 10^{-1} Pa, the predominant ionization mechanisms of sputtered neutrals are electron impact and Penning mode [1-4, 6].

For the preliminary SNMS analysis, the samples chosen, GaAs and AgAu(Au:60 wt%), are composed of two components, one of which is highly sensitive to SIMS analysis and the other insensitive. Two kinds of mass spectra for positive ions from AgAu sample at V_s = 0 and 500 V are shown in Fig.2 in order to clarify the post-ionized neutrals (Ag^+, Au^+ and Cu^+ ions).

Hence, the Cu peak at V_s = 500 V originates from the Cu sample holder. It is noticeable that the Au peak can be detected in the same range as the Ag peak, which is usually impossible in SIMS. Similar results are also obtained for the GaAs sample. They are tabulated in Table 1. These two analyses indicate that the elemental sensitivities in SNMS are in a narrow range, less than an order of magnitude.

Table 1 Comparison of relative elemental sensitivity between SNMS and SIMS

	As^+/Ga^+	Au^+/Ag^+
SNMS (Ar)*	5×10^{-1}	9×10^{-1}
SIMS (O_2^+)**	1×10^{-3}	5×10^{-3}
(Ar^+)**	7×10^{-4}	

*Plasma gas; **Primary ion beam

Two problems are encountered in the present SNMS system. One is the appearance of misleading peaks in mass spectra which originate from the plasma wall, as noticed from the spectra at V_s = 0 V in Fig.2. The plasma potential

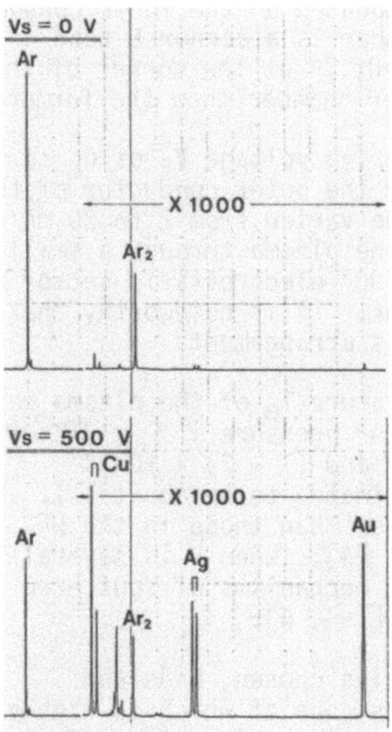

Fig.1 Schematic diagram of SNMS system

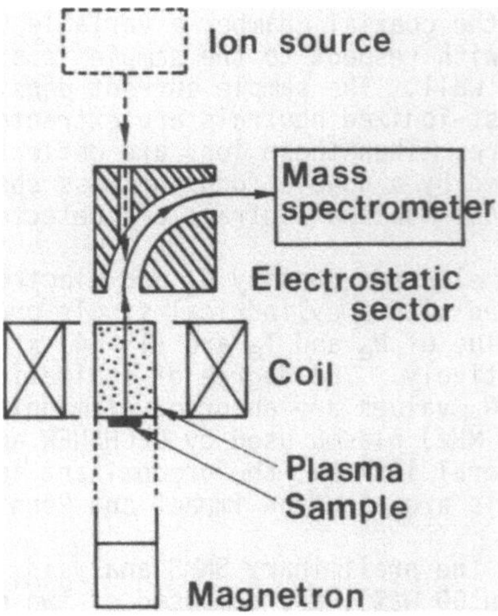

Fig.2 Typical mass spectra of positive ions from AgAu (Au:60 wt%) sample set on Cu holder at V_s = 0 and 500 V

is more positive than the plasma wall potential by a floating potential of some 10 V. Therefore, the ions in the plasma bombard the plasma wall to bring about sputtering. The other is the residual gas effect. The noise intensities resulting from the residual gas species in the mass spectrum are lower than the Ar^+ intensity by two or three orders of magnitude, but comparable to the signal intensities of sample elements. This affects the SN ratio and lowers the detection limit of SNMS. The equivalent pressure to the sputtered particle density is roughly estimated as some 10^{-4} Pa, which is on the same order as the background pressure of the plasma. It is especially necessary to lower the background pressure in order to improve the SN ratio.

References

1. J.W. Coburn and E. Kay, Appl. Phys. Lett. 18, 435 (1971).
2. J.W. Coburn, E. Taglauer and E. Kay, J. Appl. Phys. 45, 1779 (1974).
3. H. Oechsner and W. Gerhard, Phys, Lett. 40A, 211 (1972).
4. H. Oechsner and E. Stumpe, Appl. Phys. 14, 43 (1977).
5. T. Ishitani, N. Sakudo, H. Tamura and I. Kanomata, Phys. Lett. 67A, 375 (1978).
6. B.L. Bentz, C.G. Bruhn and W.W. Harrison, Int. J. Mass Spectrom. Ion Phys. 28, 409 (1978).

Investigation of Monolayers at High Primary Ion Current Densities

K.D. Klöppel
Physikalische Chemie, Gesamthochschule Siegen
5900 Siegen 21, Postfact 21 02 09, BRD

W. Seidel
Physikalische Chemie, Universität Gießen
6300 Gießen, Heinrich Buff Ring 58, BRD

Today SIMS is almost routine for bulk analysis and depth profile determinations. Its application to investigate monolayers was first shown by BENNINGHOVEN and co-workers [1]. They used the so-called static SIMS with very low primary ion (PI) densities and low partial pressures of oxygen. By strongly reducing the intensity of the PI the lifetime of a monolayer is extended to hours. At the same time perturbations by sputtering processes are reduced. Such perturbations cannot be avoided in principle on account of continuous impact of primary ions. Therefore, it was assumed that high densities of primary ions will lead to untimely destruction of the material under investigation, especially if larger, i.e., organic, molecules on a surface are to be analyzed.

We will show in this paper that surface reactions, e.g., the build up of an "oxide monolayer" on a metal surface, can be followed even at PI-currents in the A/cm^2 region, if short PI-pulses are employed.

Our instrument consists of an ion discharge source with magnetic field for production of the primary ions and a target chamber with a quadrupole mass spectrometer for the analysis of the secondary ions [2]. The positive ions are measured with a 14-stage Cu/Be-multiplier combined with a fast preamplifier. The shortest possible time to record a spectrum in the mass range between 0 and 200 amu is a tenth of a second.

The formation of oxide layers can be followed in two different ways. The first method is an indirect one in which the decomposition is investigated on oxide layers which have been grown under various conditions of oxygen pressure and exposure time. Because the secondary ion (SI) emission is assumed to be proportional to the oxygen coverage of the surface, it is possible to determine the formation quality of an oxide layer from the SI intensities at the beginning of any decomposition measurement. In Fig.1 the logarithmic Cu^+-SI intensities are plotted versus the sputtering times for differently oxidized copper surfaces. The lowest line shows that it is possible to follow the removal of a monolayer by such decomposition measurements.

When the Cu^+-SI intensities for the onset of those decomposition measurements are plotted against the oxygen exposure dose in Langmuir, the curve in Fig.2 is obtained. From this curve a dose of 200 Langmuir is seen to be sufficient to reach saturation of the Cu^+-emission. For a CuO^+-emitting layer the saturation is about 9000 Langmuir.

The second way to look at a copper oxide layer being formed is a more

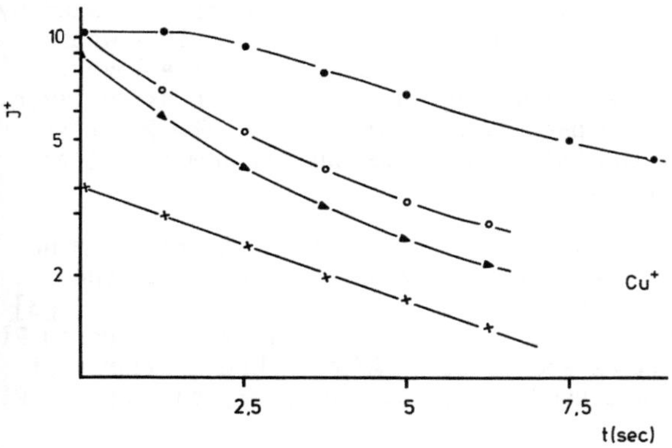

Fig.1 Change of Cu^+-SI emission during sputtering of copper oxide layers generated by different oxygen exposure

x 2 min/2 · 10^{-7} Torr
▶ 3 min/5 · 10^{-7} Torr
o 3 min/8 · 10^{-6} Torr
● 3 min/8 · 10^{-6} Torr

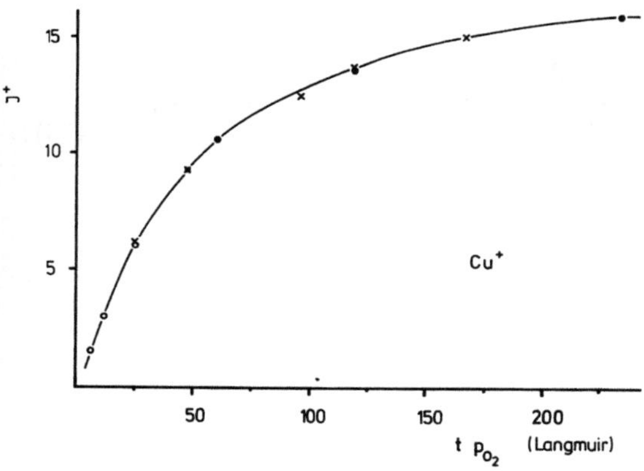

Fig.2 Change of Cu^+-SI emission from a copper surface as a function of oxygen exposure dose

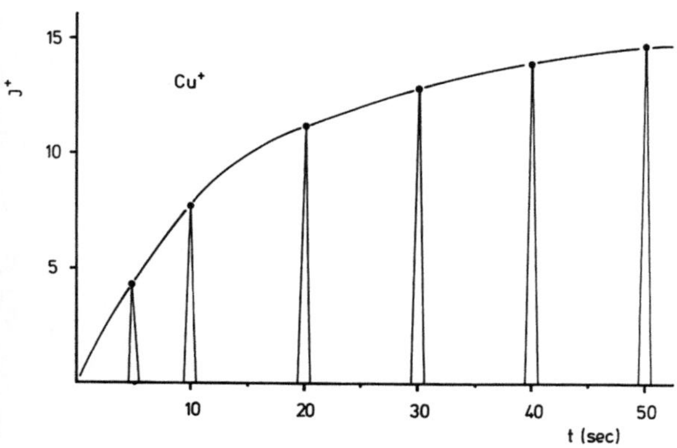

Fig.3 Oxidation of a copper surface investigated by the SIMS pulse technique: Change of Cu^+-SI emission as a function of O_2-exposure time

direct one. The primary ion beam consists here of short pulses generated by using the beam monitor as a chopper. The PI pulses cause SI pulses whose intensities are characteristic for the O_2 coverage reached in the layer. Fig.3 shows Cu^+-SI intensities obtained during the growth of an oxide layer on a copper surface under 4.10^{-4} Pa partial pressure of O_2. The PI pulse time is about 10^{-1} sec. Again, the Cu^+-emission reaches saturation at an exposure dose of about 200 Langmuir.

By applying this pulsed beam method to analysis of thin layers of amino acids on solid targets we obtain the mass spectra of alanine and serine. The spectra are similar to those first reported by BENNINGHOVEN et al., [3] using static SIMS. Best results are obtained from graphite targets using PI-energies of 2 keV. The PI-densities are about 4 $\mu A/cm^2$. When mixtures of graphite powder and amino acids are used, it is not necessary to chop the PI-beam for spectra measurement.

References

1. A. Benninghoven, Z. Physik 230, 403 (1970).
2. K.D. Klöppel, W. Seidel, J. Mass Spectrom. Ion Phys., in press.
3. A. Benninghoven, D. Jaspers, W. Sichtermann, Appl. Phys. 11, 35 (1976).

VII. Geology

SIMS Measurement of Mg Isotopic Ratio in a Chondrite

J. Okano and H. Nishimura
Institute of Geological Sciences
College of General Education
Osaka University
1-1 Machikaneyama-cho, Toyonaka
Osaka 560, Japan

1. Introduction

The Allende carbonaceous chondrite has been reported to contain primordial white inclusions with evidences for the nucleosynthesis during the early history of the solar system [2,3]. ^{26}Al (half life - 7.2 x 10^5y) possibly remained in Al-rich white inclusions in the early time and ^{26}Mg would be enriched there due to the decay of ^{26}Al. GRAY and COMPSTON [1] reported the ^{26}Mg excess of 0.4% in an inclusion of Allende. Successive discoveries of the Mg isotopic anomalies have been done by many workers [2-5] in the white inclusions of Allende. BRADLEY, et al., described that the excess of ^{26}Mg was found to be 40% for an anorthite crystal with the ion microprobe technique [4]. SHIMIZU, et al., also reported 13% excess of ^{26}Mg in an anorthite inclusion with the ion microprobe [5]. Those excesses were concluded to be due to the in situ decay of ^{26}Al.

In the present work, the distributions of Mg isotopic ratios were examined for two inclusions of Allende by line profile analysis with SIMS.

2. Experimental

A homemade SIMS with a hollow cathode primary ion source [6,7] was used. The sample surface was bombarded by 8 keV oxygen primary ion beam. The probe size was about 100 μm, and the current, 3×10^{-7}A. The oxygen pressure was about 2.5×10^{-5} torr in the sample chamber.

Two inclusions of Allende (AL1 and AL2) were analyzed for Mg isotopic ratios of $^{25}Mg/^{24}Mg$ and $^{26}Mg/^{24}Mg$ along probed lines across the inclusions every 0.1 mm. $^{27}Al/^{24}Mg$ ratio was also measured. A terrestrial forsterite sample from Ehime Pref., Japan was intermittently analyzed as a laboratory standard. The interferences of molecular and doubly-charged ions ($^{12}C_2^+$, $^{48}Ti^{++}$, $^{50}Ti^{++}$, $^{52}Cr^{++}$, $^{48}Ca^{++}$, $^{13}C_2^+$) and hydride ions to the mass range of 24-26 were carefully examined. Their interferences to the subject peak heights were less than 0.9×10^{-3} and neglected.

3. Results and Discussion

The results are shown in Figs.1,2,3 and 4. In Fig.1 are shown the deviations of raw values of $^{25}Mg/^{24}Mg$ and $^{26}Mg/^{24}Mg$ ratios from the reference values. Δ_{25} (ordinate) and Δ_{26} (abscissa) are calculated according to

$$\Delta_n = \left[\frac{(^nMg/^{24}Mg)meas}{(^nMg/^{24}Mg)ref} - 1\right] \times 1000 \quad (n = 25,26) \qquad (1)$$

where $(^{25}Mg/^{24}Mg)ref = 0.12484$ and $(^{26}Mg/^{24}Mg)ref = 0.13626$, were the mean values for the terrestrial forsterite sample. In Fig.1, the normal fractionation line through the reference point with the slope of 1/2 is illustrated.

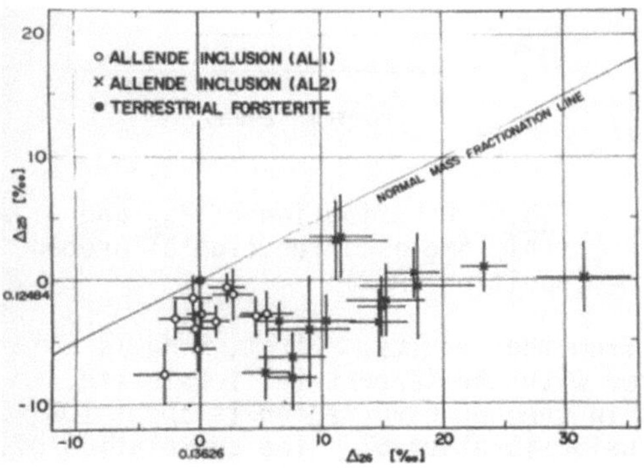

Fig.1 The deviation Δ_{25}, as a function of Δ_{26}. Δ_{25} and Δ_{26} were calculated from (1). A closed circle in the figure indicates the reference point for the terrestrial forsterite

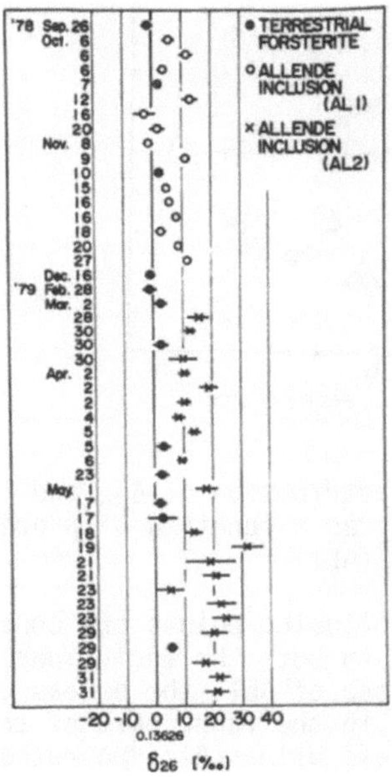

Fig.2 The deviation, δ_{26}, of normalized $^{26}Mg/^{24}Mg$ for AL1 and AL2. The measured data for the terrestrial forsterite are also plotted

The results for the inclusions after the correction for the mass fractionation are shown in Fig.2. The data are plotted in the chronological order. The abscissa represents δ_{26} calculated from

$$\delta_{26} = \left[\frac{(^{26}Mg/^{24}Mg)meas/(1-\alpha)^2}{(^{26}Mg/^{24}Mg)ref}\right] - 1 \quad \times 1000 \qquad (2)$$

$$1-\alpha = \frac{(^{25}Mg/^{24}Mg)meas}{(^{25}Mg/^{24}Mg)ref} \qquad (3)$$

The distributions of δ_{26}'s for two inclusions, AL1 and AL2, are shown in Figs.3 and 4. The abscissa represents the probed position on the sample surface. $^{27}Al/^{24}Mg$ ratios are also plotted in the upper parts of the figures. The sketches of the inclusions are shown in those figures. The error bars are the standard deviations.

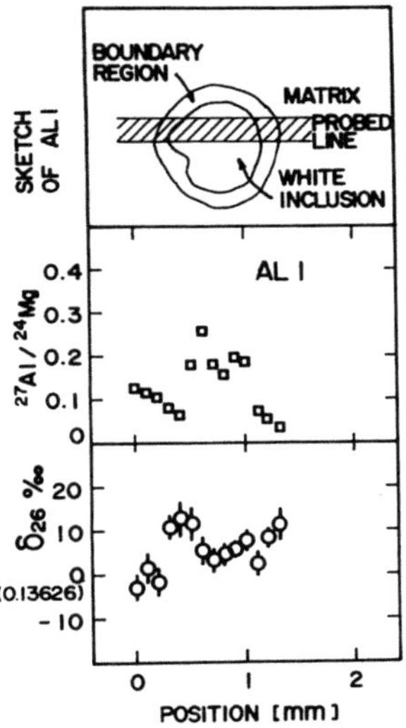

Fig.3 Distribution of δ_{26} and $^{27}Al/^{24}Mg$ as a function of probed position (AL1)

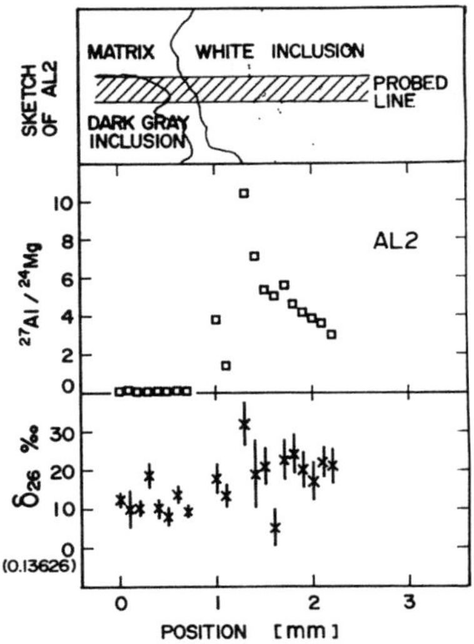

Fig.4 Distribution of δ_{26} and $^{27}Al/^{24}Mg$ as a function of probed position (AL2)

The following points are concluded from the results. First, ^{26}Mg is enriched in both the inclusions compared with the terrestrial forsterite. In the case of AL1, the excess of ^{26}Mg in the boundary region is about 10% and that in the inner part of the inclusion is about 5%. The correlation of ^{26}Mg excess with $^{27}Al/^{24}Mg$ ratio is weak. Second, for AL2 inclusion, the ^{26}Mg excess is high in the high Al/Mg portion, and the excess is about 20%. This excess seems to correlate with Al concentration and likely resulted from the decay of ^{26}Al.

Acknowledgements

The authors are indebted to Prof. N. Takaoka of Yamagata University for his permission to use the samples of Allende. They also express their thanks to the head of the Institute of Geological Sciences for his offer of the terrestrial forsterite sample, whose locality is Akaishi mine, Uma gun, Ehime, Japan.

References

1. C.M. Gray, W. Compston, Nature 251, 495 (1974)
2. G.J. Wasserburg, T. Lee, D.A. Papanastassiou, Geophys. Res. Lett. 4, 299 (1977)
3. R.N. Clayton, T. Mayeda, Geophys. Res. Lett. 4, 295 (1977)
4. J.G. Bradley, J.C. Huneke, G.J. Wasserburg, J. Geophys. Res. 83, 244 (1978
5. N. Shimizu, M.P. Semet, C.J. Allegre, Geochim. Cosmochim. Acta 42, 1321 (1978)
6. H. Nishimura, J. Okano, Jap. J. Appl. Phys. 8, 1335 (1969)
7. H. Nishimura, J. Okano, Mass Spectroscopy 23, 9 (1975)

Negative Molecular Ion Analysis of Inorganic Sulfur-Oxygen Salts by SIMS

J. Ganjei, R. Colton, and J. Murday
Chemistry Division
Naval Research Laboratory
Washington, D.C. 20375

An important feature of secondary ion mass spectra is the presence of polyatomic ions. Due to the low mass resolutions of most SIMS instruments, these ions were initially regarded as a serious mass interference problem for SIMS elemental analysis. Currently, several research groups are exploring the analytical potential of molecular secondary ions for providing bonding structural information. This paper describes an investigation of the relationship between secondary ion emission and anion stoichiometry in inorganic salts. Three simple sulfur-oxygen anions, SO_4^{-2}, SO_3^{-2}, and $S_2O_3^{-2}$ were examined. A clear relationship between the anion stoichiometry and the negative molecular secondary ion emission pattern was established.

The experiment was performed under dynamic conditions with 1 Kv Ar^+ primary ions at 1-2 A/cm^2 beam density. The salts were embedded as powders in indium foil. Secondary ion measurement was accomplished with an Extranuclear quadrupole mass spectrometer previously described [1]. Significant ion signals were detected only when sample charging was reduced by flooding the surface with electrons from a neutralized filament. The possibility of beam damage was monitored by XPS, present in the same UHV chamber (base pressure 1×10^{-9} torr).

Initially, both the positive and negative secondary ion mass spectra of the salts were recorded. The (+) secondary ions included the alkali metal ions (present as the cation and impurities) and related (MO^+, M_2O^+) oxide species. These molecular oxide ions did not exhibit any apparent relationship to anion stoichiometry. In contrast, the (-) spectra were dominated by the SO_x^- ions as previously reported by BENNINGHOVEN, et al., [2] for sulfate anions. The major peaks consisted of the 16n (n = 1, 2...6) series; mass 17, OH^-; mass 19, F^-; mass 26, CN^-; and mass 42, CNO^-. These spectra were carefully recorded under the same experimental conditions for Na_2SO_4, Li_2SO_4, Na_2SO_3, and $Na_2S_2O_3$. An initial period of presputtering was necessary to remove an overlayer of carbon contamination introduced during sample preparation.

The results of these analyses are presented in Table 1 where the columns represent the 16n series normalized to the mass 32 intensity. The relative standard deviation of five separate measurements of Na_2SO_4 is also given in order to show that ratio changes above 15% are significant. The last two rows headed by N* represent the number of combinations available to form SO_x from S with four oxygens (SO_4) or three oxygens (SO_3).

There are several obvious features in the salts' spectra. First, the sulfate anion emits substantially more high mass ions of mass 80 and 96,

Table 1 Negative secondary ion intensities from $S_xO_4^{-2}$ anions

	m/z					
Salt	16	32	48	64	80	96
	O	O_2^-,S^-	SO^-	S_2^-,SO_2^-	SO_3^-	SO_4^-
Li_2SO_4	12	1.0	.56	.31	.12	.021
Na_2SO_4	9.4±1.1	1.0	.46±.02	.32±.02	.14±.02	.029±.004
Na_2SO_3	6.2	1.0	.37	.17	.037	.006
$Na_2S_2O_3$	2.8	1.0	.31	.23	.023	.004
N* SO_4	-	1	4	6	4	1
N* SO_3	-	1	3	3	1	-

N* equals the number of combinations available to form SO_x from SO_n

reflecting its solid bulk compositional structure. Second, the variation of 16/32 intensity ratio reflects the changes in proportion to the O/S atomic ratio in the anions. Third, while some differences exist, the Li_2SO_4 emission pattern closely resembles Na_2SO_4. Due to the above differences, one can distinguish between the anions on the basis of the relative intensities of the (-) molecular secondary ions (fingerprint pattern).

Another interesting relationship is the correlation of the above mentioned N* and the relative secondary ion intensities of Na_2SO_4 and Na_2SO_3. In this case, the differences between the ion intensity ratios 80/32, 64/32, and 48/32 in the two anions are accurately predicted (within 10%) by the statistical factor of SO_x formation from SO_N. Thus, the 80/32 ratio in Na_2SO_4 is four times the ratio in Na_2SO_3.

This treatment was not extended to $Na_2S_2O_3$due to the complications of $^{64}S_2^-$and $^{80}S_2O^-$ mass interferences. N* does not predict the relative ratios of the different mass fragments of one anion due to the added factors of ion yield and instrument transmission. However, its accuracy in comparing SO_4^{-2} to SO_3^{-2} does present the possibility that MO_x^-/M^- relative ion intensities can be predicted rather than simply recorded as a fingerprint pattern. At this point further analyses with other central atom anions are necessary to evaluate the statistical factor significance of combination.

Sample damage during ion bombardment was a major concern since the experiment was performed under dynamic conditions. There existed a distinct possibility that the damaged sampling volume was composed of metal oxides and/or sulfides bearing no relationship to the original salt structure. However, XPS analysis of samples before and after bombardment showed only a small enrichment of cation surface concentration. The measured values for Na1s, O1s, NaKVV Auger and S2p electron binding energies agreed with the literature values for all three salts (including the two S_2p peaks in $S_2O_3^{-2}$). Other experimental factors which will be discussed are the effects of primary ion energy and secondary ion energy filtering on the relative secondary ion intensities.

Future projects include an investigation of transition metal salts and other central atom anion series.

References

1. R.J. Colton, J.S. Murday, J.R. Wyatt, and J.J. DeCorpo, Surf. Sci., in press.
2. A. Benninghoven, Z. Naturforsch, 24a, 859 (1969).

SIMS Measurements of Tracer Diffusivity of Oxygen in Titanium Dioxides and Sulfur in a Calcium Sulfide

M. Someno and M. Kobayashi
Department of Metallurgical Engineering
Tokyo Institute of Technology
Ookayama 2-12-1, Meguroku, Tokyo 152, Japan

1. Introduction

SIMS has the great advantage of in-depth analysis of stable isotopes. This makes it possible to determine the tracer diffusivity, ranging from 10^{-15} to 10^{-19} m^2/sec, directly from a penetration curve of a stable isotope. By using this method, oxygen tracer diffusivity has been measured in ZrO_2 [1], UO_2 [2,3] and Al_2O_3 [4]. The purposes of this report are (1) to examine the effect of Cr_2O_3-doping on the oxygen tracer diffusivity in n-type semiconductor rutile (TiO_2), and (2) to identify predominant charge carriers in ionic conductor calcium sulfide by the sulfur tracer diffusivity measured by SIMS.

2. Experimental

Pure and 0.08 mol% Cr_2O_3-doped rutile single crystals were diffusion-annealed in a quartz chamber under about 5 kPa of an oxygen gas mixture of 21.3 at% ^{18}O in ^{16}O. Sintered dense calcium sulfide discs and 10 mg of 30-50% dilute ORNL stable isotope S-34 (original content, 94.33 at%) were put into a vacuum-sealed quartz tube and diffusion-annealed. Sulfur vapor pressure in the tube was 80-100 kPa during the diffusion-annealing.

The samples were analyzed with a Hitachi ion microanalyzer using Ar^+ primary ion beam of 1 mm diameter. Electrons with 3 keV were showered on the calcium sulfide sample surface to prevent charge build-up. After each analysis, the diameter and the depth of the ion-etched crater were measured with a commercially available stylus device (Taly-Step). A flat-bottomed crater was obtained by defocusing the primary ion beam on the sample surface. The etching speed was about 1 A/sec.

3. Results and Discussion

A typical result of the depth profile measurement for the oxygen isotopes in the rutile sample is shown in Fig.1. The measured intensity of m/e = 18(I_{18}) had a background in the tailing region. The background is considered to consist mainly of $H_2{}^{16}O^-$. The ^{18}O concentration ($I_{18}/I_{16}+I_{18}$) at the surface was determined by linear extrapolation from the 0.3 µm depth, and was smaller than the concentration in the gas phase (0.213). This difference indicates that the ^{18}O transfer from the gas phase to the solid phase was controlled not only by diffusion in the solid but also by the surface exchange reaction. Assuming the reaction of first order, the solution of Fick's second law in a semi-infinite solid is expressed by

$$C/Co = erfc(x/2\sqrt{Dt}) - exp(hx + h^2Dt)\ erfc(x/2\sqrt{Dt} + h\sqrt{Dt}) \tag{1}$$

for the following conditions

$$-D(\partial C/\partial x)_s = k(Co - Cs),\ C = 0 \text{ for } t = 0, \text{ and } Co = 0.213. \tag{2}$$

In (1), Co is the fraction of ^{18}O in the gas phase, x is the depth, D is the tracer diffusivity, t is the diffusion time, k is the rate constant of the exchange reaction, and h = k/D. In Fig.2, the calculated values of D are plotted against the reciprocal temperature. The best fitting equations for the measured D values for the // c-direction diffusion can be expressed as

$$D\ (m^2/sec) = 3.4 \times 10^{-7} exp[-251(kJ/mol)/RT]\ \text{(pure), and} \tag{3}$$

$$D\ (m^2/sec) = 2.8 \times 10^{-8} exp[-204(kJ/mol)/RT]\ (Cr_2O_3\text{-doped}). \tag{4}$$

At 1073K, the diffusivity of oxygen for the c-direction was 1.5 times greater than that for // c-direction in the same sample. From Fig.2, it is found that the tracer diffusivity of oxygen increased with Cr_2O_3-doping. The increment is considered to be due to interaction of oxygen atoms with Cr atoms in the titanium lattice sites [5].

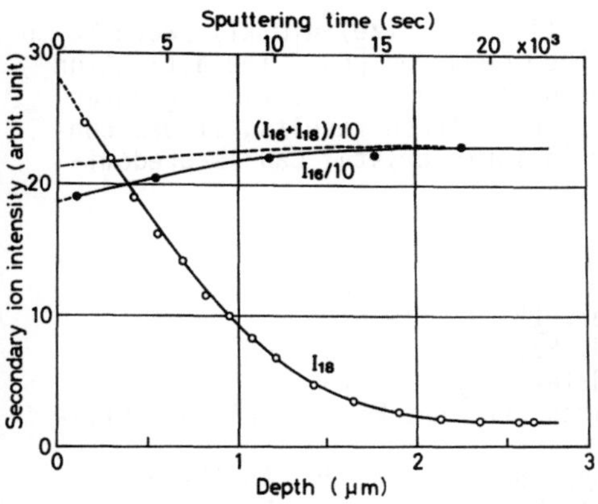

Fig.1 In-depth intensity profile of negative secondary ions of m/e = 16(I_{16}) and m/e = 18 (I_{18}) from a pure rutile sample diffusion-annealed at 1353K for 1.16×10^4 sec

The depth profile for the sulfur isotopes in calcium sulfide was measured using CaS^+ ions because of their much higher sensitivity than S^+ ions. No difference of the ^{34}S concentration was observed between at the surface and in the gas phase. In this case, the solution of Fick's second law is expressed by

$$C/Cs = erfc(x/2\sqrt{Dt}). \tag{5}$$

In Fig.2, the calculated values of D are plotted. The best fitting equation for the measured D can be expressed as

$$D\ (m^2/sec) = 1.48 \times 10^{-6} exp[-243(kJ/mol)/RT]. \tag{6}$$

The electrical conductivities, calculated from the measured sulfur tracer

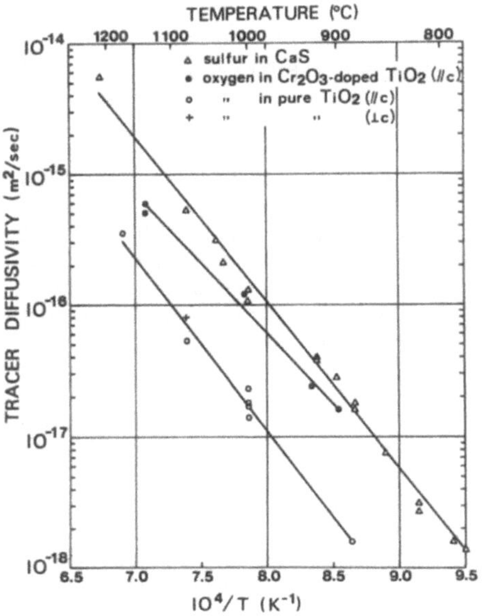

Fig.2 Tracer diffusivity as a function of reciprocal temperature for oxygen in pure rutile, for oxygen in 0.08 mol% Cr_2O_3-doped rutile, and for sulfur in calcium sulfide

diffusivities and the Nerst-Einstein relation ($\sigma = (ze)^2 n_i D/kT$), were 10 to 100 times smaller than those which had been measured [6]. The activation energy (243kJ/mol) for the sulfur tracer diffusion was 1.5 times greater than that for the electrical conduction. From these results, it was concluded that the electric charge carrier in ionic crystal CaS is predominantly Ca^{2+}.

References

1. B. Cox, J.P. Pemsler, J. Nucl. Mater. 28, 73 (1968)
2. P. Contamin, G. Slodzina, C.R. Acad. Sc. Paris, 267C, 805 (1968)
3. J.F. Marin, P. Contamin, J. Nucl. Mater., 30, 16 (1969)
4. D.J. Reed, A.J. Garrat-Reed, Abstracts 3rd Ann. Meeting FACSS, Philadelphia, No. 388 (1976)
5. M. Ikebe, Y. Miyako, M. Date, J. Phys. Soc. Japan, 26, 43 (1969)
6. H. Nakamura, K. Gunji, J. Japan Inst. Metals, 42, 635 (1978

SIMS Analysis of TiO_2, Including Depth Profiling[1]

J.L. Peña[2]
Laboratory for Surface Studies and Department of Physics
University of Wisconsin, Milwaukee, Wisconsin 53201

M.H. Farias and F. Sanchez-Sinencio[3]
Departamento de fisica, Centro de Investigacion del
Instituto Politecnico Nacional,
Apartado Postal 14-740 Mexico 14, D.F.

Titanium dioxide shows a blue coloration after being placed in contact with a sodium amalgam at room temperature [1]. Sodium and hydrogen diffuse during treatment. Water decomposition by an electrochemical cell with an n-TiO_2 electrode (treated in a H_2 gas atmosphere) was first demonstrated by FUJISHIMA and HONDA [2]. The characterization of the surface structure and chemical composition of the active TiO_2 surface is needed in order to understand the mechanisms of the photocatalytic reaction and the coloration process [3,4].

Titanium dioxide specimens, cut from an ingot of a single crystal rutile, were ground and polished to a thickness of 1 mm and area of ~2 cm^2 perpendicular to the c axis. TiO_2A samples were placed in contact with sodium-amalgam and left for 3 hrs. at a temperature of 400°K. TiO_2HI was implanted with 100 keV H^+ and shows blue coloration. TiO_2HAT was heated at 1300°K in a H_2 gas atmosphere (p = 1 atm.) for 5 minutes.

The experiments were performed with a Balzers SIMS [5]. The spectra were taken in the positive mode with primary Ar^+ ions with an energy of 2.1 keV (Ep) and current density (Ip) 1×10^{-7} Amp-cm^{-2}. The depth profiling is measured using Ep = 1 keV, and Ip = 2.5×10^{-5} Amp-cm^{-2}. Such conditions produce a sputtering rate of ~50Å min^{-1}. The spectrum of the TiO_2 surface showed clearly the Ti-isotopes. The spectrum shows the existence of TiO_x (Ti_mO_n, m + n = x) clusters, with $0 \le x \le 6$. When $3 \le x \le 6$, one might expect to find Ti_2O_2, and Ti_2O_3, each of which would contain nine peaks. We find only groups containing five peaks. TiO and TiO_2 clusters can be related to TiO_xH^-formation because they do not maintain the Ti-isotopic ratio. The spectrum of TiO_2A has the same peaks as those related to Ti and TiO_2. There are peaks that exceed those of TiO_2 in height, for example: ^1H, ^7Li, ^{23}Na, ^{40}Ca, $^{56}(CaO)$, ^{56}Fe, ^{88}Sr, $^{116}(SrSi)$, ^{138} Ba and $^{154}(BaO)$. The impurities in the sample itself are almost the same as in TiO_2 [6]. Most of these impurities are found in the Na-amalgam [7], which is therefore a source in the diffusion processes. The formation of bronzes [8] can be noted in the spectrum. The TiO_2HAT spectrum shows that hydrogen diffusion causes changes in the crystal surface due to the production of $^{49}(TiH)$.

The depth profiling analysis did not reveal any differences in the ^1H signal between TiO_2 and the samples with hydrogen diffusion. Fig.1 shows the $^{17}(OH)$ profile for TiO_2HI. To a first approximation, we expected the ^1H profile to be a Gaussian curve with maximum peak around 1µm. However, it was

---

[1]Work partially supported by PNCB-CONACYT (Mexico) and SEPAFIN (Mexico).
[2]Permanent address: F.C. Fisico-Mathematicas, UANL (Monterrey N.L. Mexico).
[3]Also in Escuela Superior de Fisica y Matematicas del IPN.

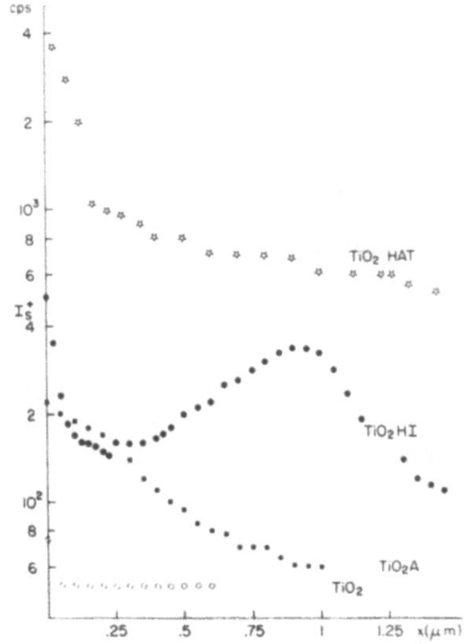

Fig.1 SIMS profile of ^{17}(OH) in TiO_2 (virgin) TiO_2A, TiO_2HI, and TiO_2HAT. Sputtering is done with 1 keV Ar^+, $Ip=2.5x10^{-5}$ $Amp\text{-}cm^{-2}$

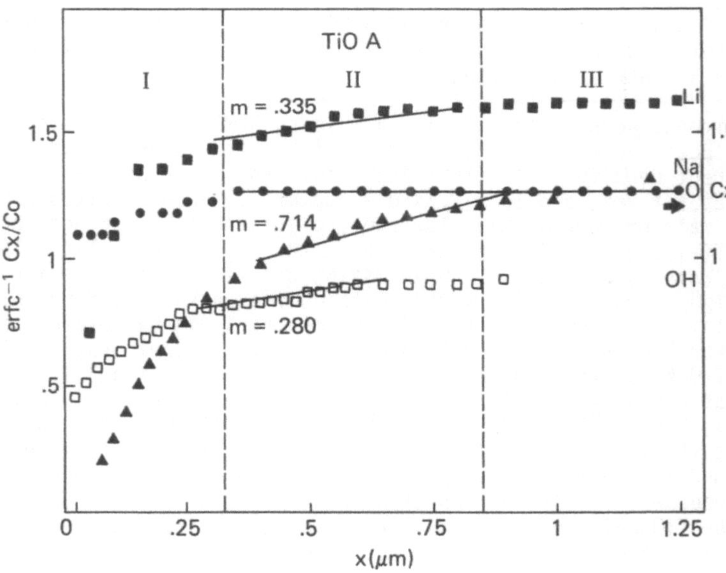

Fig.2 Plot of $erfc^{-1}$ (C_x/Co) versus depth when H, Li, and Na are diffused into TiO_2A. m is the slope for each straight line. The right vertical axis shows the oxygen relative concentration

the ^{17}(OH) that showed these characteristics (maximum at ~1μm). It was known that this type of chemical implantation occurs in TiO_2-rutile [9]. The result is accounted for by the formation of (OH) through the reaction of O and H inside the crystal. This result suggests a new method of using SIMS for the detection of H by means of monitoring the $^{17}(OH)^+$ signal. Fig.1 shows the ^{17}(OH) profiles for TiO_2, TiO_2A and TiO_2HAT. ^{23}Na and ^{12}C profiles in TiO_2, ^{23}Na and ^{7}Li in TiO_2A, and C in TiO_2HAT were obtained. They show typical decay behavior. From the results obtained, it can be concluded that diffusing samples exhibit higher counts of the elements and compounds which are dealt with in these works. The profiles can be used for atomic diffusion studies.

To calculate the diffusion coefficients (D) suppose that the number of secondary ions at depth x is proportional to the concentration in the sample. Since the diffusion process is only concerned with relative concentrations at different depths, the proportional factor disappears, and D can be calculated using the detected number of secondary ions. The one-dimensional model of diffusion (D constant) is used, so the solution is [10]: $C_x/Co = \mathrm{erfc}\,[x/(4Dt)^{\frac{1}{2}}]$. If a plot of $\mathrm{erfc}^{-1}\,[C_x/Co]$ against x is made, it should be linear, and have a slope of $(4Dt)^{-\frac{1}{2}}$. Fig.2 shows the plot for ^{17}(OH), ^{23}Na, and ^{7}Li in TiO_2A. There are three different slopes for each of them. It is well known [11] that, for large penetration distances, short-circuiting diffusion along defects, distorts the ideal picture, so the apparent D value will be higher. Close to the surface, the slope is larger and the apparent D value is abnormally low. This phenomena has been called the near-surface effect (NSE) (region I). The NSE is related to the formation of a different compound. That can be observed in the behavior of the oxygen relative concentration (ORC) (right vertical axis). ORC changes occur only in region I, afterwards it is constant. The ORC increase indicates a reduction in region I. Consequently, region II gives the correct value of D. The D for hydrogen was calculated using the ^{17}(OH) data. The values are $D_H = 3\mathrm{x}10^{-12}$ cm^2-sec^{-1}, $D_{Li} = 2.1\mathrm{x}10^{-12}$ cm^2-sec^{-1}, $D_{Na} = 4.5\mathrm{x}10^{-13}$ cm^2-sec^{-1}. In the case of TiO_2HAT the graph was plotted for ^{12}C and ^{17}(OH) (Fig.3). ORC changes in all the depth analysed and shows reduction, so the analysis is done in the NSE region. The values are $D_H = 5.9\mathrm{x}10^{-10}$ cm^2-sec^{-1}, and $D_C = 4.5\mathrm{x}10^{-11}$ cm^2-sec^{-1}.

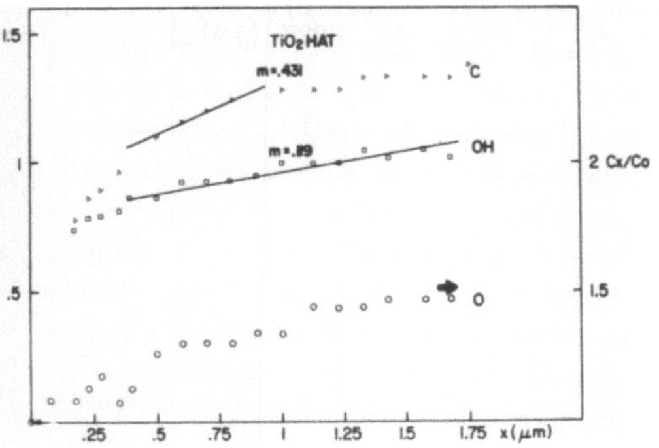

Fig.3 Plot of erfc^{-1} (C_x/Co) versus x when H and C are diffused into TiO_2HAT. The right vertical axis shows the oxygen relative concentration

In summary, surface analysis and depth profiles using the SIMS technique were made for the following samples: TiO_2, TiO_2HI, TiO_2HAT, and TiO_2A. TiO_2HI showed chemical implantation. This result supports a new method involving using SIMS for the detection of hydrogen by means of monitoring the ^{17}(OH) signal. Furthermore, the method was used to get H-diffusion coefficients for TiO_2A and TiO_2HAT. Carbon contamination was detected in TiO_2HAT. The Na and Li diffusion coefficients in TiO_2A were obtained. The coloration has been associated with a reduction process. JOHNSON has associated the high-temperature coloration with atomic impurity diffusion [12]. Coloration could also be due to the formation of titanium bronze compounds at high temperature [8]. In the present work, TiO_2HAT shows: carbon contamination, reduction and (OH) formation; TiO_2A shows: formation of titanium bronze compounds, impurities diffusion, and (OH) formation; TiO_2HI shows: (OH) formation. All the different samples were colored even after sputtering and a strong reduction of impurities (e.g., Li, C, Na, Ca, etc.). The only point in common was the (OH) formation, which can reasonably be expected to result in a broad free carrier absorption in the visible and near IR in addition to the IR OH stretching absorbances. Consequently, this is a possible interpretation for the coloration process.

References

1. J. Gonzales H., J. Gonzales-Basurto, F. Sanchez-Sinencio, J.S. Helman and A. Reyes-Flotte, J. Appl. Phys. 49(8), 4509 (1978).
2. A. Fujishima and K. Honda, Nature (Lond.) 238, 37 (1972).
3. Victor E. Henrich, G. Dressellhauss and H.J. Zeiger, Phys. Rev. Lett. 36, 1335 (1976).
4. W.J. Lo, Y.W. Chung and G.A. Somarjai, Surf. Sci. 71, 199 (1978).
5. W.K. Huber, H. Selhofer and A. Benninghoven, J. Vac. Sci. Technol. 9, 482 (1972).
6. G.J. Hill, J. Appl. Phys. (J. Phys. D) 1, Ser. 2, 1151 (1968).
7. N.H. Forman, Standard Methods of Chemical Analysis, Vol. 1, Academic Press (1969).
8. S. Hirano, M. Ismail and S. Somiga, Mater. Res. Bull. 11, 1023 (1976).
9. B. Siskind, D.M. Gruen and R. Varma, J. Vac. Sci. Technol. 14, 537 (1977).
10. J. Crank, The Mathematics of Diffusion, 2nd ed., Oxford Univ. Press, London (1975).
11. D.K. Reimann and J.P. Stark, Acta, Met. 18, 63 (1970).
12. W.D. Ohlsen and O.W. Johnson, J. Appl. Phys. 44, 1927 (1973).

Ion Microprobe Analysis of Small Heavy Metal Particles and Their Compounds

J. Gavrilovic
Walter C. McCrone Associates, Inc.
Chicago, IL 60616

1. Introduction

Increased levels of air pollution in certain areas in the past 20-40 years have generated a strong demand for full characterization and identification of airborne microparticles in order to determine and hopefully eliminate their source. Among the most harmful pollutants are microparticles containing heavy metals and their compounds. Lead, along with its compounds, is probably the most common and most widely investigated airborne contaminant.

The most frequently used tools for chemical characterizations of microparticles are the Scanning Electron Microscope and Electron Microprobe [1]. These two instruments equipped with energy dispersive spectrometers will provide chemical analysis of a microparticle down to submicrometer size in a matter of minutes for most elements above 0.3 to 1% by weight [2]. Chemical analysis of microparticles below that level, however, requires an enormous increase in effort and counting time.

In such cases the ion microprobe analyzer offers several significant advantages. It will detect all elements, including hydrogen, with detection limits ranging from 0.1 to 0.001% (in a reasonable analysis time). It will also measure isotope ratios and detect thin films on the particles [3]. Trace element capability becomes a very significant factor in determination of the exact nature and origin of certain microparticles and pollutants.

2. Sample Preparation and Mounting

Perhaps the most important step in identification of microparticles with an ion microprobe is their preparation for analysis and mounting on a substrate.

Microparticles may be analyzed "in situ" but in virtually all cases it is advantageous to physically remove them and place first on a glass slide for microscopical examination and then on a special substrate for analysis. A substrate for ion microprobe analysis of microparticles has to satisfy the following conditions:

-- should be flat with a mirror-like surface
-- has to be chemically pure and not very reactive
-- free of inclusions
-- electrically conductive
-- should have relatively low secondary ion yield
-- should have a simple mass spectrum, i.e., few isotopes

Two substrates which satisfy most of these conditions and are in use at McCrone Associates are high purity gold and vitrified graphite. Gold is preferred because of its very high reflectance which facilitates microscopical inspection of the smallest microparticles. Gold beads of 99.9999% purity are placed between specially cleaned glass or quartz slides in the mounting press. Under gradually increasing pressure, gold planchettes of 5-10 mm diameter are produced with mirror-like clean surfaces. For small particle mounting and analysis, a small rectangular grid is inscribed on the gold using fine pointed tungsten needles.

Microparticles are then lifted off the samples using tungsten needles with tips electrochemically etched to a 1 µm point. The particle is transferred onto the gold substrate and fixed with a drop of very dilute collodion [4]. Electrically conductive particles in most cases do not require collodion for mounting. They will remain on the gold plate provided they are carefully handled. After first contact with the high energy ion beam, the sputtered and possibly resputtered ions from the particle and the substrate efficiently bond them together as shown in Fig.1.

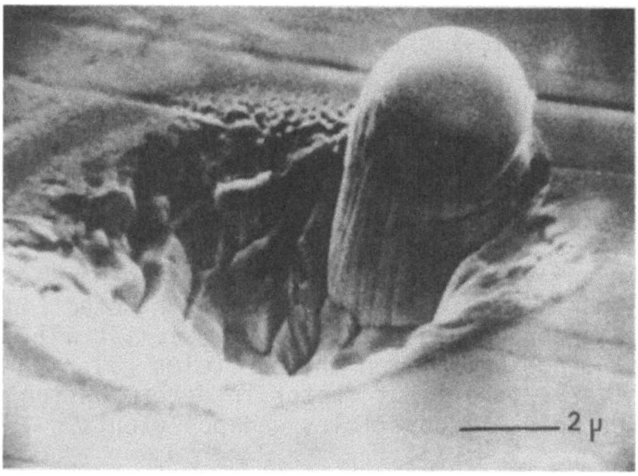

Fig.1 A spherical microparticle of fly ash on a gold substrate that has been bombarded by an off-center argon beam. Note the resputtered gold on the lower side of the particle which effectively bonds it to the substrate. Note shadowing effect and etching of the substrate that shows inverse Gaussian curve.

3. Analysis

Microparticles, mounted on a substrate with low secondary ion yield such as high purity gold, are analyzed by subjecting them to a partially focused beam of high energy oxygen, nitrogen, argon or cesium ions.

It has been established that all (including non-conductive) microparticles can be analyzed using a positive primary ion beam. The main reasons seem to be low density of the ion beam, small size of the particles, and possible resputtering of the metal substrate on the particle.

Small particles can be analyzed with either a scanning or stationary ion beam that is at least twice, and sometimes five times larger than the projected area of the particle.

A 2 µm particle under a 10 µm beam will receive, if properly centered, approximately one-third of the total intensity of the ion beam. The full scan for mass range 1-260 will take approximately five minutes and will display most elements present in a concentration of 0.1% or higher. For lower concentrations, ranging from 1000 down to 100 ppm, scanning times become prohibitively long.

4. Results and Interpretation

In the first 10 to 20 seconds only about 20-50 nm of particle surface is removed. In many cases this layer is surface contamination, such as organic thin films and sub-micron size dust particles. After preliminary etching, the microparticle will be clean and subsequent analysis will produce a full mass spectrum.

A mass spectrum of a 5 µm microparticle from a leaded fuel car exhaust (shown in Fig.2), was recorded in 15 secs.

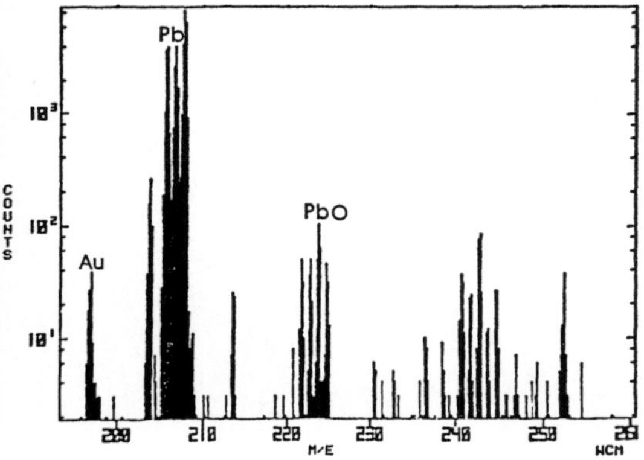

Fig.2 Lead signal from a small particle collected from a car exhaust

Particle was almost pure lead oxide. A trace amount of lead in a glass - 500 ppm required 125 secs. scanning time to produce readily identifiable lead signal (Fig.3).

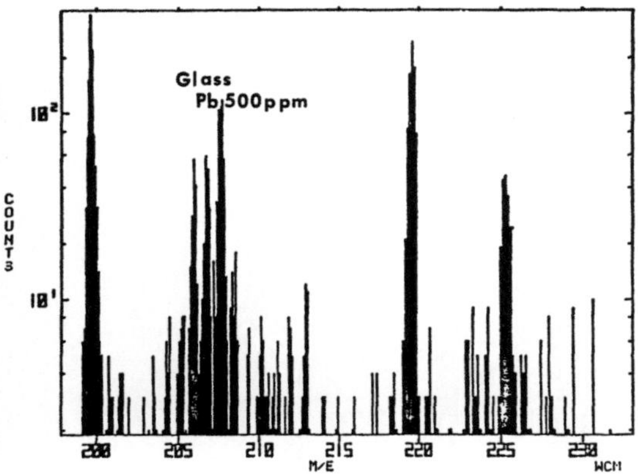

Fig.3 Mass spectrum of a small particle showing lead isotopes recorded at 500 ppm by weight

Fluorides of heavy metals are differentiated easily from oxides and metals by their characteristic fragmentation in a narrow spectral range.

All of these results were produced in an ARL microprobe that was under the control of a Data General computer. The program for analysis was developed at GE Vallecitos Nuclear Center and is particularly suitable for small particles.

References

1. W.C. McCrone, J. Delly, "The Particle Atlas Two," Ann Arbor Science Publishers, Ann Arbor, MI.
2. M. Bayard, "Application of the Electron Microprobe to the Analysis of Free Particulates" - C.A. Andersen, Ed., Wiley, NY, 1973.
3. C.A. Andersen, "Analytical Methods and Applications of the Ion Microprobe Mass Analyzer" - C.A. Andersen, Ed., Wiley, NY, 1973, Chapter 17.
4. A. Teetsov, J. Brown, "Some Techniques for Handling Particles in SEM Studies," Proc. Scanning Electron Microscopy, 1976 (Part III), IIT Research Institute, Chicago, IL.

VIII. Panel Discussion

Applications of Particle Accelerator-Assisted Ultra-High Sensitivity SIMS

I.L. Kofsky
Panel Discussion Chairman

G.H. Morrison, K.H. Purser, C.A. Evans, Jr., N. Shimizu
Panel Members

The panel discussion on particle accelerator-assisted ultra-high sensitivity SIMS covered the principles of this novel new technique in more detail than its application. Members of the audience were intensely interested in the principles and viability of the methodology. There were no disappointments as the exciting details were revealed. Although there are two particle accelerator approaches [1-4] being explored at this time, the tandem electrostatic accelerator [1,2] received more panel discussion attention than the cyclotron [3,4] due to the interests and expertise of the panel members.

Block diagrams of the two techniques are given in Fig.1. The tandem accelerator, developed by the Toronto-General Ionex-Rochester collaborators (TIR), method utilizes conventional magnetic-sector mass spectrometers before and after the accelerator to provide the necessary m/e separation. Because ion stripping is employed in the tandem accelerator only negative secondary ions can be analyzed. By contrast, the cyclotron method can analyze both positive and negative secondary ions. Both methods utilize nuclear particle detectors to determine E and dE/dx and thereby further discriminate against unwanted interferences.

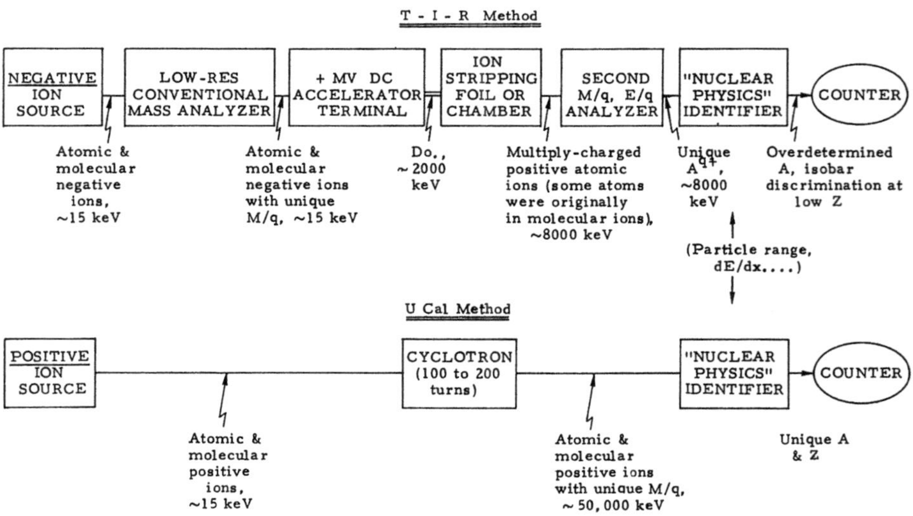

Fig.1 Particle accelerator assisted SIMS: tandem accelerator (TIR) and cyclotron (University of California) methods

A simple schematic diagram of the TIR (Toronto-General Ionex-Rochester) tandem accelerator instrument is given in Fig.2. In this method, negative ions are produced under primary ion (e.g., Cs^+) bombardment and accelerated into a low-mass-resolution high-transmission sector mass spectrometer. At defining aperture #1 the mass-to-charge ratio (m/e) of the desired atom or molecule is selected. In the first section of the tandem accelerator ion energies are increased from a few keV to a few MeV. These negative ions then pass through a metal foil or gas stripping chamber where several electrons are removed from each particle. In +2, +3, and +4 charge states molecules are unstable and disassociate; their resulting fragments have m/e ratios and energies far different than the atomic ions of interest. The

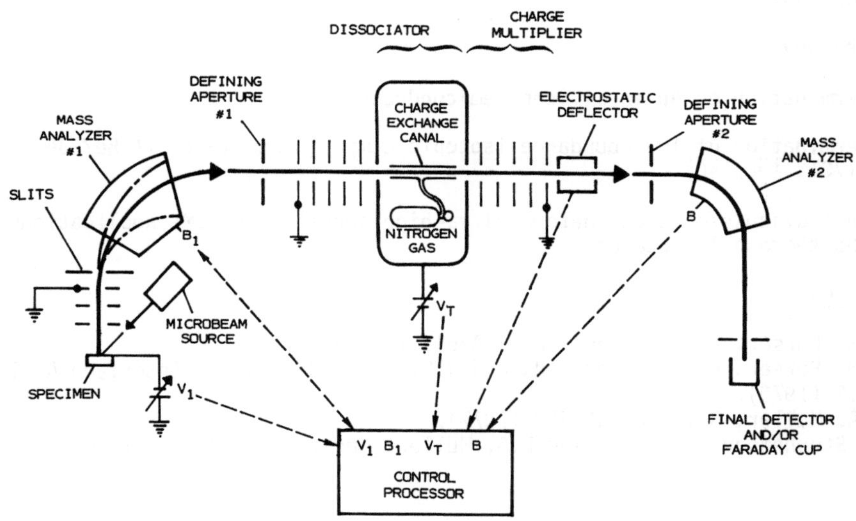

Fig.2 Tandem accelerator (TIR) assisted SIMS

resulting positively charged ions are accelerated by the few MV potential on the terminal (e.g., +3 ions are accelerated to ~9 MeV) and those ions with selected charge state and energy will pass through defining aperture #2 and into the second mass analyzer, which performs another m/e separation. By the time the ions pass through the detector defining aperture molecular ion interferences have been eliminated and scattered ion interferences are very low. To provide additional information and to discriminate against isobaric interferences (e.g., $^{14}N^{+3}$ from $^{14}C^{+3}$) a E, dE/dx detector is used instead of the Faraday cup shown in Fig.2. In addition to very high abundance sensitivity, this instrument is capable of high transmission efficiency; for example, the ratio of C ions detected to C atoms sputtered from a carbon sample is ~0.03.

A very short discussion of applications was held between panelists and audience. Applications mentioned include:

- Carbon-14 age dating

- Determination of As and P in silicon semiconductors

- Determination of impurities in lead salts, in silicon, and in HgCdTe
- Determination of trace transition elements in glass (fiber and laser optics)
- Determination of alpha emitting elements in semiconductor encapsulating materials
- Determination of trace inorganic elements in cells (biomedicine) and in body serums, hair, and nails (physiology and diagnosis)
- Geochronology
- Petrology
- Determination of Au in silicon semiconductors
- Determination of low abundance isotopic compositions (e.g., $^{10}Be/^{9}Be$, $^{36}Cl/Cl$, $^{13}C/^{12}C$)

The panel discussion ended before ultra-high sensitivity SIMS applications could be thoroughly treated.

References

1. K.H. Purser, et al., Rev. Phys. Appl. 12 1487 (1977).
2. K.H. Purser, A.E. Titherland, and J.C. Rucklidge, Surf. Interface Anal., 1 12 (1979).
3. R.A. Muller, Science 196 489 (1977).
4. E. Stephenson, T. Mast, and R.A. Muller, Nucl. Inst. Meth., in press.

IX. Biology

Biomedical Applications of Secondary Ion Emission Micro-Analysis

P. GALLE
Département de Biophysique
Faculté de Médecine de Créteil
6, rue du Général Sarrail
94000 Créteil, France

1. Introduction

Proposed by R. CASTAING and G. SLODZIAN in 1962, a new microanalytical method *"Secondary Ion Emission Microanalysis"* has been applied to the study of some variety of biological tissues. It appears that this method offers many new possibilities in biological research ; the problems of specimen preparation and some difficulties in the interpretation of microanalytical data will be briefly discussed and some typical applications will be presented.

2. Material and Methods

2.1 Preparation of Biological Specimens

Two conditions are required to obtain a good image : a flat surface of the specimen and a good electrical conductivity. The problems of specimen preparation are different for soft tissues and hard tissues.

2.2 Preparation of Hard Tissues (Bones, Teeth...)

Ground sections of about 100 µm thickness are achieved with a specially built microtome using as cutting device a fine emery coated wire of 0.1 mm diameter. These transverse sections are polished to a thickness of about 30 µm with a 0.5 to 3 µm diamond powder. The selected polished ground sections are fixed on a specimen holder of the ion microanalyzer. Under vacuum, an aluminum grid is deposited on the polished surface. The purpose of the grid is the elimination of the electrical charges accumulated under the primary ion beam at the surface of the non-conductive sample. Bars are 10 µm wide, 2 to 3 µm thick and the distance between them is 150 µm. The ion analysis is performed between the square meshes of the grid.

2.3 Preparation of Soft Tissues

Experience shows that the addition of a metallic grid is not necessary if the section of such a tissue is thin enough [9,11]. The maximum thickness of the section depends on the embedding material : frozen or paraffin sections less than 10 µm thick and fixed on a gold specimen holder can be studied without any metallic grid ; for epon or araldite sections, of very low conductivity, the maximum thickness of the section on the same gold specimen holder must be less than 1 µm. For thicker sections, the addition of a metallic grid is necessary.

Biomedical Applications

Many results obtained on soft tissues [1,3,9,10,11,12,24,25,26,27,28] and hard tissues [8,12,13,14,15,16,17,18] have already been published ; compared to other microanalytical methods two characteristics of secondary ion emission microanalysis appear to be particularly interesting in biomedical research : 1) the very high sensitivity for many elements making possible for the first time the study of elements at low or even at trace concentration in soft tissues [9,28]; and 2) the very good resolution in depth which allows us to obtain good images of distribution of elements at the surface of bulk samples of hard tissues [15,16,17]. A local isotopic analysis, a study of very light elements such as lithium, fluorine or beryllium are still other interesting possibilities [1,12].

We shall present here some typical applications on soft tissues, isolated cells and hard tissues.

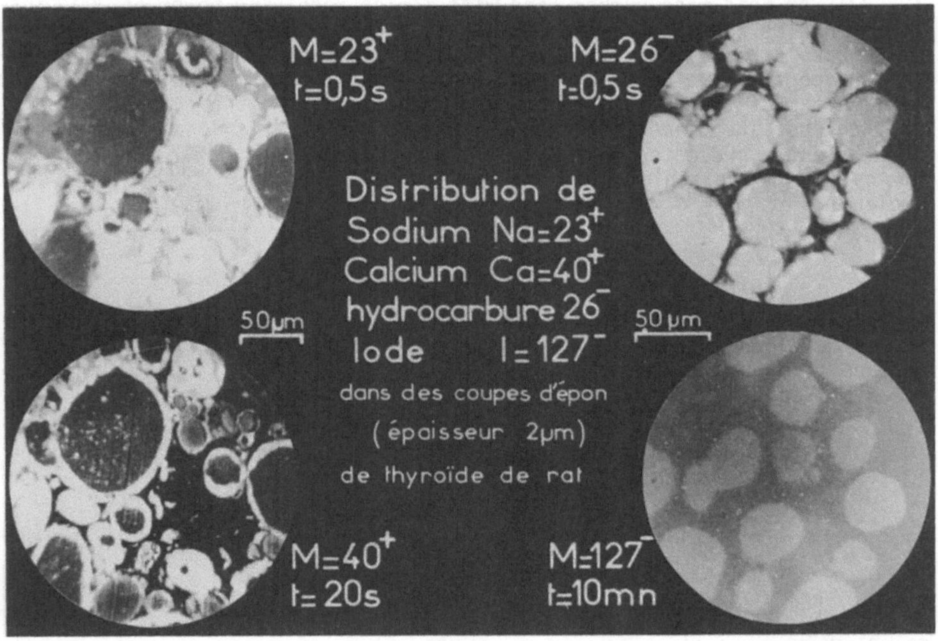

Fig. 1 - Distribution of $^{23}Na^{+}$, $^{40}Ca^{+}$, $^{26}CN^{-}$ (or $^{26}C_2H_2^{-}$) and $^{127}I^{-}$ at the surface of an epon section of a rat thyroid

Figure 1 shows the distribution of $^{40}Ca^{+}$, $^{23}Na^{+}$, $^{26}C_2H_2^{-}$ or $^{26}CN^{-}$ and $^{127}I^{-}$ at the surface of an epon section of a rat thyroid (primary ions : O_2^{+}). This figure shows that the sensitivity for a given element is very dependent on the "ion emission yield" N_i/N_o where N_i is the number of atoms of an element etched in an ionized form and N_o is the total number of atoms of the same element etched during the same time. Under these working conditions (primary ions O_2^{+}), the ratio N_i/N_o for Na^{+} is very high (probably very close to 1) and the sensitivity for this element is very good. Although of a very low concentration in this epon section (most of the diffusible elements have been removed from this epon section during dehydration), the sodium image was obtained in 0.5 sec. On the other hand, the time required to obtain the image of a negative ion such as $^{127}I^{-}$ is more than 1000 times longer (600 sec) than for $^{23}Na^{+}$ although iodine is present in this epon section of thyroid gland in a much higher concentration than sodium; the low ion emission yield for $^{127}I^-$ under these conditions explains the very low sensitivity for this element. The ion emission yield of a given element is very dependant on the chemical nature of this element, its chemical bonds in the specimen and the working conditions, for example, the ratio N_i^-/N_o for negative halogen ions such as $^{127}I^-$ should be very much enhanced using a gas flow of cesium at the surface of the specimen. This fundamental problem of the ionization process is discussed in different papers [2,5,6,19,21]. The same Fig.1 shows that the organic elements may easily be localized when the sample is bombarded with O_2^+ through an examination of a negative polyatomic ion at mass 26, corresponding simultaneously to $^{26}CN^-$ and $^{26}C_2H_2^-$.

Figure 2 shows the distribution of mass 56 at the surface of a frog nucleated red blood cell. This image has been obtained with transfer optics (22,23). At this mass number, and with a mass resolving power of 300, interference is possible between $^{56}Fe^+$ (true mass = 55.934932) and $^{56}CaO^+$ (true mass = 55.957504) ; in order to separate these two masses high resolution spectra have been obtained with the electrostatic analyzer from two small areas of the cell corresponding respectively to the cytoplasm and the nucleus. The diameter of the areas selected here is less than 5 μm. These spectra presented on Fig.3 show a high ratio of $^{56}Fe^+/^{56}CaO^+$ in the cytoplasm containing hemoglobin and a low ratio $^{56}Fe^+/^{56}CaO^+$ in the nucleus.

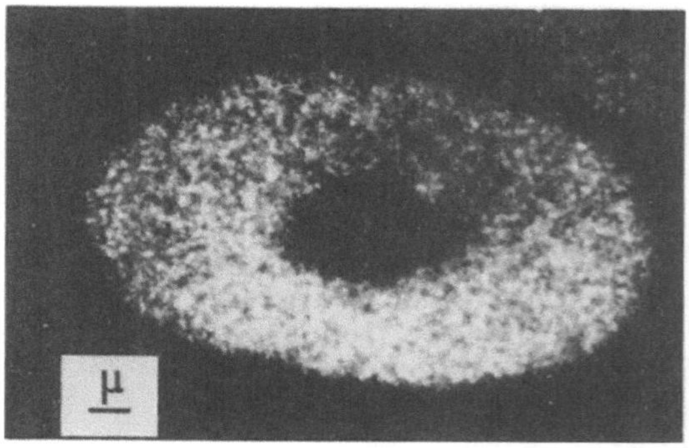

Fig.2 Distribution of mass 56 , ^{56}Fe or $^{56}CaO^+$ at the surface of a frog nucleated red blood cell

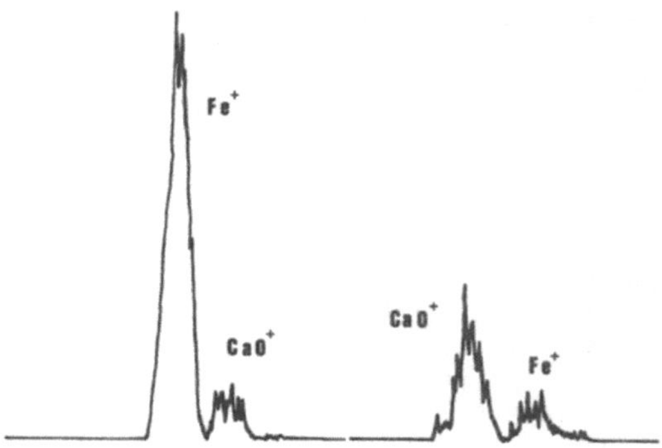

Fig.3 High resolution spectrum at mass 56 with the electrostatic analyzer. At the left side, the two peaks are obtained from the cytoplasmic area of the red blood cell of Fig.2; at the right side, the two peaks are obtained from the nucleus area

Figure 4 shows the distribution of $^{40}Ca^+$ at the polished surface of normal human tooth enamel; the calcium distribution is homogeneous.

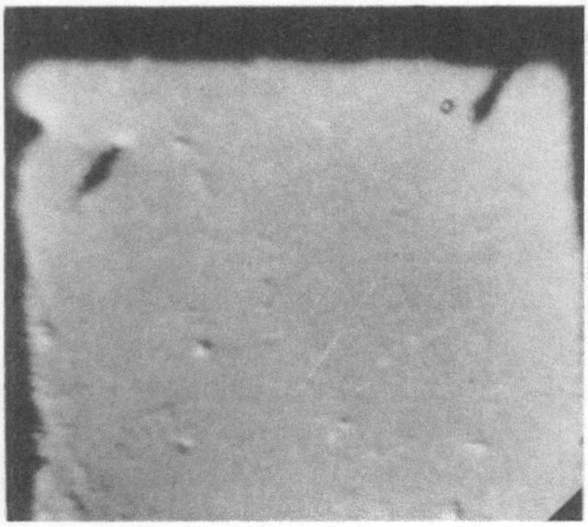

Fig.4 - Normal human tooth enamel : homogeneous distribution of $^{40}Ca^+$ at the micronic level

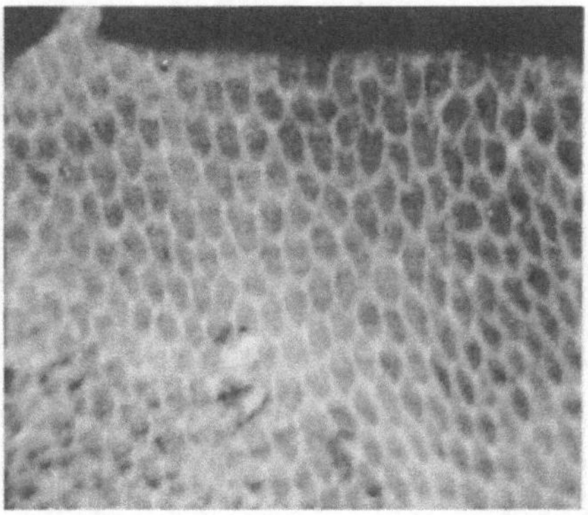

Fig.5 - Human incipient carious lesion: the distribution of $^{40}Ca^+$ is no longer homogeneous in the enamel; destruction of enamel prism is clearly visible

Figure 5 shows an incipient carious lesion of human tooth enamel ($^{40}Ca^+$ distribution); one can easily observe the disappearance of the rod core leaving a honey-combed network of interrod material.

Figure 6 is an assembly of six micrographs ($^{40}Ca^+$ distribution) which shows the area of destruction of enamel in a human incipient carious lesion. The edge of the caries is in the upper part of the image. Normal enamel can be observed in the lower left hand corner.

Figure 7 shows the distribution of six elements at the polished surface of of a 200 million year old polymetallic submarine nodule. Skeleton of a fossil radiolaria can be easily identified. Exposure time required to obtain an image varies from 1/250 sec to 15 sec, depending on the concentration and the ionization yield of the element. Assuming that the sputtering rate is of the order of 10 angstroms/sec under these conditions (primary beam density), it can be easily deduced that the total time to obtain these 6 microanalytical images is less than 20 sec. and necessitates less than 200 angstroms of material. The ease of obtaining such microanalytical images rapidly is one of the characteristics of this method.

In conclusion, it can be noted that most of the results presented here could not be obtained by any other microanalytical method. The most evident advantages of secondary ion emission microscopy are the very high sensitivity for many elements, the very good resolution in depth, the possibility to study the lightest elements, three of these : lithium, beryllium and fluorine being interesting in biomedical research, and the ease to obtain microanalytical images rapidly . Most of the difficulties encountered in the early applications of this method to biological research (charging effect, emission of polyatomic ions) have been practically solved. This method appears very well adapted for the study of diffusible ions ; however, for this particular application, the problem of preparation of the tissue remains unsatisfactory.

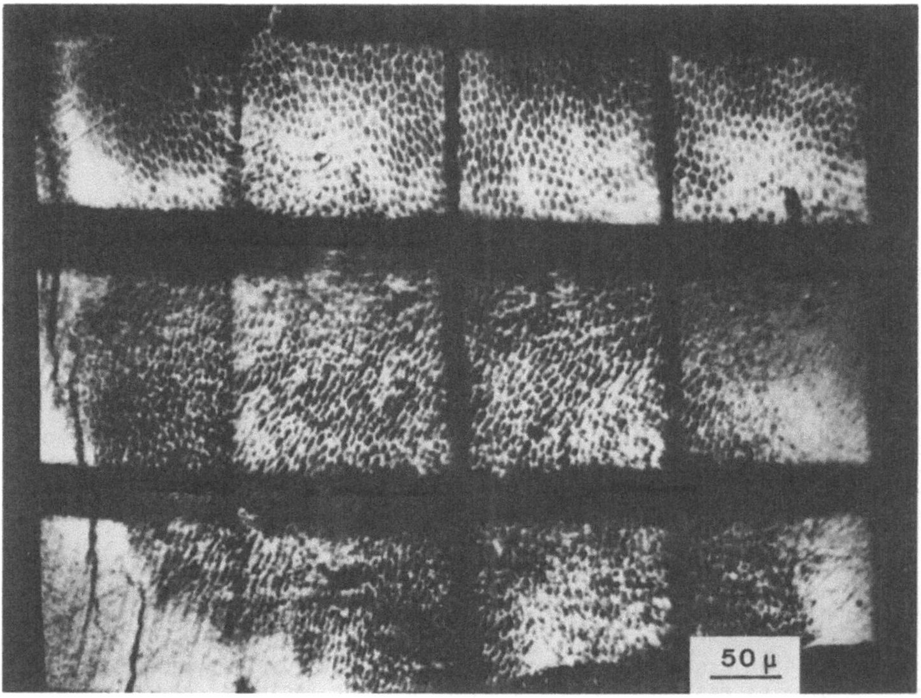

Fig.6 Human incipient carious lesion of the enamel : $^{40}Ca^{+}$ distribution ; the edge of the cary is in the upper part of the image ; subnormal enamel is visible in the lower left hand corner ; calcium distribution at the surface of the carious lesion is visible between the square meshes of the aluminum grid

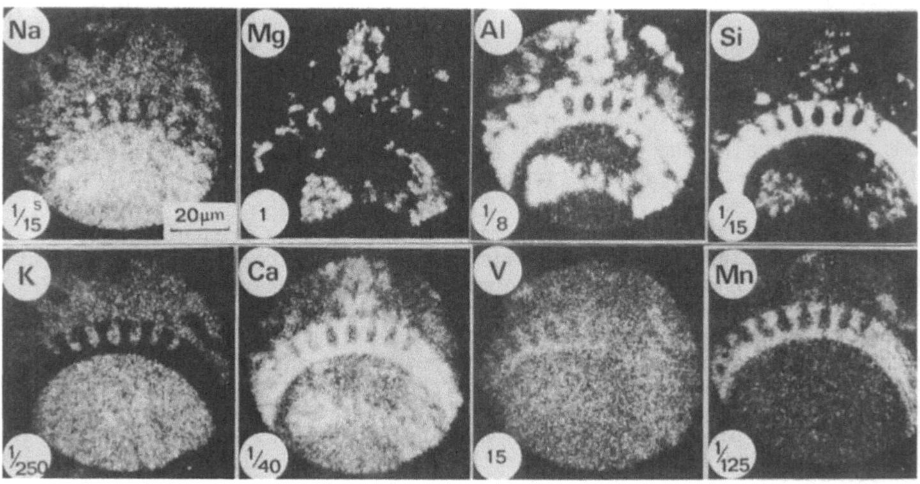

Fig.7 Distribution of Na, Mg, Si, K, Ca, V and Mn at the surface of a polished section of a 200 million years submarine polymetallic nodule; skeleton of a radiolaria is clearly visible

References

1. J.L. Abraham, VIIIth Intern. Conf. X-ray Optics and Microanalysis, Boston (1977), 11.
2. C.A. Andersen and J.R. Hinthorne, Science, 175, 853 (1973).
3. M.B. Bellhorn and R.K. Lewis, Exp. Eye Res., 22, 505 (1976).
4. M.B. Bellhorn and D.M. File, VIIIth Intern. Conf. X-ray Optics and Microanalysis, Boston (1977), 137.
5. M. Berheim and G. Slodzian, J. Microsc. Spectrosc. Electron., 2, 291 (1977).
6. G. Blaise and G. Slodzian, J. Phs. (Paris), 35, 237 (1974).
7. R. Castaing and G. Slodzian, J. Microscopie (Paris) 1, 395 (1962).
8. J.P. Cuif and R. Lefevre, C.R. Acad. Sc. Paris, 278, 2263 (1974).
9. P. Galle, Ann. Phys. Biol. Med., 4, 84 (1970).
10. P. Galle, G. Blaise, and G. Slodzian, IVth Nat. Conf. Electron Microprobe Analysis, A.A. Chodos ed., Calif. Inst, Tech., 36 (1969).
11. P. Galle, Microprobe Analysis as Applied Cells and Tissues, Acad. Press (1974), 89.
12. P. Galle, J.P. Berry, and R. Lefevre, Scanning Electron Microscopy (1979/11), SEM Inc. 703.
13. R. Lefevre, Intern. Colloquium on Crystallography of Apatites of Biological Interest, C.N.R.S. ed. Paris, 1973, report No. 230.
14. R. Lefevre, J. Micr. Biol. Cell, 22, 335 (1975).
15. R. Lefevre and R.M. Frank, J. Biol. Buccale, 4, 29 (1976).
16. R. Lefevre, R.M. Frank and J.C. Voegel, Calc. Tiss. Res. 19, 251 (1976).
17. R. Lefevre, VIIIth Intern. Conf. X-ray Optics and Microanalysis, Boston (1977), 143.
18. R. Lefevre, J. Marcoux, and J.P. Cuif, J. Microsc. Spectrosc. Electron. 3, 469 (1978).
19. G.H. Morisson and G. Slodzian, Analytical Chem. 47, 932 A (1975).
20. G. Slodzian, Thesis Sc. Paris (1963).
21. G. Slodzian, Surface Sci. 48, 161 (1965).
22. G. Slodzian and A. Figueras, VIIIth Intern. Conf. on X-ray Optics and Microanalysis, Boston, 127 (1977).
23. G. Slodzian, J. Microsc. Spectrosc. Electr., 3, 447 (1978).
24. M. Truchet, J. Microscopie 17, 103 (1973).
25. M. Truchet, Thesis Sc. 3ème cycle Paris (1974).
26. M. Truchet, J. Microscopie Biol. Cell 22, 465 (1975).
27. M. Truchet, J. Microsc. Spectrosc. Electron. 2, 77 (1977).
28. M. Truchet and S. Trottier, J. Microsc. Spectrosc. Electron. 3, 463 (1978).

Diffusible Ion Localization in Biological Tissue by Ion Microscopy

K.M. Stika and G.H. Morrison
Department of Chemistry
Cornell University
Ithaca, NY 14853

Abstract

The ion microscope is a particularly sensitive analytical tool for the localization of ions such as sodium, potassium and calcium which are highly diffusible and of considerable interest in many biological systems. Histochemical information obtained from ion micrographs of samples processed by conventional methods of tissue preparation (chemical fixation, dehydration, embedding in plastic and thin sectioning) agrees well, in many cases, with microanalysis reported using other methods. However, the possibility of ion loss and redistribution introduced by the preparation method introduces some question as to the validity of ion distribution observed for ions not strongly bound to the biological matrix.

The necessity of low temperature preparative procedures for diffusible ion localization at the 1 µm level, which is the spatial resolution of the CAMECA IMS-300 direct imaging ion microanalyzer, will be discussed. Methods of cryo-fixation, low temperature sectioning and freeze-drying will be presented with a discussion of their effect on the resulting ion image. Application of cryotechniques to specific biological systems in pathology, pharmacology and botany will also be discussed.

Determination of Isotope Ratios of Calcium and Iron in Human Blood by Secondary Ion Mass Spectrometry

K. Wittmaack, F. Schulz[a)] and E. Werner[b)]

Gesellschaft für Strahlen- und Umweltforschung mbH (GSF)
D-8042 Neuherberg, W. Germany

This study relates to the problem of the fractional intestinal absorption of iron and calcium. Iron is a necessary substrate for hemoglobin formation whereas calcium controls a variety of physiological functions and bone homeostasis Deficiency of these elements is quite common even in industrial countries. It is desirable, therefore, to account for any deficiency of these elements, whether temporal (e. g. during menstruation or after an accident) or permanent (due to misfunction of the respective organ), by oral application of a suitable drug. Previous quantitative studies of intestinal absorption have been performed by means of radioactive tracers. In course of (and subsequent to) such an experiment the patient was exposed to ionizing radiation emitted from the incorporated source. To avoid this side--effect, the feasibility of non-radioactive tracer techniques has to be investigated.

In principle, the extent of intestinal absorption of a certain element can be deduced from the changes in concentration of one isotope of that element, introduced by incorporation of isotopically enriched quantities of the tracer material. In order to evaluate the response of oral application quantitatively, a direct comparison with the result of simultaneous intraveneous application is necessary. This implies the use of two tracers of the same element, enriched in two different isotopes. In view of these requirements the neutron activation technique did not look promising, either because of the lack of two suitable isotopes (iron) or because of inproper activation cross sections and life times (calcium). Therefore, we have investigated the potential of Secondary Ion Mass Spectrometry (SIMS) for the determination of isotope ratios of calcium and iron in human blood.

The SIMS experiments were performed in the DIDA ion microprobe described elsewhere [1]. In this study sample erosion was achieved by bombardment with either Ar^+ or O_2^+ ions at an energy of 10 keV. The primary ion beam was defocussed to cover a sample area ~1 mm in diameter. The beam current was varied between 10 nA and 1 μA, depending upon the secondary ion intensities observed with the samplesof different preparation. The quadrupole mass filter was tuned to provide mass independent transmission in the region of interest.

---

a) Presently on leave at New York University, New York
b) GSF, Abteilung für Biophysikalische Strahlenforschung, D-6000 Frankfurt/M., FR Germany

In the first experiment a thin layer of human blood was deposited on a high purity germanium substrate and dried in air. This sample was subsequently introduced into the target chamber. SIMS analysis was performed at a base pressure of $\sim 2x10^{-6}$ Pa. Similar to most (but not all) of the samples investigated, the air-dried layer exhibited an electrical resistance low enough to prevent build-up of a positive surface charge. Therefore, compensation of the primary ion charge by means of simultaneous electron bombardment [2] was usually not required. With air-dried blood, however, we observed signal inhibition at the very beginning of exposure to the primary ion beam. An almost step-like increase in SIMS intensity was recorded after a bombardment fluence of $\sim 2x10^{16}$ O_2^+/cm^2 ($\sim 1x10^{16}$ Ar^+/cm^2). Subsequently, the $^{56}Fe^+$ intensity (which was studied in some detail) exhibited a fluence dependence very similar to the behaviour of metal ions emitted from oxygen bombarded metals [3, 4], i. e. the intensity passed through a pronounced "surface" peak followed by a minimum and then increased again to finally attain a steady state level comparable in height to the surface peak (bombardment fluence $\sim 10^{17}$ O_2^+/cm^2). Under Ar^+ impact the steady state $^{56}Fe^+$ signal was about a factor of three smaller than under O_2^+ impact.

Most important in the present context are the following two observations. (i) The intensity of the ions of interest emitted from air-dried blood was rather low. E. g., the steady state signal of $^{56}Fe^+$ in the peak of the secondary ion energy distribution was only $\sim 10^4$ counts/µC. (ii) The mass spectrum exhibited relatively large intensities of molecular ions composed of H, C, N and O atoms. These ions fequently interfered with the metal ions of interest. E. g., a very prominent line appeared at mass number 46 [u] (probably NO_2^+). Fortunately, the molecular ions are characterized by an energy spectrum which tails off much more rapidly towards high energies than the spectrum of atomic ions. Therefore one can discriminate against molecular interference by tuning the energy window of the mass spectrometer to accept only ions of sufficiently high energy [5]. This procedure, however, affects also the intensities of atomic ions (cf. Fig.1). A loss by one to two orders of magnitude is typical. Accordingly, the intensity of $^{58}Fe^+$ from air-dried blood, under conditions of discrimination against molecular interference, was only of the order of 10 counts/µC. Clearly, this was much too low for the purpose in consideration.

On the basis of these results we have investigated various methods of sample preparation. The present status of our experiments is illustrated by Figs.1(a) and (b) which show energy spectra of secondary ions observed either in the calcium or the iron regime. The results depicted in Fig.1(a) relate to a sample prepared by dry-ashing of blood plasma in a platinum cup (24 h at 600°C) followed by oxalate precipitation. A thin film of this solution was deposited on a high purity copper substrate and dried in air. The secondary ion intensities observed at mass numbers 42, 43, 44, and 48 and at energies $\gtrsim 50$ eV agree with tabulated abundances to within typically $\pm$ 2%. This is illustrated by crosses which mark the natural abundance at an energy where the intensity of $^{44}Ca^+$ corresponds to (10^3 counts/60 nA) x (abundance (%)). Due to molecular interference, the intensity at mass 46 is still a factor of five too high at that energy.

Figure 1(b) relates to a sample which was prepared from natural blood, dry ashed as above, followed by sulfide precipitation. The most interesting result of Fig.1(b) is that the intensity at mass 57 decreases less rapidly than the intensities of $^{54}Fe^+$ and $^{58}Fe^+$ which exhibit the correct

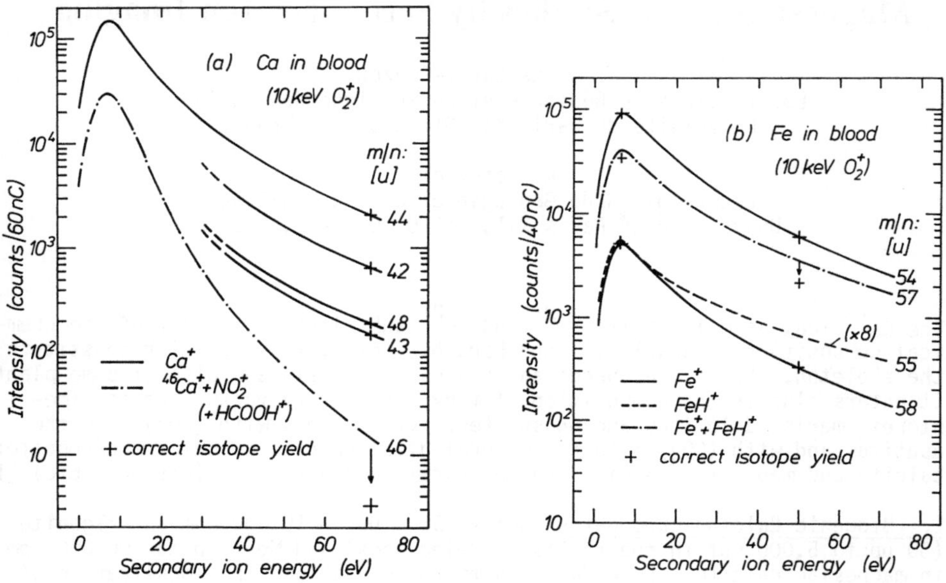

Fig.1 Energy spectra of secondary ions emitted from chemically processed blood samples. (a) Calcium regime, (b) iron regime. The crosses mark the intensities to be expected without molecular interference

intensity ratio throughout the whole energy spectrum (negligible molecular interference). Comparison of the spectra of $^{54}Fe^{1}H^+$ and $^{58}Fe^+$ reveals that *the energy distribution of FeH$^+$ tails off more slowly than the distribution of Fe$^+$*. This effect has not been observed previously. It may be specific to metal hydride molecules. Notice that because of the large mass mismatch only very little energy can be transferred from H to Fe and vice versa.

In conclusion we have shown that, suitable sample preparation provided, relative concentration of selected isotopes of calcium and iron in human blood can be measured with sufficient accuracy. Additional experiments are in progress to allow all isotopes of calcium and iron to be used for studies of intestinal absorption.

References

1. K. Wittmaack, Ad. Mass Spectrometry VII, 758 (1977).
2. K. Wittmaack, J. Appl. Phys. 50, 493 (1979).
3. C.A. Andersen, Int. J. Mass Spectrom. Ion Phys. 2, 61 (1969).
4. K. Wittmaack, Int. J. Mass Spectrom. Ion Phys. 17, 39 (1969).
5. K. Wittmaack, Appl. Phys. Lett. 28, 552 (1976).

Biogenic and Non-Biogenic Carbonates of Calcium and Magnesium: New Studies by Secondary Ion Imaging

J. Archambault-Guézou
Laboratoires de Géologie et de Physique des Solides
Université de Paris XI, 91405 Orsay, France

R. Lefèvre
Laboratoires de Géologie et de Microanalyse
Université de Paris XII, 94010, Creteil, France

The Ca^{2+} ion, as well as its Sr^{2+} and Mg^{2+} substitutes, is part of the chemical composition of nearly all the hard biological tissues which constitute the skeleton, shells, carapaces and scales of animals and also of some plants. It enters also in the composition of many sedimentary rocks such as limestones, marls, dolomites and phosphates. It combines with $(PO_4)^{3-}$ in the apatites and with $(CO_3)^{2-}$ in the mineralogical species of calcium carbonate: calcite and magnesian-calcite (rhombohedral) and aragonite (orthorhombic) [1].

1. Biogenic Calcium Carbonates contain Sr^{2+} up to 10,000 ppm in aragonite (50 up to 5,000 ppm in the shells of Pelecypods) and Mg^{2+} up to 300,000 ppm in magnesian calcite. Secondary ion microscopy allows the localization of aragonite and magnesian-calcite, when present on X-ray power diffraction patterns, by means of the localization of ^{88}Sr and ^{24}Mg.

1.1 In a triassic Red Algae, more than 200 million years old, a mixture of calcite and aragonite was found by means of X-ray diffraction. Ion microscopy has shown a succession of $^{24}Mg^+$ rich layers and $^{88}Sr^+$ - $^{23}Na^+$ rich layers [2]. This zonation was confirmed to be a reality by means of Laser Microprobe based on Raman Effect [3]: magnesian-calcite and aragonite are layered.

1.2 In the Cockle (*Cerastoderma glaucum*) (Pl. I), all the calcium carbonate of the shell is aragonite. Its crystallographic structure, as known, is complex: in the external area of the outer crossed lamellar layer, the first order lamellae are aligned concentrically and consist in sheets of aragonite needles arranged with their long axes normal to the surface of the sheet (Fig.1,4,12,13). In the inner area of the same layer the constituent lath-like second order lamellae are inclined in opposing directions in adjacent first order lamellae. On secondary ion micrographs two types of zonations are observed [4]: simultaneous variations of $^{40}Ca^+$ and $^{23}Na^+$ (Fig.9,10) and independent variations of $^{40}Ca^+$, $^{23}Na^+$, $^{24}Mg^+$ and $^{88}Sr^+$. These two families of zones are found to be independent and often orthogonal. The first, which reproduces the microstructure, is probably artifactual and induced by surface state or crystalline-lattice effect under ion sputtering, whereas the second is a chemical reality resulting of chemical increments related to the growth of the shell.

1.3 In the Mussel (*Mytilus galloprovincialis*) (Pl. II) the inner part of the shell is made of aragonite (mother of pearl) and the outer part of low-magnesian calcite (finely prismatic layer) (Fig.7,8,13). These two parts are well distinguished on ion micrographs of $^{24}Mg^+$, $^{88}Sr^+$ and $^{23}Na^+$ (Fig.3, 4,5). A sub-zonation is present in the outer part related to growth patterns

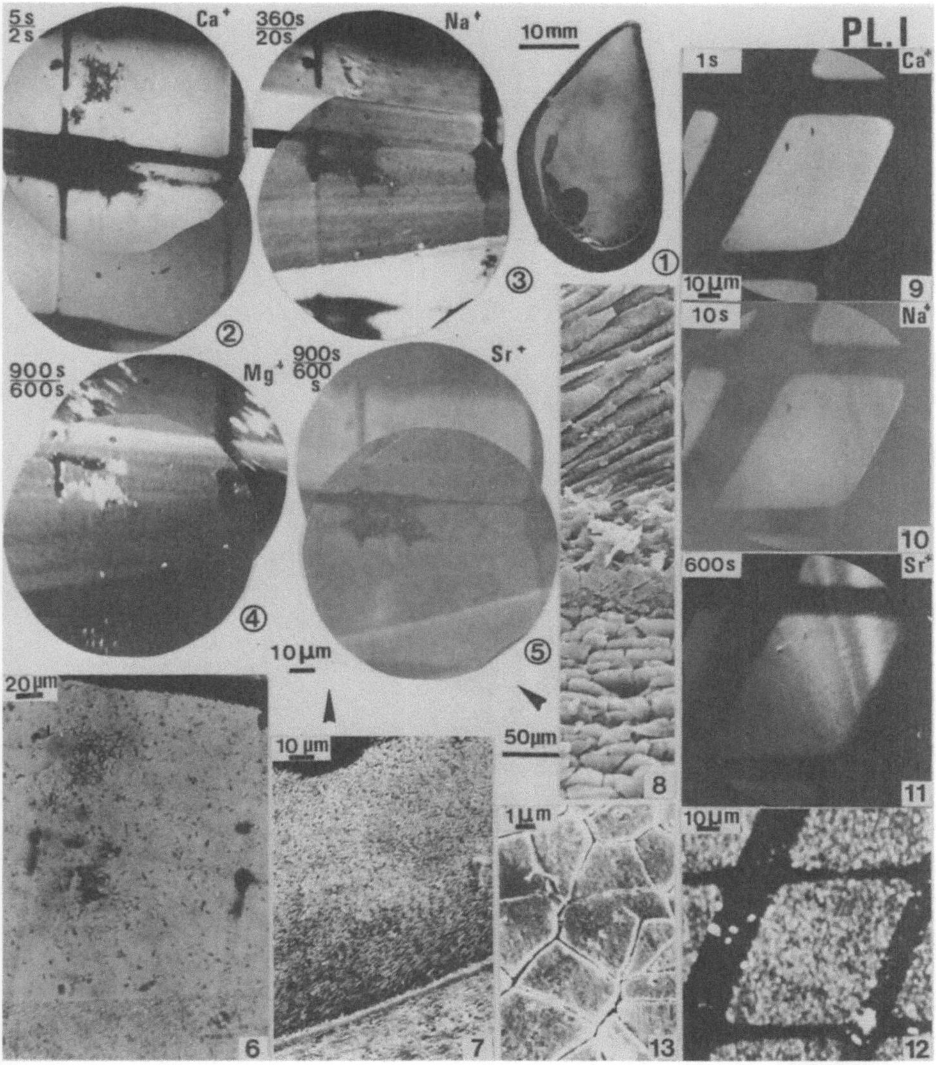

Plate I Biomineralization of *Cerastoderma glaucum*, Brug.: Fig.1 Crossed-lamellar structure. Fig.2 sp. from Camargue, France. Fig.3 Radial section. Fig.4 Radial section, pallial, light microscope, crossed Nichols. Fig.5,7,8 Distribution of $^{40}Ca^+$, $^{23}Na^+$, $^{88}Sr^+$ by secondary ion microscopy of the same area, viewing field: 150 µm; primary beam: O_2^+; secondary positive ions; CAMECA IMS 300 ion microscope with electrostatic filter and experimental transfer optics (Orsay). Fig.6 SEM, similar area. Fig.9,10 Secondary ion micrographs of $^{40}Ca^+$ and $^{23}Na^+$ distribution in the inner area of outer layer (radial section, pallial margin). Fig.11 SEM similar area. Fig.12 Radial section of the outer crossed-lamellar layer showing structural continuity of outer and middle areas. Fig.13 Crossed-lamellar structure: radial section of pallial margin of the outer layer

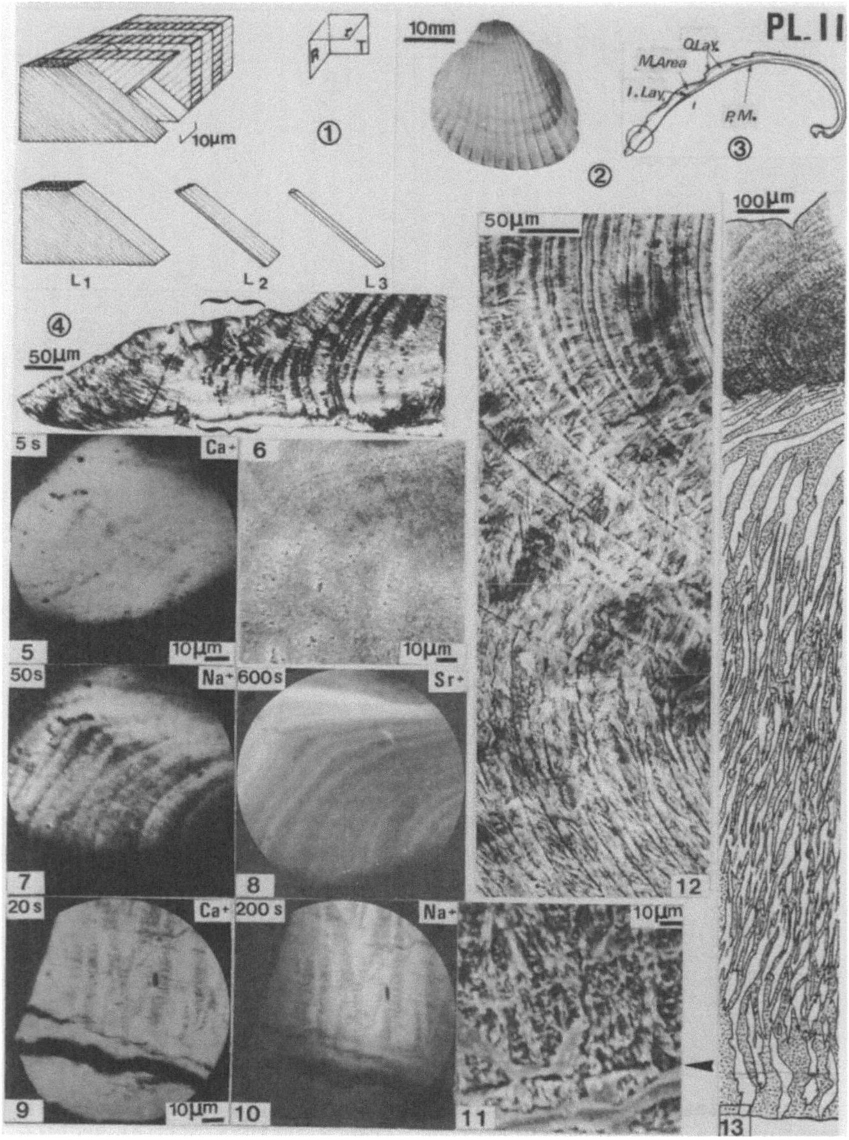

Plate II Biomineralization of *Mytilus galloprovincialis*, Lmk: Fig.1 sp. from Camargue, France. Fig.2-5 Secondary ion micrographs of radial section showing distribution of $^{40}Ca^+$, $^{23}Na^+$, $^{24}Mg^+$, $^{88}Sr^+$ in the outer prismatic calcitic layer and inner nacreous aragonitic layer, with dividing sheet of prismatic aragonitic myostracum, viewing field: 250 μm; primary beam: O_2^+, secondary positive ions; CAMECA IMS 300 ion microscope with electrostatic filter (Créteil). Fig.6 light microscope, similar area. Fig.7 SEM of analyzed area. Fig.8 Detail of the structure, ibid. Fig.9-11 Secondary ion micrographs ($^{40}Ca^+$, $^{23}Na^+$, $^{88}Sr^+$) of tangential section in the inner nacreous layer. Fig.12 SEM, similar area. Fig.13 View of (001) faces of nacreous tablets, similar area

(Fig.3,4). In the inner part, a sub-zonation is also present (Fig.9,10,11,12) related to growth bands and possibly to surface effects under ion sputtering or imperfect polishing.

1.4 The first kind of zone, probably artifactual, is also observed in the carapace of Crabs, which is made of calcite grown on a helical arrangement of chitinoproteic fibrils [4,5].

2. Non-Biogenic Calcium Carbonates arise directly by chemical precipitation from dissolved carbonates present in sea- or fresh-waters, or indirectly by stabilization of metastable biogenic or non-biogenic calcium carbonates. Secondary ion micrographs of these non-biogenic calcium carbonates, for example: dolomite and low magnesian calcite, sometimes show chemical zonations. These zonations are not artifactual as demonstrated by the complementary distribution of the elements and by quantitative electron microprobe analysis [6].

Acknowledgements

This work is supported by the French "Centre National de la Recherche Scientifique," Paris and the "Bureau de Recherches Géologiques et Minières," Orléans. Instrumental assistance for ion micrographs, R. Dennebouy and A. Allain-Quettier.

References

1. R. Lefevre, Secondary Ion Microscopy of Biomineralizations, Scanning, Special issue on Secondary Ion Mass Spectrometry, in press (1979).
2. J.-P. Cuif and R. Lefevre, C.R. Acad. Sci., Paris, 278D, 2263 (1974).
3. R. Lefevre, J. Barbillat, J.-P. Cuif, P. Dhamelincourt, and J. Laureyns, C.R. Acad. Sci., Paris, 288D, 19 (1979).
4. J. Archambault-Guezou, Y. Bouligand, M.-M. Giraud, and R. Lefevre, Biol. Cell., 35, 37a (1979).
5. M.-M. Giraud, Biol. Cell., 35, 37a (1979).
6. A. M'Rabet, H. Bizouard, and R. Lefevre, 7° Réunion Annuelle des Sciences de la Terre, Lyon, 338 (1979).

Localization of Elements in Botanical Materials by Secondary Ion Mass Spectrometry

A.R. Spurr
Department of Vegetable Crops
University of California
Davis, CA 95616

Pierre Galle
Laboratoire de Biophysique
Université Paris-Val de Marne
94-Créteil, France

1. Introduction, Materials, and Methods

Secondary ion mass spectrometry (SIMS) was evaluated for the distribution of selected elements in plant tissue including material grown at high levels of salinity and at toxic levels of boron. Seedlings of *Cucumis sativus*, *C. myriocarpus*, and *Pisum sativum* were grown in 0.5 Hoagland's colution culture [1] and with supplemental NaCl at 70 meq/ℓ Na. Some cultures of *P. sativum* seedlings were also supplemented with 1.2% $La(NO_3)_3$, 0.5% H_3BO_3, or with 20 meq/ℓ LiCl for 24 h before collection. Root segments, 0.5 cm long at 5 cm from the root tip, and leaf strips, 1 mm wide, of the control and the treated 10-day-old seedlings were fixed in 2% glutaraldehyde, dehydrated in acetone, and embedded in low viscosity epoxy resin [2]. The La, B, and Li material and their respective controls were processed by dehydration in toluene containing Al_2O_3 (acid; Brockman, activity grade 1) [3] and 0.5% dimethoxypropane (DMP) [4]. Sections 1 or 2 µm thick were mounted on 99.9999% Ag discs, 0.1 mm x 15 mm, and placed in a 60°C oven until shortly before analysis. Whole mounts of fresh leaf discs of *P. sativum* were mounted on Ag holders then dried at 60°C and carbon coated in a grid pattern. All other samples were uncoated. The analyses were made with a Cameca IMS 300 ion probe using a $^{16}O_2^+$ primary beam having a field diameter of 250 µm. The primary beam current during observations was about 1 µA and accelerated at 10 kV. Photographs of the elemental images were taken at X100 and mass spectra were prepared at different sensitivities depending upon the elements of interest.

2. Results and Discussion

Figure 1 illustrates a typical mass spectrum from a leaf of *C. sativus*. Ions with a mass/charge (m/e) ratio less than $^{23}Na^+$ were usually not included in the spectra. The intensities of the singly charged species, $^{23}Na^+$, $^{39}K^+$, and $^{40}Ca^+$, were usually relatively high. Note also the variety of polyatomic ions in the spectrum. These are a characteristic feature of organic materials [5] and of the epoxy resin embedding material in the sections and did not interfere with the elements of primary interest in this study. In experiments in which seedlings of *P. sativum* were grown at toxic levels of boron, insufficient B remained in the tissue of the roots and leaves to provide sufficient intensity for ion images. Also, it may have been more advantageous to use a primary beam of $^{16}O^-$ for boron. Mass spectra taken at high sensitivity resulted in considerable variability in the ratio of $^{10}B^+$ to $^{11}B^+$ for different specimens. However, the intensity of $^{11}B^+$ was always greater, and in some cases in approximately the correct ratio of 1:4. In similar preparations in which LiCl was introduced into the culture solution a clear separation of $^6Li^+$ and $^7Li^+$ in the mass spectra was achieved. The mass spectrum of

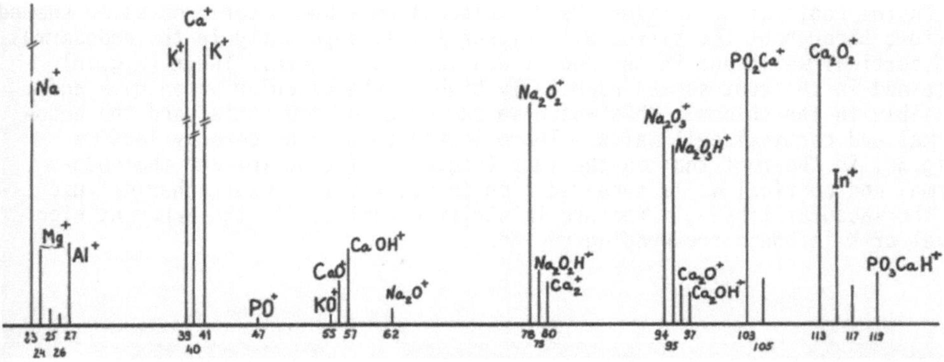

Fig.1 Secondary ion mass spectrum using a $^{16}O_2^+$ primary ion beam on a leaf section of C. sativum. Note some polyatomic ions

P. sativum roots grown in lanthanum had a strong peak for $^{139}La^+$, a very strong peak at LaO^+ mass 155, and moderate intensity at $LaOH^+$ mass 156. The very weak peak at mass 138 was presumed to be the isotope, $^{138}La^+$, and this is in keeping with its low natural abundance of 0.089%.

Rather weakly contrasted Na images were obtained (Fig.2a) as a consequence of the loss of sodium during processing. The remaining Na was mostly in the vascular bundles and in the outer walls of the upper and lower epidermis of the leaf. Potassium (Fig.2b) provided more intense ion images although much was undoubtedly lost in processing. Both epidermal layers of the leaf were intensely contrasted reflecting a high level of K in these regions. The vascular bundles also retained high levels of K, but it was more diffusely present in the nuclei, the cell walls, and in the cytoplasm closely associated with the walls. Calcium ion images (Fig.2c) were usually very bright and especially intense in the cell walls. This may be indicative of the retention of much of the Ca during processing and its accumulation in cell wall materials. The vascular bundles and nuclei also had intense images for Ca.

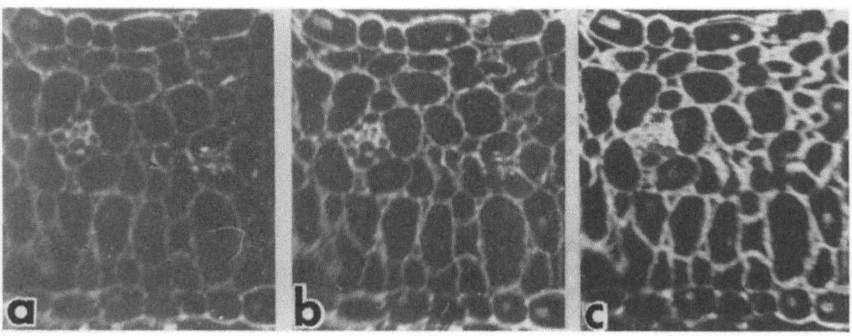

Fig.2a-c Secondary ion images for $^{23}Na^+$ (a), $^{39}K^+$ (b), and $^{40}Ca^+$ (c) from a leaf of C. myriocarpus grown at control levels of NaCl. X 280

In the roots of C. sativus the Na (Fig.3a) retained after processing seemed diffuse throughout the tissue with higher levels apparently in the endodermal and cortical walls and in the nuclei and vascular tissue. The K (Fig.3b) retained in the root seemed especially high in the vascular parenchyma and possibly in the phloem. This was also the case for the nuclei and the endodermal and cortical cell walls. There seemed to be considerably less Ca (Fig.3c) in the root than in the leaf (Fig.2c). The nuclei and the endodermal and cortical walls appeared rich in Ca, while it was rather diffuse in the vascular tissue, a feature in striking contrast to the apparent high level of K in the corresponding region.

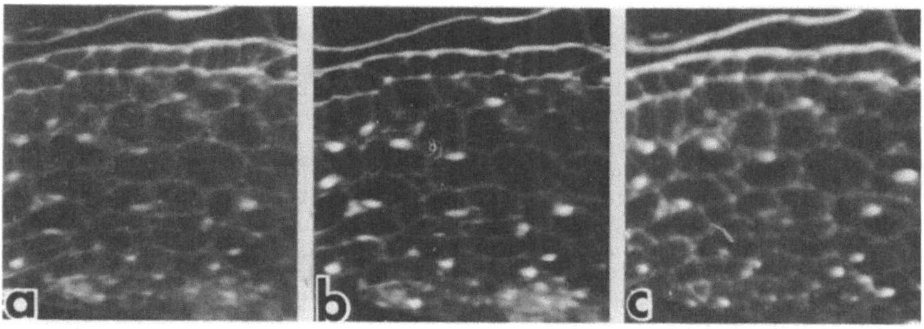

Fig.3a-c Secondary ion images for $^{23}Na^+$ (a), ^{39}K (b), and ^{40}Ca (c) from the root of C. sativus grown at 70 meq/ℓ NaCl. X 280

Lanthanum was introduced into the solution cultures of P. sativum to test the penetration and distribution into intact roots of the trivalent cation, La^{+++}. The inward movement of La appeared to be sharply limited at the endodermis (Fig.4) with the cortex brightly contrasted for La and with its exclusion from the darkly contrasted stele. The La apparently moved through the cell wall free space of the root epidermis and cortex and was prevented from moving through the endodermis by the Casparian strips. These results are in

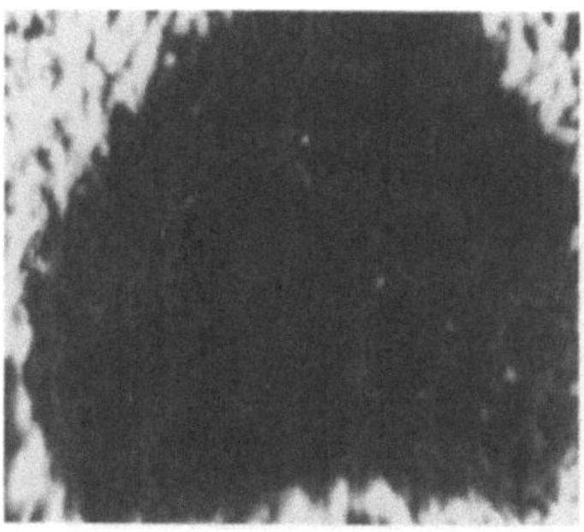

Fig.4 Secondary ion image for $^{139}La^+$ from the root of P. sativum. The light contrasted region at the periphery is indicative of La in the cortex. X 300

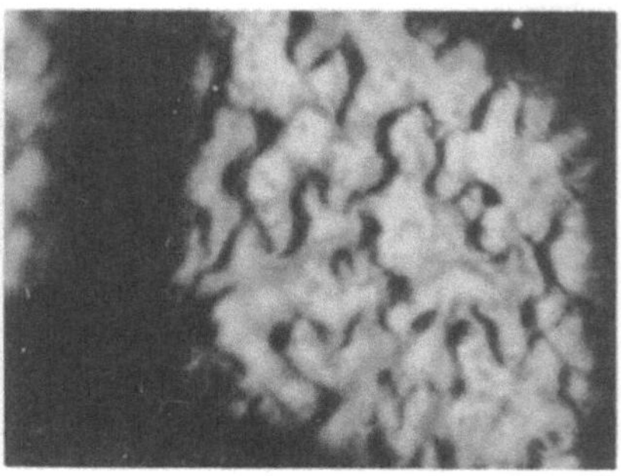

Fig.5 Secondary ion images for $^{39}K^+$ from whole mount of leaf of P. sativum. Note undulating outlines of epidermal cells and carbon coating in a grid pattern. X 300

keeping with the concept that La, at least at the more mature levels of the root, is apparently unable to traverse the plasma membrane of living cells [6,7] and that the free-space pathway for solutes is blocked by the Casparian strip [8].

The distribution of K in the dried leaf discs of P. sativum was of interest because of its role in stomatal regulation. The ion image (Fig.5) did not provide conclusive information on K distribution. The undulating darkly contrasted outlines of individual cells may have been due to irregularities in the surface of the dried preparations.

A point of particular interest demonstrated by this study is that low Z elements such as Li and B can be detected in botanical samples. Investigations by SIMS of these and other light elements of physiological importance are now more feasible.

References

1. D.R. Hoagland and D.I. Arnon, California Agric. Exp. Sta. Cir. 347, 1 (1950).
2. A.R. Spurr, J. Ultrastr. Res. 26, 31 (1969).
3. A Läuchli, A.R. Spurr, and R.W. Wittkopp, Planta 95, 341 (1970).
4. L.L. Muller and T.J. Jacks, J. Histochem. Cytochem. 23, 107 (1975).
5. M.B. Bellhorn and R.K. Lewis, Exp. Eye Res. 22, 505 (1976).
6. J.P. Revel and M.J. Karnovsky, J. Cell Biol. 33, C7 (1967).
7. W.W. Thomson, K.A. Platt, and N. Campbell, Cytobios 8, 57 (1973).
8. T.W. Tanton and S.H. Crowdy, J. Exp. Bot. 23, 600 (1972).

Secondary Ion Emission Microanalysis of the Pigments Associated with the Eye: Preliminary Data

C. Chassard-Bouchaud†
Laboratoire de Zoologie, Universite Pierre et Marie Curie
4, place Jussieu, 75230 Paris Cedex 05, France
M. Truchet†
Laboratoire d'Histophysiologie fondamentale et appliquee
Universite Pierre et Marie Curie, 12 rue Cuvier, 75005 Paris

†Laboratoire de Biophysique, Universite Paris-Val de Marne
Faculte de Medecine, 94000 Creteil, France*

It is assumed that the majority of natural pigments found in animals are present in vivo as metallic complex salts containing various sorts of metals and ligands [1]. Of interest here is whether these metals can be detected in fixed tissues and if there are relationships with environmental factors.

We selected the distal and proximal retinal pigments (ommochromes associated with proteins) of Crustacea compound eyes. In order to detect the possible influence of environmental factors [2,3], different species were collected from the same location, while the same species was samples from different regions. All the animals were fixed in Carnoy. We also analyzed non pathological human eye fixed in glutaraldehyde and compared melanin containing structures (ciliary body and iris) with a non pigmented one (cornea).

SIMS analysis was performed always under the same conditions with a CAMECA SMI 300 primary ion beam, O_2^+, 7.5 µA; 200 µm for the diaphragm of the lens; 20 V in lateral energy; mass resolution of 300 on images and complete spectra, 2000 for the spectra of isolated masses obtained with the electrostatic filter to avoid interferences [4,5]. To take into account the variable amounts of matter in the analyzed areas, the intensities were normalized to carbon ($^{12}C^+$). Among the elements which were strongly disturbed by polyatomic ions, Mg, Al, Fe, Co and Cu were analyzed by using a high mass resolution.

The results are summarized in Table 1 for Ca, Sr, Ag and Ba; Table 2 gives the percentages of metal emission at the masses 24^+, 27^+, 56^+, 59^+, and 63^+ and the normalized intensities of emission for the metal alone. Figs.1-3 give the distribution of Ca, Sr and Al in a crustacean eye.

The normalization of the intensities to the intensity of carbon permits one to compare the emission of a given metal from one sample to another, but only between similarly fixed ones, because the difference of fixation may introduce a difference in the carbon emissivity. On the other hand, it is interesting to note that despite the differences of fixations, the same elements are detected in almost all the samples, including the unexpected Ag. It is also interesting to note that Mg is always the main ion at 24^+, whereas the amounts of Al, Fe and Cu are quite variable inside masses 27^+, 56^+ and 63^+, from one sample to another. However, after normalization of the metal emission to carbon, this variability also occurs for Mg, demonstrating the importance of both high mass resolution and normalization of intensities.

*The analyses were performed with the financial support of CNRS and INSERM and the technical assistance was provided by Annick Allain-Quettier.

TABLE 1

			Origin	Ca+ (40)	Sr+ (88)	Ag+ (107)	Ba+ (138)
CRUSTACEA EYES ommochrome	marine sp.	Crangon crangon	The english Channel (Le Havre)	1212	6.6	0.4	0.8
		Crangon crangon	The english Channel (Roscoff)	266	1.35	0.15	0
		Crangon crangon	North Sea (Helgoland)	519	2.25	0	0.2
		Zoea larva	North Sea (Helgoland)	400	4.5	0.4	0.6
		Zoea larva	Mediterranean (Villefranche s/mer)	1252	8.75	0	0.8
		Mysis larva	Mediterranean (Villefranche s/mer)	213	0.8	0.5	0
		Eupagurus prideauxi	Mediterranean (Banyuls s/mer)	485	2.85	0.25	0.7
		Eupagurus prideauxi	The english Channel (Roscoff)	289	1.7	0	0
	fresh water sp.	Bythotrephes longimanus 1	Lake of Geneva (Thonon les Bains)	97	0.3	0.15	0.1
		Bythotrephes longimanus 2	Lake of Geneva (Thonon les Bains)	99	0.7	0.2	0.07
HUMAN EYE	melanin	Iris	Paris	56	0.05	0.4	0.058
		Ciliary body	"	57	0.07	0.2	0.03
		Cornea	"	27	0.04	0.06	0.01

TABLE 2

	Mg		Al		Fe		Co		Cu	
	% at 24+	I/C*	% at 27+	I/C*	% at 56+	I/C*	% at 59+	I/C*	% at 63+	I/C*
C.crangon (Helgoland)	72	0.8	28	0.21	-	-	0	0	-	-
C.crangon (Le Havre)	86	1.86	25	0.25	0.5	0.01	0	0	-	-
B.longimanus 1 (Thonon les Bains)	92	12	2	0.05	-	-	0	0	10	0.08
Ciliary body	94	2.16	65	1.23	48	1.8	0	0	71	0.1
Cornea	87	8	34	0.26	6	0.04	0	0	36	0.024

(*absolute intensities normalized to $^{12}C^{+}$)

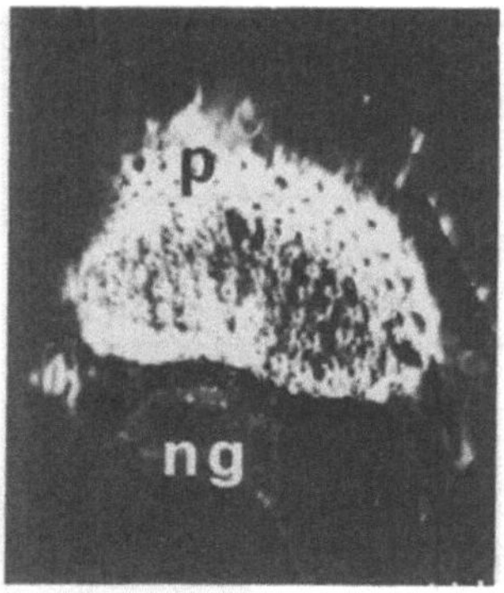

Fig.1 $^{40}Ca^+$

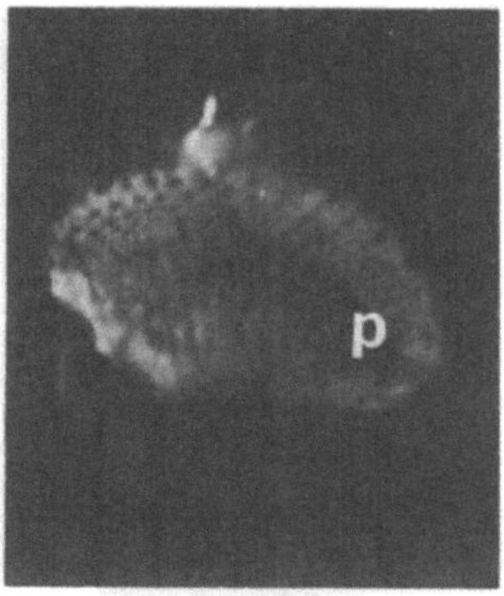

Fig.2 $^{88}Sr^+$

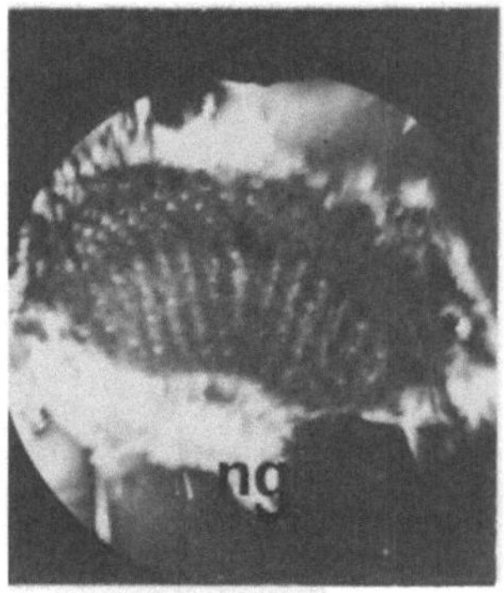

Fig.3 $^{27}Al^+$

Fig.1,2,3 Distribution of the elements in the eye of *Bythotrephes longimanus* (Crustacea): n g = nervous ganglion, p = pigment (ommochrome)

The strong emission of Ca in ommochromes is in good agreement with biochemical data [6] and electron probe results [7]; Sr and Ba are known to accumulate with Ca and were already detected in pigmented structures [8]. Here, Sr intensities are well correlated with that of Ca, while this is not the case for Ba. Such a discrimination between both elements was also described [9]; however, the weak intensities for Ba in our samples are in contrast with the stronger emission recorded in mammalian retinas [10]. Our images show that Sr and Ca are similarly located in pigments of crustacean eyes while Al is in the nervous ganglion. However, these data have to be compared with similarly treated samples because the fixation may disturb the *in vivo* localization. In marine Crustacea, Ca and Sr are more intense than in freshwater species; in all the samples the intensities of the other metals are very low and the variability of the results makes impossible, at the moment, the determination of other relationships with environmental factors. On the other hand, in the human eye, Al, Fe and Cu are more intense in melanin containing structures than in non pigmented ones; it is the contrary for Mg.

Thus, our results demonstrate that interesting data can be obtained but that further investigations are required to draw firm conclusions.

References

1. H. Kikkawa, Z. Ogita, and S. Fujito, Proc. Jap. Acad. 30 (1), 30 (1954).
2. R. Martoja, et al., J. Micr. Biol. Cell, 22 (2-3), 441 (1975).
3. C. Chassard-Bouchaud and G. Balvay, Micr. Acta, sup. 2, 185 (1978).
4. M. Truchet, J. Micr. Biol. Cell., 24 (1), 1.
5. M. Truchet, Micr. Acta, sup. 2, 355 (1978).
6. R. Martoja, C.R. Acad. Sc. Paris, 273 D, 368 (1971).
7. C. Chassard-Bouchaud, C.R. Acad. Sc. Paris, 274 D, 2511 (1972).
8. Z.M. Bacq, in Handbuch der Exp. Pharmacol. (Springer-Verlag, 1963).
9. A.C. Brannon and K.R. Rao, Comp. Biochem. Physiol., 63A, 261 (1979).
10. M.B. Bellhorn and R.K. Lewis, Exp. Eye Res., 22, 505 (1976).

Comparison of Spectra of Biochemical Compounds and Tissue Preparations

M.S. Burns
Department of Ophthalmology
Albert Einstein College of Medicine, Bronx, N.Y.

D.M. File
Electron Optics Lab, Naval Weapons Support Center
Crane, Indiana

A. Quettier and P. Galle
Department of Biophysics
University of Paris, Creteil, France

Abstract

Questions of importance in assessing the potential of SIMS for analysis of unknown biological samples are: 1) Do biologically important macromolecules have characteristic spectra? 2) Is the diverse chemical nature of a tissue area apparent from the mass spectrum? 3) What is the practical detectability of predictable levels of elements in biological samples?

In order to study the first question we have investigated the mass spectrum of major classes of biochemical compounds found in tissues. These include protein, lipids, deoxyribonucleic acid, ribonucleic acid, polysaccharides and amino acids. Operating conditions using the CAMECA IMS-300 were positive oxygen primary beam at current densities of 10^{-5} A/cm^2 and detection of positive secondary ions. Spectra of the pure chemicals showed some characteristic mass peaks for each class of compound. Deuterated derivatives were used to decipher the contribution of various hydrocarbons to the mass spectrum. High resolution mass spectrum showed agreement with these assignments.

The second question was approached by investigation of the mass spectrum of specific tissue areas with known differences in chemical composition. Only minor differences in spectra were observed. The spectra of different tissue areas in lyophilized tissue preparations are dominated by the intensity of the water soluble ions, sodium and potassium, and other differences appear minor.

The third question was approached by calculating the minimum detectable limits of elements in biological tissue with reference to detection limits in empirical standards. These limits are compared to levels found in biological tissue such as whole blood, as an example of the practical usefulness of SIMS for biological tissue analysis.

These results indicate that a considerable degree of understanding of both the biological problem and of the instrumental operating parameters will be necessary for SIMS analysis of unknown biological samples.

X. Combined Techniques

High Sensitivity SIMS Using DC Accelerators

K.H. Purser
General Ionex Corporation
19 Graf Road
Newburyport, Massachusetts 01950 USA

Abstract

Experiments have been conducted which show that secondary ion mass spectrometry (SIMS), together with the addition of a rather low voltage (2 MV) DC accelerator, can become an extremely powerful new technique in many fields of research related to the characterization of surface and bulk solids and with high spatial resolution. The technology is available today to build an instrument that would have high lateral resolution (~ 1 μm) coupled with a secondary ion transmission coefficient from target to detector >10% with no mass interferences from molecular ions. Molecular ions can be fragmented and analyzed separately. We believe that this combination of secondary ion mass spectrometers and accelerators represents a most significant advance in analytical techniques.

Combined SIMS, AES, and XPS Investigations of Oxygen-Covered 3d Transition Metal Surfaces

O.Ganschow, L.Wiedmann and A.Benninghoven
Physikalisches Institut der Universität Münster
Schloßplatz 7, D-4400 Münster, W. Germany

The interaction of some 3d metals with oxygen in the monolayer range under UHV conditions has been investigated by quasisimultaneous static SIMS, static AES, and XPS. The combination of these methods is particularly useful for this purpose, because each of the stages of metal-oxygen interaction causes significant changes of specific SIMS, AES, or XPS signals.

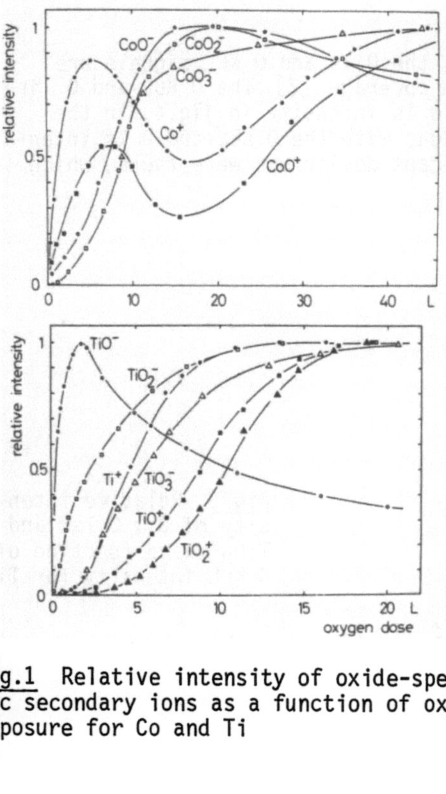

Fig.1 Relative intensity of oxide-specific secondary ions as a function of oxygen exposure for Co and Ti

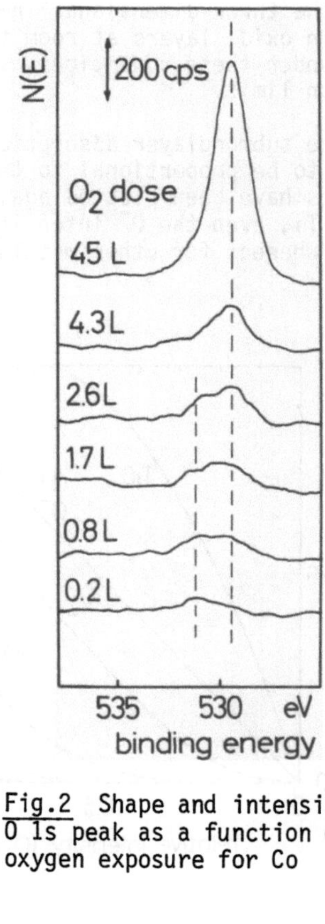

Fig.2 Shape and intensity of O 1s peak as a function of oxygen exposure for Co

The intensity vs. oxygen exposure curves of the "oxide-specific" secondary ions $MeO_n^{\pm}$ lead to the distinction of two different types of behaviour (Fig. 1). The "normal" sequence of appearance (e.g. for Ti and Fe) is determined by the valency model of PLOG et al. [1]. For some metals (e.g. Co and Ni), an additional increase of positive secondary ion yield at low oxygen doses is found. In the case of Co, this increase can be correlated with the existence of a shifted O 1s level in XPS (Fig.2), and therefore indicates a special oxygen chemisorption stage.

Whereas the intensity changes of $MeO_n^{\pm}$ ions indicate the formation of a two-dimensional "surface oxide" [1], the most significant changes of AES signals, especially from valence band transitions, take place in the submonolayer and monolayer oxygen adsorption stage. According to Fig.3, where the O^-, O 1s, and $TiO_n^{\pm}$ intensity is plotted against the O KLL intensity, most of the $TiO_n^{\pm}$ intensity changes occur at exposures for which the O KLL (as well as the Ti LVV and MVV) signals have almost reached saturation. This behaviour is also found for other 3d metals.

Chemical shifts of substrate peaks in XPS are only observed at oxygen exposures exceeding the saturation dose for the AES and SIMS signals, and indicate the three-dimensional in-depth growth of the oxidized layer. Ni forms very thin oxide layers at room temperature and oxygen exposures up to some 10^4 L. Under these conditions, shifted Ni lines were below the experimental detection limit.

In the submonolayer adsorption range, the O 1s and O KLL signals are assumed to be proportional to the oxygen coverage [2]. The O KLL and O^- intensities have been plotted against the O 1s intensity in Fig.4. In the case of Ti, even the O^- intensity coincides with the O KLL and O 1s intensities, whereas for other metals significant deviations were found, which

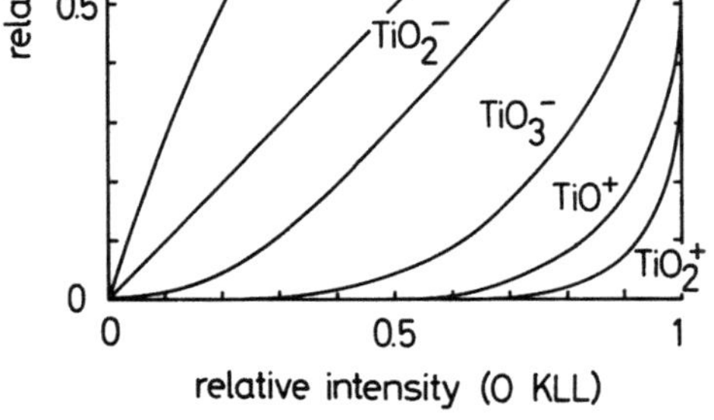

Fig.3 Relative intensity of O^-, O 1s, and $TiO_n^{\pm}$ as a function of O KLL intensity for Ti

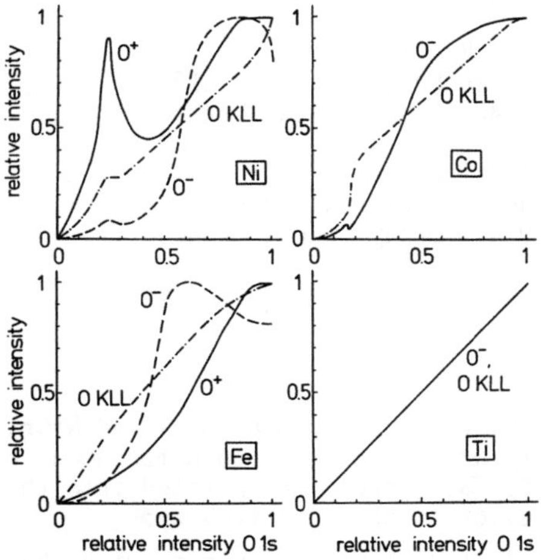

Fig.4 Relative intensity of O^- and O KLL as a function of O 1s intensity for some 3d transition metals

indicate that the O^- signal can usually not even approximately be taken as a measure for oxygen coverage. On the other hand, for Co and Ni slight differences were even found between the O 1s and O KLL intensities, which might be due to the complicated oxygen chemisorption behaviour of these metals.

References

1 C.Plog, L.Wiedmann and A.Benninghoven, Surface Sci. 67 (1977), 565
2 C.R.Brundle, J.Vac.Sci.Technol. 11 (1974), 212

Oxidation of Nickel Base Alloys Flooded with Oxygen Under Ionic Bombardment

J.C. Pivin and C. Roques-Carmes
Laboratoire de Métallurgie Physique, Bât. 413
Université Paris-Sud, 91405 Orsay Cedex, France

G. Slodzian
Laboratoire de Physique du Solide, Bât. 510
Université Paris-Sud, 91405 Orsay Cedex, France

1. Introduction

Two steps have been observed for the fixation of oxygen on Ni and NiCr, NiFe, NiFeCr alloys, when flooded with oxygen under argon bombardment [1,2,3]. First, oxygen is chemisorbed on the surface, then incorporated into the bulk at a critical coverage θc. This behavior is characterized by:

i) variations of monoatomic and polyatomic emission yields with the oxygen pressure (or the exposure in static SIMS);
ii) comparison of mass spectrograms with those of standard oxides;
iii) "depth profiles" which showed that the thickness of the oxidized layer increases abruptly at θc.

Further experiments have shown that this behavior is general for Ni and Co base alloys (with Ti, V, Cr, Fe, Co, Ni, Si, Al, Be solutes). Emission yields of solvent species Ni^+, NiO^+, Ni_2^+ (or Co^+, CoO^+, Co_2^+) and solute species (M^+, MO^+, M_2^+) change discontinuously at θc. The present paper shows evidences for these phenomena on NiM alloys. The discontinuities ΔI (M^+) and ΔI(Ni^+) may be positive or negative according to the nature of the solute M, its concentration and the energy of the selected ions.

2. Experiments

The alloys were melted in a plasma furnace; their grain size was about 1 mm and secondary ions emitted by one grain could be selected. All experiments were carried out in a CAMECA SMI 300 microanalyzer. The target was sputtered by 5.5 keV Ar^+ ions; the primary ion density on the surface was about 3 μA/mm^2. Under these bombardment conditions, measurements of the emission yields were recorded when a dynamic equilibrium was established for each pressure. Generally, [0,25 eV] secondary ions were selected for the measurements.

Figure 1 shows the variations of the emission yields for alloys containing comparable concentrations of Ti, V, Cr, Fe and Co. For NiFe and NiCo alloys the discontinuous change ΔI of monoatomic emissions at θc is always positive. For Ti, V, Cr and Si, Al (refer to Figs.2 and 3) it may be positive for solutes of low abundance and is negative for more abundant ones.

For the same alloy the sign of ΔI may be positive or negative according to the energy of secondary ions (refer to Fig.4 for a NiCr alloy).

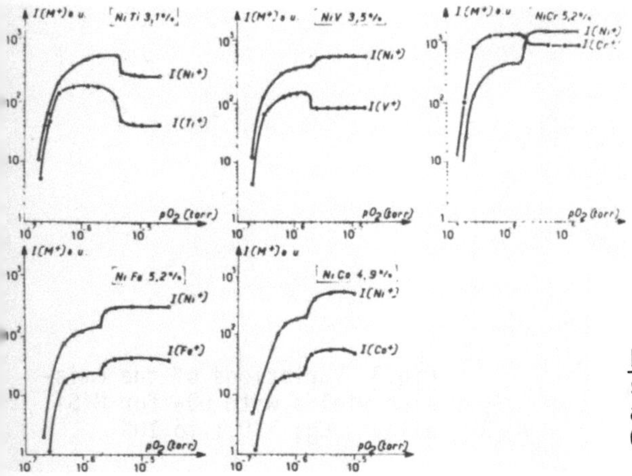

Fig.1 Variations of the emission yields with pO_2 for NiM alloys, at a given concentration C_M = 3 to 5% [4]

3. Discussion and Conclusions

The sputtering yield S_{NiM} was measured on both sides of the transition for several alloys: in all cases it falls by a factor of about 1.8 at θc. Thus, larger decreases of the M^+ emission yields, observed at c for the same alloys, must be attributed either to changes of the ionization probabilities P^{M}_{NiM} and P^{Ni}_{NiM} of particles sputtered as monoatomic species or to modifications of the relative yields of monoatomic and polyatomic species. The two phenomena may also combine their effects. Moreover, for a given grain size of an alloy, P^{Ni}_{NiM} may increase or decrease according to the energy of the selected ions.

Note that small differences are observed in the M^+ emission yields from one grain to another before θc: They are certainly due to differences in the sputtering yields of these grains. But the Ni^+, M^+ intensities emitted by saturated surfaces (above θc) remain nearly identical whatever the surface orientation may be.

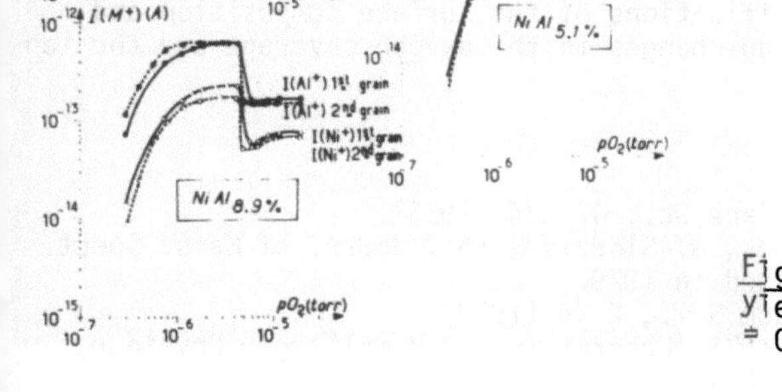

Fig.2 Variations of the emission yields with pO_2 for NiAl alloys, C_{Al} ≐ 0.1 to 10%

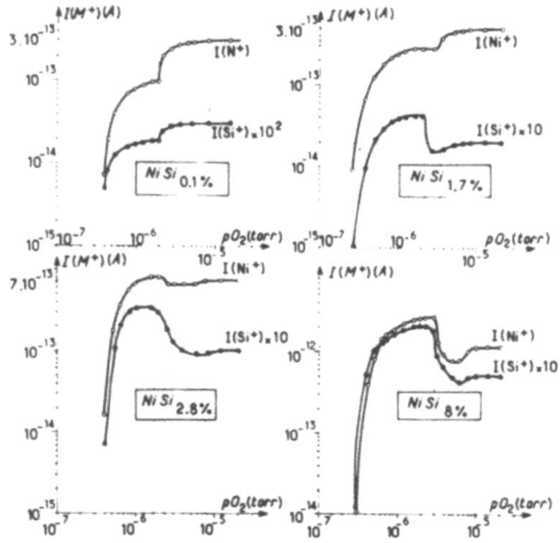

Fig.3 Variations of the emission yields with pO_2 for NiSi alloys, C_{Si} = 0.1 to 10%

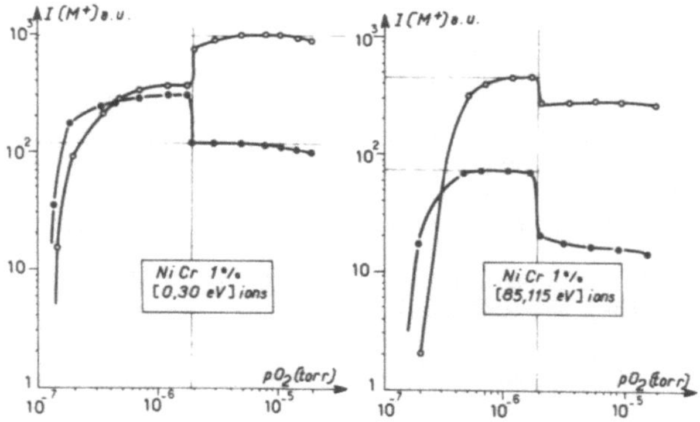

Fig.4 Influence of secondary ions energy for an NiCr 1% alloy

o = Ni^+; ● = Cr^+

Othersie, the Ni^+ emission yield of saturated surfaces is enhanced by additions of Ti, V, Cr, Al, Si in the nickel matrix, this being particularly clear in Fig.2 and 3. This problem is discussed in another paper. The ionization probability of nickel changes with the nature of the neighbors encountered during the ejection of the nickel atom and the velocity of the ejection. Variations of the Ni^+ emission yield at c with the nature of the solute and the energy of the selected ions must be due to the same process as those of P_M^{Ni} with C_M on saturated surfaces. Further experiments are at present being carried out using electron spectrometry to correlate variations of the emission yields with modifications of the surface composition and electronic structure accompanying changes in the oxygen coverage and the ion bombardment density.

References

1. G. Blaise, M. Bernheim, Surface Sc., 47 324 (1975).
2. J. C. Pivin, C. Roques-Carmes, G. Slodzian, Int. Journ. of Mass. Spect. and Ion Phys., to be published in 1979.
3. K. H. Rieder, Applied Surface Sci., 2 74 (1978).
4. G. Bellegarde, Thèse Orsay 1978 (Laboratoire de Métallurgie Physique, Université Paris-Sud, 91405 Orsay).

Matrix Effect Studies by Comparative SNMS and SIMS of Oxidized Ce, Gd and Ta Surfaces

H. Oechsner, W. Rühe, H. Schoof, and E. Stumpe
Physikal. Institut TU Clausthal,
D-3392 Clausthal-Zellerfeld and Sonderfor-schungsbereich 126,
Göttingen-Clausthal, W. Germany

The main problem for the quantitation of SIMS is the non-linear dependence of the secondary ion intensities on the chemical surface composition. It is the aim of the present paper

i) to demonstrate quantitatively the influence of such matrix effects for oxidized metal surfaces, and
ii) to supply experimental information on the variation of the ionization coefficients with the surface oxygen concentration. Both kinds of information are obtained by an in-situ comparison of SIMS and SNMS for oxidized polycrystalline samples of Ce, Gd and Ta.

SNMS (Sputtered Neutral Mass Spectrometry) was introduced as a novel method for surface analysis a few years ago [1]. In SNMS, sputtered neutral particles are postionized when traversing a hot maxwellian electron gas constituting the electron component of a low pressure high frequency plasma maintained by electron cyclotron wave resonance. Postionization coefficients α° in the order of 10^{-2} are obtained, whereas the interaction with heavy plasma particles is negligibly small.

The experimental arrangement is schematically shown in Fig.1. For SIMS the hf plasma is shut down and the electrical diaphragm in front of the target is opened for secondary ions. The specimens were bombarded at 45° with 4 keV Ar^+ ions at current densities around 10^{-5} Acm^{-2}. Before each SIMS or SNMS run the samples were under continuous ion bombardment exposed to such oxygen pressures $p(O_2)$ for which the SNMS and the positive SIMS spectra became independent of $p(O_2)$. For more details refer to [2].

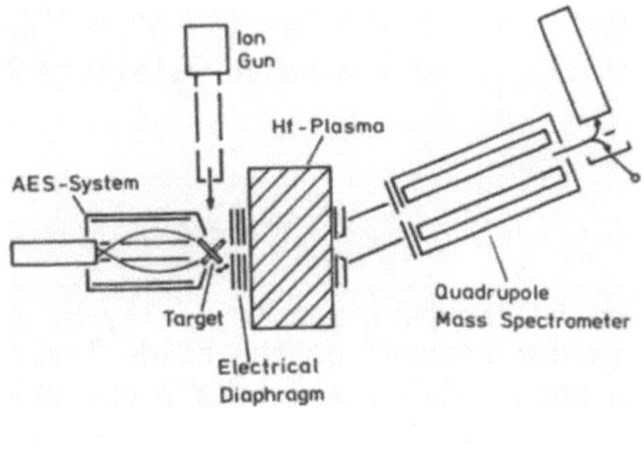

Fig.1 Schematic diagram of the SNMS apparatus with additional AES system

The SNMS spectra show that the flux of ejected neutrals consists almost quantitatively of metal atoms Me and monoxide molecules MeO [3]. The highest peaks in the positive SIMS spectra referred to these species, too. Hence, the intensity ratios $Q^\circ = I(MeO^\circ)/I(Me^\circ)$ of postionized neutrals, and $Q^+ = I(MeO^+)/I(Me^+)$ of positive secondary ions were used to characterize the composition of both kinds of particle fluxes from the samples. Corresponding results are shown in Fig.2 and 3. Q° and Q^+ are always plotted versus the total dose of the bombarding Ar^+ ions. Zero ion dose refers to the throttling down of the O_2 inlet into the system. The O_2 pressure dropped down below 10^{-8} torr before the second data point along the dose axis was reached. The actual surface oxygen concentrations c_O^S are obtained either by in-situ AES or from the SNMS-signals $I(MeO^\circ)$ using the model for MeO formation developed in [3].

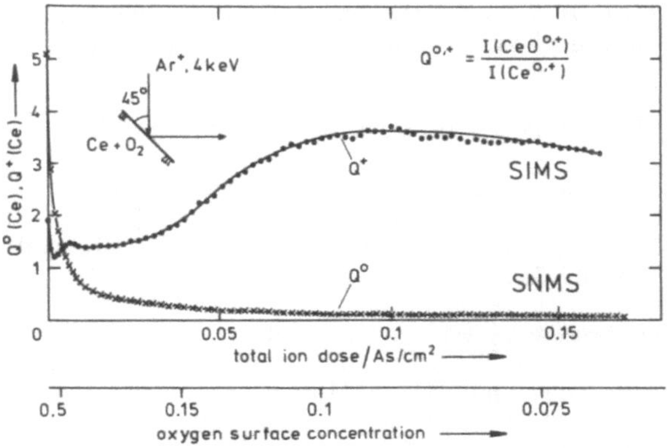

Fig.2 MeO/Me intensity ratios Q° and Q^+ from SNMS and positive SIMS spectra for oxidized polycrystalline Ce

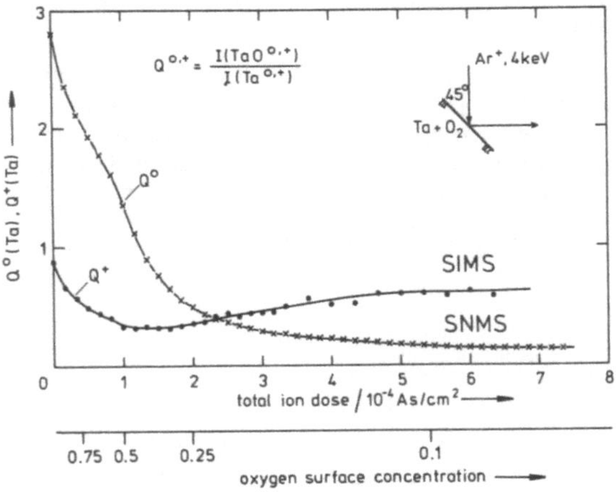

Fig.3 Q° and Q^+ for oxidized polycrystalline Ta

The SNMS ratios Q° were always found to decrease monotonically by 1 to 2 orders of magnitude indicating a continuous removal of the oxide layer. The SIMS ratios Q^+ vary only little, since both $I(MeO^+)$ and $I(Me^+)$ decrease with decreasing oxygen surface concentration c_O^S. In the low bombarding dose regime the reduction of MeO^+ formation exceeds that of Me^+. This behavior is reversed for further reduction of c_O^S.

The decoupling between the emission and the ionization process, realized by SNMS, permits measurement of the ionization coefficients $\alpha_X^\pm$ for secondary ion emission. For SNMS and for SIMS the detected intensity I of a species X is given by

$$I(X^{\pm,o}) = I_p \eta_X \, Y_X \alpha_X^{\pm,o} \qquad (1)$$

with I_p primary ion current, η_X geometry and transmission factor for X, Y_X partial sputtering yield of X. Eq. (1) is valid for SNMS when $\alpha_X^{\pm} \ll 1$. The SNMS postionization probability α_X^o for a certain species X is a constant of the apparatus. For alternate in-situ SNMS and SIMS we obtain

$$\alpha_X^{\pm}/\alpha_X^o = I(X^{\pm})/I(X^o) \qquad (2)$$

Corresponding results describing the variation of α_{Me}^{+} with the surface concentration of oxygen are presented in Fig.4 for Ce^+, Gd^+ and Ta^+. For Ta^+ additional results for α^+ at nitrided Ta surfaces are included.

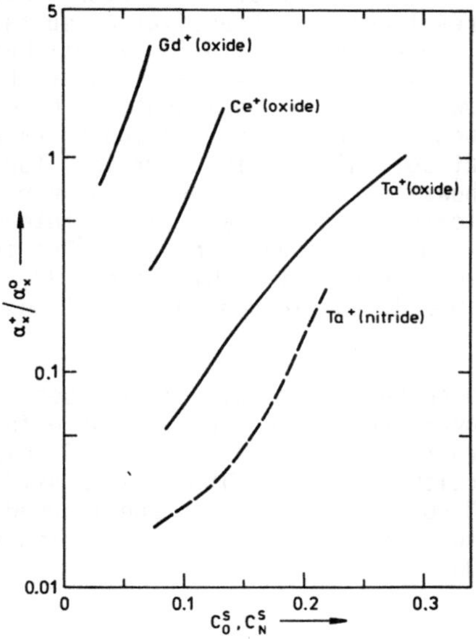

Fig.4 Ratio of ionization coefficients α^+ and constant postionization factors α^o for Me atoms as a function of oxygen or nitrogen surface concentration c_O^S and c_N^S

The α^o values for Gd and Ce are almost equal [2]. Hence, we obtain α^+ Gd to be above α_{Ce}^{+} by approximately one order of magnitude for c_O^S of several at.%. The α_{Ta}^{+} values for oxidized and nitrided Ta surfaces can be compared immediately in view of constant α_{Ta}^{o}. For c_O^S and c_N^S between 10 and 20 at.%, α_{Ta}^{+} for oxygen covered Ta is larger by a factor of 2-3 than α_{Ta}^{+} for the nitrided surface.

References

1. H. Oechsner, W. Gerhard, Phys. Lett. 40A, 211 (1972).
2. H. Oechsner, E. Stumpe, Appl. Phys. 14, 43 (1977).
3. H. Oechsner, W. Rühe, E. Stumpe, Surface Sci. 85, 289 (1979).
4. H. Oechsner, H. Schoof, E. Stumpe, Surf. Sci. 76, 343 (1978).

XPS/LEED/SIMS Study of the Ni(100)/O_2 System: Origin of the Two O(1s) Features

C.R. Brundle and H. Hopster*
IBM Research Laboratory
San Jose, California 95193

1. Introduction

Numerous previous papers generally agree that the room temperature adsorption of oxygen on Ni/O_2 at low pressures leads to a "passivated" surface with an oxide 2-3 atomic layers thick. However, despite individual LEED [1], XPS [2-4], and SIMS [5,6] studies, it is still not clear as to what the mechanism of the oxidation stage following initial chemisorption, or what is the exact nature of the oxide product. In this paper we concentrate on the latter aspect. LEED has shown that some NiO domains certainly form on single crystal surfaces. XPS often shows *two* O(1s) features, however, indicating the presence of two types of atomic oxygen. The relative intensities of the two features vary greatly from study to study. The two O(1s) features have been ascribed to chemisorbed O atoms on top of an NiO layer [3]; to the formation of "Ni_2O_3" in addition to NiO [2]; and in non-UHV situations to $Ni(OH)_2$ [7].

In the present study we are able to correlate the different types of information from LEED, XPS, and SIMS over the entire exposure range for Ni(100)/O_2. LEED observes only the ordered structures formed; XPS follows the *total* oxygen uptake and the conversion of surface Ni metal to oxide, but is unable to unambiguously identify "Ni_2O_3" or distinguish chemisorbed O from OH; and SIMS has the capability both to detect the onset of oxidation from chemisorption and distinguish O from OH.

2. Experimental

All experiments were performed in a VG Scientific combination LEED/Auger/XPS/UPS/SIMS system which is described elsewhere [8]. The relevant information here is (i) the base pressure is ~7×10^{-11} Torr; (ii) the XPS emission detection angle can be varied, thereby varying the effective sampling depth; and (iii) SIMS is performed in the static mode on both positive and negative ions. The Ni(100) crystal is oriented within 1° and cleaned by sputter-heating combinations.

3. Results and Discussions

A typical O(1s) total intensity versus exposure curve for an oxygen pressure of 5×10^{-9} Torr is given in Fig.1. The shape of the O(1s) structure at 45°

---

*Permanent address: Institut für Grenzflâchenforschung und Valkuumphysik, Jülich, West Germany.

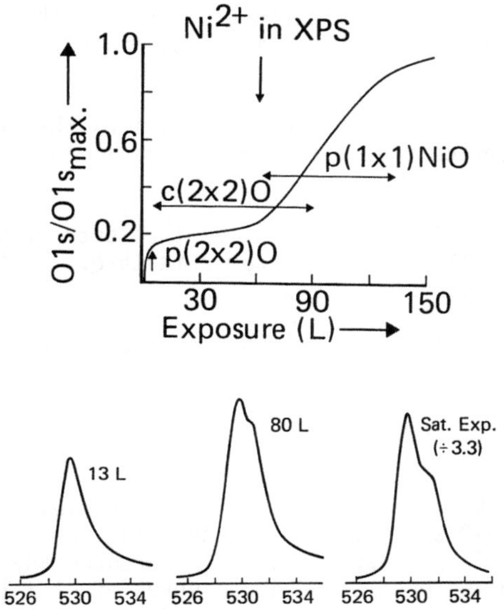

Fig.1 Relative oxygen coverage as a function of oxygen exposure at 300K. The exposure extent of the observed LEED patterns are marked, as is the onset of NiO as detected by XPS. The shape of the O1s spectrum at various exposures is shown below

emission angle as a function of exposure is shown in the panels above. At 15° the higher Binding Energy feature becomes relatively more intense. The LEED sequences are marked on the figure as is the onset position of Ni^{2+} as evidenced from the Ni2p XPS spectrum. Fig.2 shows a plot of the high B.E./low B.E. O(1s) ratio against the observed O^-/OH^- SIMS intensities. Plots of the intensities of any of the major positive or negative SIMS ion intensities versus coverage show sharp breaks (decreases of rate of growth) at the Ni^{2+} onset position identified by XPS.

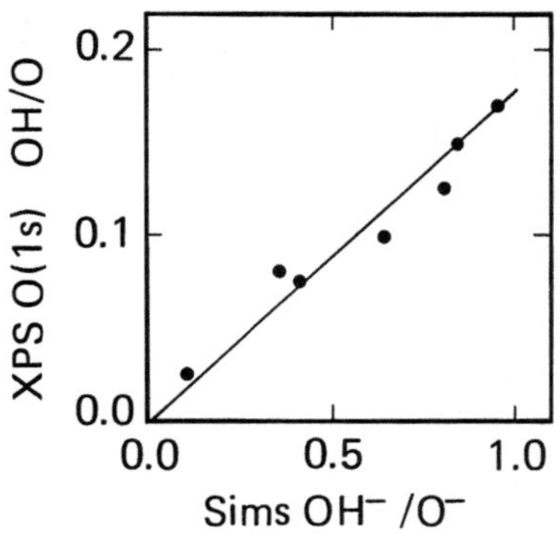

Fig.2 Plot of the ratio of the high/low B.E. O1s peak as a function of SIMS OH^-/O^- relative intensities

Several conclusions can be drawn from the above sets of correlated data:

(1) LEED, XPS, and SIMS all identify the onset of NiO formation towards the end of the stable C(2x2) chemisorbed O regime, approximately co-incident with the increase in sticking probability.

(2) There is a clear correlation between the higher B.E. O(1s) and the SIMS OH^- intensities, thus demonstrating that this O(1s) feature represents OH.

(3) The fact that the slope of Fig. 2 is less than unity indicates that the OH is confined to nearer the surface than the oxide. This is correlated by the O(1s) angular behavior.

(4) The assignment of the higher B.E. O(1s) to OH implies that both C(2x2) O and ordered NiO have the *same* O(1s) B.E., the *lower* feature. This has been a source of controversy for some time.

We have subsequently demonstrated that the OH arises from the reaction:

$$H_2O + O \rightarrow 2\,OH$$

or $$2H_2O + O_2 \rightarrow 4\,OH$$

and that this takes place during the C(2x2) O conversion to NiO nucleation period by co-adsorption of residual H_2O and O_2. Simply replacing the gas phase O_2 with H_2O during this stage of the reaction will *not* produce any OH. The important point is that this happens even at base pressures of $<1 \times 10^{-10}$ Torr. If the partial pressure of the residual H_2O is reduced by performing the reaction at *high* oxygen pressure (10-6 Torr or greater), the OH content of the oxide drops to nearly zero. If H_2O is deliberately introduced along with O_2, the final OH content can be increased above that shown here.

References

1. P.H. Holloway and J.B. Hudson, Surf. Sci. 43, 123 (1974).
2. K.S. Kim and N. Winograd, Surf, Sci. 43, 625 (1974).
3. C.R. Brundle and A. Carley, Chem. Phys. Lett. 31, 423 (1975).
4. P.R. Norton, R.L. Tapping, and J.W. Goodale, Surf. Sci. 65, 13 (1977).
5. T. Fleisch, W. Winograd, and W.N. Delgass, Surf. Sci., to be published.
6. A. Benninghoven, V.H. Müller, M. Schemmer, and P. Beckmann, Appl. Phys. 16 367 (1978).
7. J. Haber, J. Stoch, and L. Ungier, J. Electron. Spectr. 9, 459 (1976).
8. C.R. Brundle, IBM Journal Res. & Dev. 22, 235 (1978).

Combined SIMS, AES and XPS Study of $Cd_xHg_{1-x}Te$

O. Ganschow, H.M. Nitz, L. Wiedmann and A. Benninghoven
Physikalisches Institut der Universität Münster
Schloßplatz 7, D-4400 Münster, Germany F.R.

The ternary compound $Cd_xHg_{1-x}Te$ is of great importance in infrared technology. The band gap of this material depends critically on the concentration x. Therefore, a detailed knowledge of the surface and bulk composition of the material is necessary for reliable control of the desired properties.

The surface composition of $Cd_xHg_{1-x}Te$ with known bulk concentration x = 0.2 was investigated in an UHV system by quasisimultaneous static SIMS, static AES, and XPS. In situ scraping of the specimen was used for producing clean surfaces with known composition $Cd_{0.2}Hg_{0.8}Te$. By comparison with these surfaces, the AES and XPS signals of Cd, Hg, and Te from differently prepared surfaces could be converted into absolute elemental concentrations.

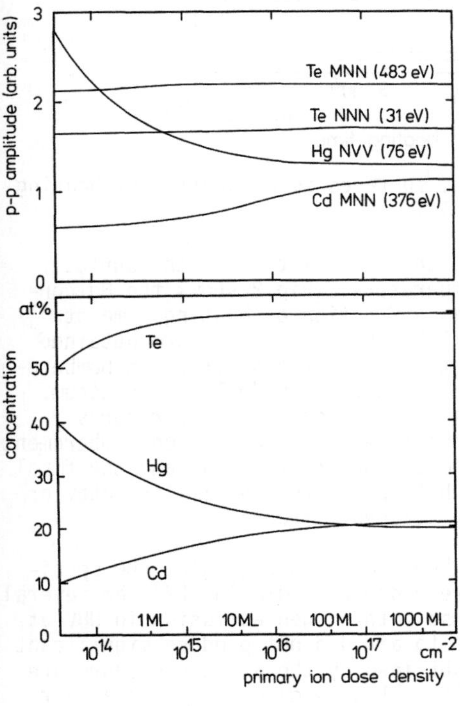

Fig.1 AES peak-to-peak amplitudes and elemental concentrations of Cd, Hg, and Te as a function of primary ion dose density during ion bombardment of an in situ scraped surface

Quantitative SIMS measurements were complicated by the fact that the Hg^+ signal showed a quadratic dependence on primary ion current density due to post-ionization of sputtered Hg atoms by the primary beam [1].

Ion bombardment of in situ scraped surfaces causes significant concentration changes due to preferential sputtering. Fig.1 shows the changes in the AES signals and in the atomic concentrations during ion bombardment (Ar^+, 5 keV, 70° angle of incidence) of a scraped surface. The equilibrium surface concentration is reached at about 10^{17} ions/cm^2 and differs from the bulk composition up to a factor of 2. Identical equilibrium surface concentrations are reached at similar ion doses irrespective of the pre-treatment of the surface.

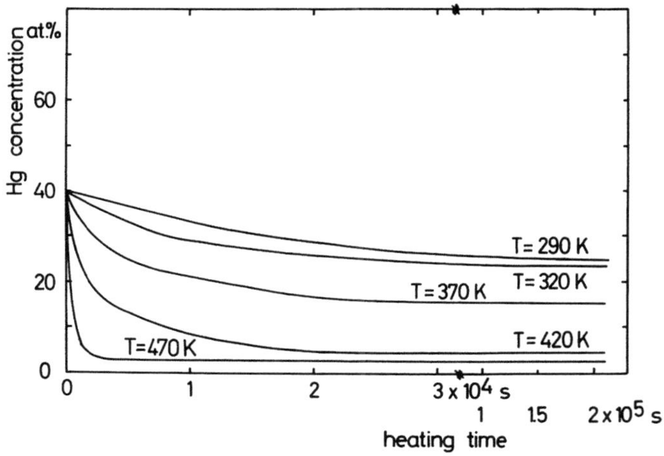

Fig.2 Hg concentration of in situ scraped surfaces as a function of heating time at temperatures between 290 K and 470 K

Heating of the specimen causes significant Hg depletion at the surface. This effect is important even at room temperature. Fig.2 shows the Hg concentration of a freshly scraped surface as a function of heating time at various temperatures between 290 K and 470 K. Similar curves are obtained when starting from the sputter equilibrium concentrations after ion bombardment. The evaporation of Hg can be detected in the residual gas spectrum. Fig.3 shows the Hg evaporation curves obtained at constant temperatures between 330 K and 470 K after reaching sputter equilibrium by ion bombardment of the surface at the same temperature. Both the initial peak and the final equilibrium evaporation rate increase with temperature. A similar behavior was found for Hg evaporation from freshly scraped surfaces.

In order to investigate the oxidation behavior of $Cd_xHg_{1-x}Te$, the specimens were oxidized in an oxygen atmosphere under UV radiation [2] for several days. This procedure was necessary because both oxygen admission in UHV at room temperature up to 10^6 L and exposure to air did not produce significant amounts of surface oxide. The surfaces obtained by the above procedure are characterized by selective oxidation of Te. This is evident by the appearance of TeO^+ and TeO_n^- (n = 1,...3) secondary ions and a chemical shift of

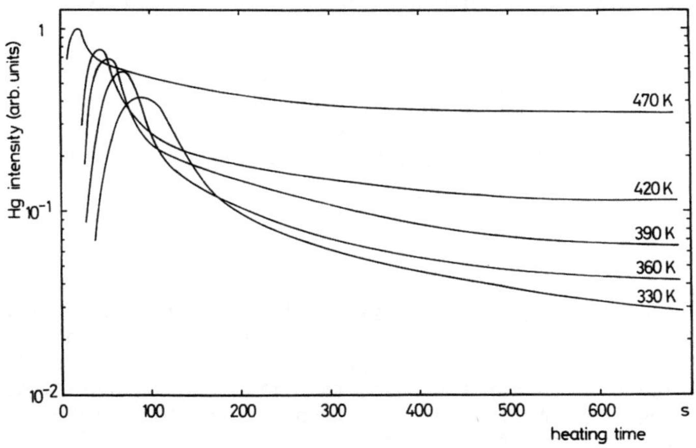

Fig.3 Hg desorption signal from ion-bombarded surfaces as a function of heating time at temperatures between 330 K and 470 K

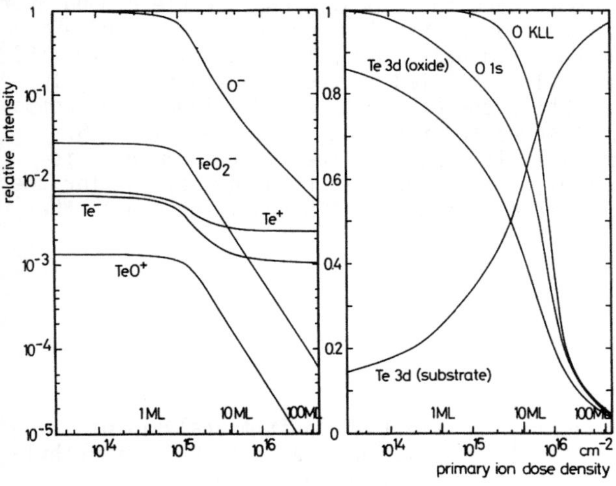

Fig.4 Intensity of typical SIMS, XPS, and AES signals as a function of primary ion dose density during depth profiling of an oxidized surface

the Te 3d lines in XPS, whereas no $CdO_n^{\pm}$ or $HgO_n^{\pm}$ ions and no shift of Cd or Hg lines could be detected. Fig.4 shows the course of typical SIMS, AES, and XPS signals during depth profiling of the oxide layer by ion bombardment. According to these curves, the thickness of the oxide is estimated to be on the order of 10 monolayers.

References

1. H.M. Nitz, O. Ganschow, L. Wiedmann and A. Benninghoven, submitted to Appl. Phys. Lett.
2. H.J. Richter and U. Solzbach, to be published in Surface Sci.

SIMS Depth Profiling of Thin and Ultrathin Films of Covalently Bonded Organic Overlayers

J.F. Evans, M. Ross, and D.M. Ullevig
Department of Chemistry
University of Minnesota
Minneapolis, MN 55455 USA

The feasibility of carrying out analyses of thin and ultrathin films of organic and organometallic species bonded to conductive carbon and semiconductive tin oxide substrates using SIMS depth profiling has been preliminarily evaluated. In certain cases thin film analyses using SIMS detection appear to be superior to Auger electron spectroscopic (AES) detection in terms of definition of the location of the overlayer-substrate interface. This is presumed to be a consequence of the increased sensitivity of the SIMS technique.

The most thoroughly studied systems have been the composites formed from the solution reaction of organosilanes (e.g., N-2-aminoethyl-3-aminopropyltrimethoxysilane) with substrate surface functional groups to yield monolayer and multilayer coverages of bound silane. Simultaneous acquisition of AES and SIMS profiles (1 keV argon bombardment) of multilayer samples (∿100-150 Å of bound silane) have given results typified by the data shown in Fig.1 for a glassy carbon substrate. While the AES profiles of silane layer components (Si, O) are monotonically decreasing functions of ion dose, the SIMS responses for m/e = 28,29 are initially constant, then decrease monotonically. This is taken as an indication that the SIMS technique allows for the definition of an interface where none is noted in AES profiling due to the

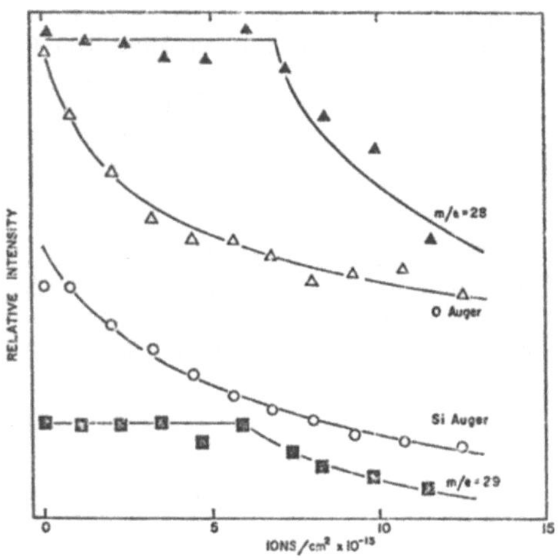

Fig.1 Thin film analysis of thick overlayer of N-2-aminoethyl-3-aminopropyltrimethoxysilane bound to glassy carbon. O and Si AES lines were monitored at 504 eV and 80-86 eV, respectively; sputtering conditions: 1 keV Ar^+, alternately focussed and defocussed

larger sampling depth of AES. If this interpretation is correct, then, from the SIMS profile, the thickness of the silane layer may be established if the sputter rate of this overlayer is known.

For monolayer or submonolayer coverages of silanes chemically bonded to doped SnO_2 (n-type), the SIMS profiles obtained for film components (m/e = 28 (Si) and m/e = 45 (SiOH)) can be fit to an exponential decay after BENNINGHOVEN [1], whereas these profiles for such films on glassy carbon show significant deviation at long times (high ion doses). These differences have been shown by AES and X-ray photoelectron spectroscopic (XPS) analysis carried out at various points during the depth profile to be a consequence of knock-in of components of the silane overlayer into the glassy carbon substrate.

All thin film analyses which rely on the removal of surface material by sputtering must be cautiously interpreted, since there can be significant physical and/or chemical rearrangement ("beam damage") occurring when the energy of the primary ion is dissipated at the surface. This is particularly problematic in SIMS depth profiling, where ion yields can be drastically changed by the chemical structure of the surface. No spectroscopic evidence (XPS or AES) has been found for chemical alteration of the silane films studied here during ion or electron beam bombardment. This is an unexpected result, since the silane monomer has a very similar structure

(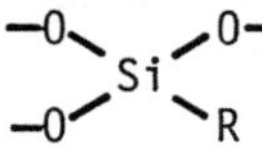 , R = organic group)

to SiO_2, which is known to undergo reduction when bombarded with either high energy ion [2] or electron beams [3-6]. Control experiments have been carried out using thin (∿ 250 Å) films of chemically vapor deposited SiO_2 on glassy carbon and tin oxide, and the results show that the aforementioned reduction of SiO_2 does occur on conductive or semiconductive substrates other than silicon. We conclude, therefore, that the sputtering of silane overlayers (of the thicknesses studied here) proceeds in a significantly better behaved manner than for SiO_2.

To determine the sputter rates of organosilanes and other bonded organic and organometallic overlayers, an apparatus has been constructed which utilizes a quartz crystal microbalance as the target in sputtering/thin film analysis experiments. One face of the quartz crystal (coated first with a silver film) is modified by the deposition of the overlayer of interest. The frequency of the crystal is then monitored as a function of ion dose during SIMS/AES or ESCA profiling. As such, the constancy or variation of sputter yield of the overlayer may be determined concurrently as depth profiling proceeds. Our initial experiments have focussed on sputter yield variations for hydrocarbon polymers pre-deposited <u>via</u> plasma polymerization or dip coating. For these materials the sputter yields are found to decrease by as much as a factor of 15-20 during removal of approximately thirty equivalent monolayers (Fig. 2). Analogous experiments are being carried out with more complex systems such as plasma polymerized organometallics and covalently attached multilayers of cyanuric chloride and silanes to ascertain whether this reduction in sputter yield (and, therefore, sputter rate) as a function of ion dose is a general phenomenon or is related to the specific chemical structure of the overlayer.

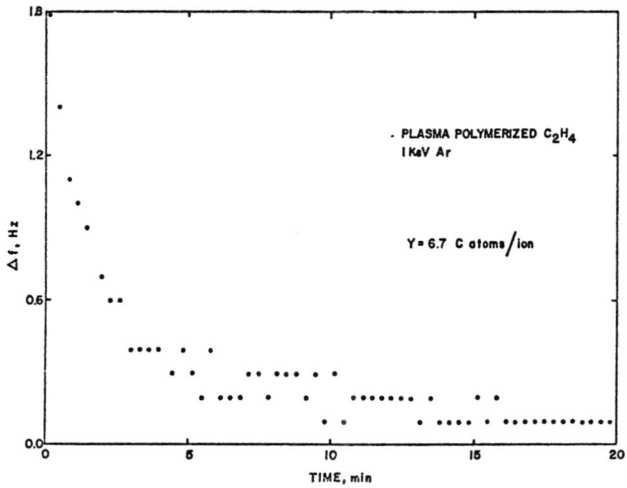

Fig.2 Frequency change for quartz crystal target (modified by deposition of plasma polymerized ethylene) as a function of sputtering time (1 keV Ar^+ current density 1 x 10^{-6} A/cm^2); -Δf is proportional to Δm according to the SAUERBREY equation [7]

References

1. A. Benninghoven, Surface Sci., 35, 427 (1973).
2. R. Holm and S. Storp, Appl. Phys., 12, 101 (1977)
3. S. Thomas, J. Appl. Phys., 45, 161 (1974).
4. J. S. Johannessen, W. E. Spicer and Y. E. Strausser, J. Appl. Phys., 47, 3028 (1976).
5. H. Koyama, K. Matsubara and M. Mouri, J. Appl. Phys., 48, 5380 (1977).
6. M. Menyhard and G. Gergely, Proc. 7th Intern. Vac. Congr. & 3rd Int. Conf. Solid Surfaces (Vienna 1977), p. 2165.
7. G. Sauerbrey, Z. Physik, 155, 206 (1959).

Alpha-Recoil and Fission Fragment Induced Desorption of Secondary Ions *

O. Becker, W. Knippelberg, D. Nederveld and K. Wien
Institut für Kernphysik
Technische Hochschule
6100 Darmstadt, West-Germany

1. Introduction

The energy loss of 100 keV particles in matter is dominated by nuclear stopping power. One would expect that mass and energy spectra of secondary particles, released from a solid surface under the bombardment with projectiles of these energies should reflect both kinds of energy losses. In the following, radioactive decay processes are used to produce the high energetic projectiles: the α-active isotope Am-241 emitting the 92 keV-revoil nucleus Np-237 and the spontaneous fission isotope Cf-252 emitting fragments with masses around 100 and energies of 80-120 Mev. Mass and energy spectra are obtained by a time-of-flight technique [1], where the accompanying nuclear radiation produces the start signal. The mass spectra were taken from Am evaporated onto a carbon foil. In this work only the ions C^+, CH^+, CH_2^+ and CH_3^+ are investigated, which are released from the carbon matrix or from contamination layers. Generally, secondary ion desorption induced by α-recoil particles shows more fragmentation than fission fragment induced desorption.

2. Experiments

Following a suggestion of SIGMUND [2], the α-active material, 0.3 μg Am-241, was evaporated onto a 20 μg/cm² carbon foil. The Americium surface is oxydized and covered by contamination layers. From energy spectra obtained for the recoiling Np-nuclei, their mean energy loss in the surface layers was determined to be 15 keV. If an α-particle is emitted in the direction of

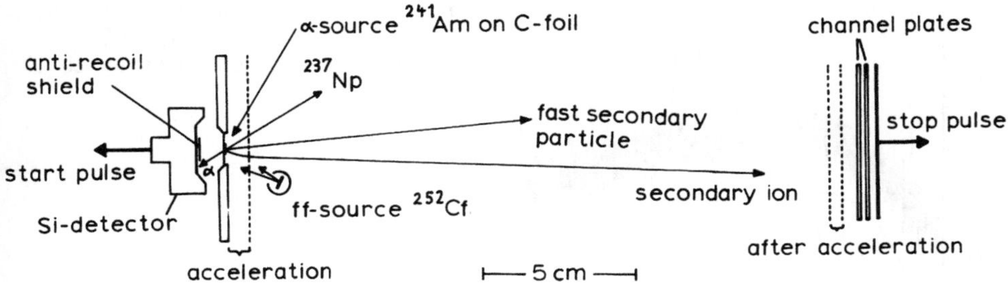

Fig.1 Scale drawing of the target and detector arrangement

*Work supported by the Deutsche Forschungsgemeinschaft.

the start detector, three kinds of heavy particles - detectable by the stop detector - emerge from the target surface: Np-recoil nuclei up to 92 keV, fast charged or neutral particles pushed off by direct impact with Np nuclei (for instance C-12 up to 17 keV) and slow ions released by atomic collision cascades [2], thermal processes [3] or other mechanisms. All of them contribute to the time-of-flight spectra shown in Fig.2b and 2c. The Np-recoil nuclei can be suppressed by a little shield between the α-source and the Si-detector hindering the detection of α-particles, which are related to Np-nuclei flying into the direction of the stop detector. A measurement without acceleration voltage was used to determine the yield of all fast particles (with $E > 1$ keV). With help of the angular distribution this yield was found to be 3.5 per α event. The yield of the slow CH^+, CH_2^+ or CH_3^+ ions is only ca. 0.0015/α. No background of fast particles is observed at fission fragment (=ff) bombardment as demonstrated in Fig.2a. The secondary ion-yields are in the order of 0.004/ff, but can reach 0.2/ff for other ions not shown in Fig.2a.

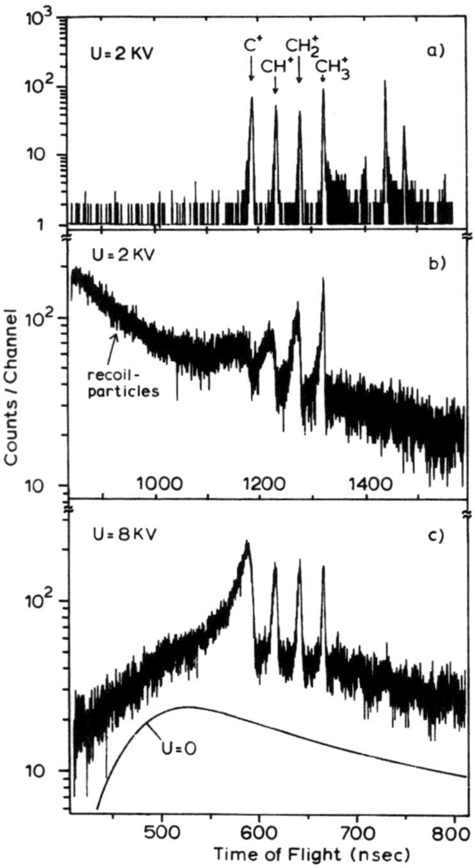

Fig.2 Time-of-flight spectra of secondary ions emitted from a contaminated carbon foil, flight distance 20.6 cm

a) Fission fragment induced desorption, acceleration voltage U = 2 kV. The Si detector was replaced by a channel plate assembly, which reduces the overall time resolution to 0.8 nsec.

b) α-recoil induced desorption, U = 2 kV, without anti-recoil shield, time resolution < 1 nsec.

c) α-recoil induced desorption, U = 8 kV, with anti-recoil shield. The solid line under the spectrum represents a measurement without acceleration voltage (mainly C-12 atoms with energies up to 17 keV)

3. Energy Distributions and Discussion

With regard to the time resolution below 1 nsec the shape of a mass line depends mainly on the axial energy component E_z of the secondary ions (z = direction of flight), if the acceleration voltage is sufficiently small. E_z-distributions derived from the 4 considered mass lines are given in Fig.3. Dashed curves indicate that these distributions are deformed by ion losses due to high radial energy components. These losses are smaller at high acceleration voltages (see Figs.2b and 2c).

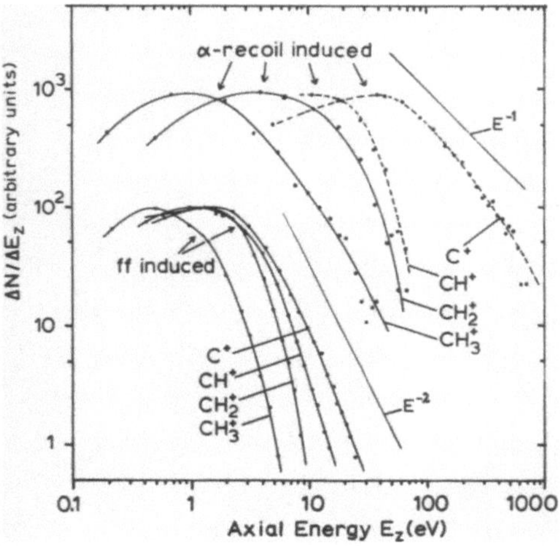

Fig.3 Axial energy distributions, each group normalized to the same height. Dashed curves indicate ion losses

Table 1 Exponents of the high energy tails in the E_z-distributions

	Desorption induced by	
	-recoil	ff
C^+		-2.2
CH^+		-3.0
CH_2^+	-1.8	-4.3
CH_3^+	-1.3	~-4

SIGMUND'S sputter theory [2] predicts an E_z^{-1} dependence for single atoms released from an elementary target. This dependence might be present in the case of -recoil desorbed C^+ ions, where however, the measurement was falsified by ion losses. Also, the exponent of the undisturbed CH_3^+ curve is close to -1, in disagreement to the expected steeper decrease [3] for molecules composed by 4 atoms. Fission fragment bombardment transfers much less translational energy to secondary ions. The exponents show the decrease with the number of molecule constituents. Comparing the results of both desorption procedures, we conclude that the emission of secondary ions by fission fragment bombardment is not induced by atomic collision cascades. On the other hand, no contradiction arose against desorption by a strong collective electronic disturbanc as proposed by KRUEGER [4].

References

1. N. Fürstenau, W. Knippelberg, F.R. Krueger, G. Weib and K. Wien, Z. Naturforschung 32a, 711 (1977).
2. P. Sigmund, Rev. Roum. Phys. 17, No. 9, 1079 (1972).
3. M. Szymonski, H. Overeijnder and A.E. DeVries, Rad. Eff. 36, 189 (1978).
4. F.R. Krueger, Z. Naturforschung 32a, 1084 (1977).

XI. Postdeadline Papers

Use of the IMS-3f High Mass Resolving Power

J.-M. Gourgout
CAMECA
103 Boulevard Saint-Denis
92403 Courbevoie, FRANCE

The first generation of ion probes was designed for microanalysis but did not allow optimization of the secondary ion collection efficiency over varying analytical areas. We have recently designed a secondary ion mass spectrometer which employs the transfer optics of Slodzian to maximize the useful ion yield from 400 x 400 µm to areas as small as a few micrometers.

The IMS-3f was first introduced in August, 1977 at the VIIIth International Conference on X-Ray Optics and Analysis. This paper will present applications of this instrument using small primary ion probes and high mass resolving power developed since that time.

The instrument employs a three lens primary ion column to produce probe diameters as small as 1 µm and with current densities of 150 mA/cm^2 (3 µm probe diamter). After acceleration to an energy of ~4500eV, the secondary ions enter the transfer optics which match the ion emission area to the acceptance of a double-focusing mass spectrometer. This allows the useful ion yield to be maximized over a wide range of analyzed areas. The double-focusing mass spectrometer gives fully stigmatic and achromatic images. Ion detection consists of an electron multiplier and a faraday cup for quantitative ion measurements and a 17 mm fluorescent screen for observing secondary ion images.

Analytical applications will be divided into two groups. First we will illustrate the instrument as a microprobe, and second, we will display the IMS-3f's high mass resolution capabilities in both the imaging and depth profiling modes.

Figure 1 was made using the "Step-Scanning" subroutine on the H-P computer. A copper mesh of 10 µm width has been deposited onto an Al base. The primary beam has been focussed to a diameter of 1.4 µm diameter. The secondary ions ($^{27}Al^+$) have been detected in synchronism with a sample displacement of 1 µm per step, the total displacement represented in the figure being 50 µm. The transfer optics in conjunction with the image projector system have been adjusted to detect ions from an area of 15 µm diameter.

Figure 2 demonstrates the ability to perform microanalytical step scans using a finely focused probe for high bombarding current density and an image plane aperture to define lateral resolution. By placing a 1.5 µm field aperture in the image plane, we can obtain lateral step-scanning resolution better than that obtainable with a gaussian shaped primary ion beam of the same diameter with tails which may extend to much greater distances. The image clearly shows the concentration of carbon in the grain boundaries of a stainless steel within a 100 µm viewed area. The step-scan analysis gives the quantitative ion intensity data over a 20 µm scan.

Figure 3 shows that with the IMS-3f we are able to quantitatively analyze inclusions in a sample as small as 5 x 5 µm using high mass resolution.

Three different analyses have been performed on three different sized grains of pyrite (FeS_2) in a matrix of $CaCO_3$. The concentration of calcium in the grain of FeS_2 is 0.1%. The goal of this analysis is to show that we could detect the true concentration of calcium in a FeS_2 grain as small as 5 µm. Three spectra at mass 56 are shown each recorded using high mass resolution. On the left we see the doublet, Fe-CaO, from a grain 50 µm x 20 µm in size. At the right is the Fe-CaO doublet from a grain of 5 x 5 µm. In each case, the Fe to CaO^+ ratio remains consistent showing that the primary ion probe does not sample the surrounding $CaCO_3$ matrix even for the smallest grain.

Figure 4 shows high mass resolution images of a sample composed of four different minerals, plagioclase, pyroxene, olivine and magnetite. Using the image of mass 56 from a 60 µm diameter analyzed area at low mass resolution it is impossible to distinguish which mineral phase contains either iron or calcium. With high mass resolution imaging, however, we can immediately see which grains contain iron or calcium. With this we can, without a doubt, recognize the minerals and proceed to quantitative analysis using the static probe mode discussed above.

Now, we demonstrate the ability of the IMS-3f to perform depth profiles at high mass resolution. The high resolution mode of analysis is accomplished by displaying the mass of interest on the fluorescent screen. Adjustment of the spectrometer slits, ion optics and exact magnetic field for the ion of analytical interest can be easily performed. The computer is set for a depth profile in the normal manner. Fig.5 shows a depth profile of a $^{31}P^+$ implant into Si after annealing using the high resolution mode of the IMS-3f. This profile was taken at 30.97376 amu with the $^{30}SiH^+$ molecule at 30.98158 being rejected (mass difference of 1/4000). This mode permits a peak-to-background improvement of >100 while maintaining sufficient $^{31}P^+$ intensity to detect a concentration $<5\times10^{15}$ at-cm^{-3}.

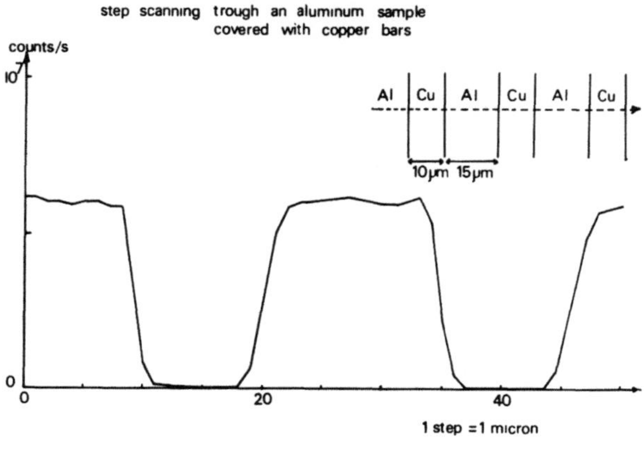

Fig.1

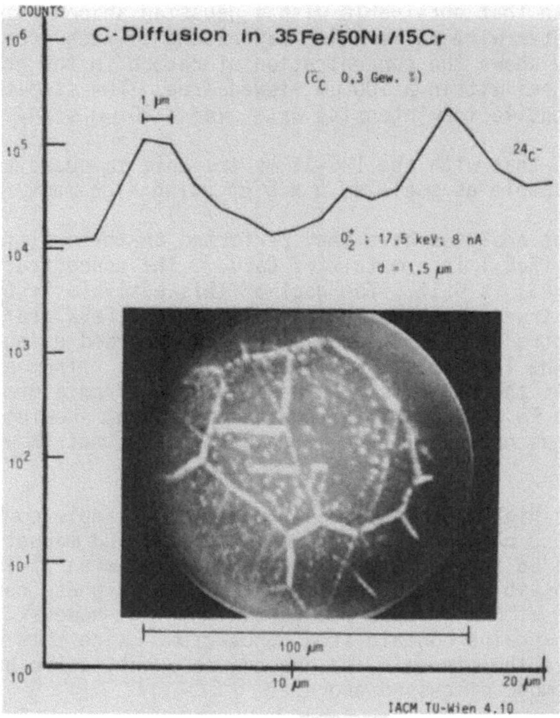
COUNTS
C-Diffusion in 35Fe/50Ni/15Cr
($\bar{c}_C$ 0,3 Gew. %)
1 µm
$^{24}C^-$
O_2^+ ; 17,5 keV; 8 nA
d = 1,5 µm
100 µm
10 µm
20 µm
IACM TU-Wien 4.10

Fig.2

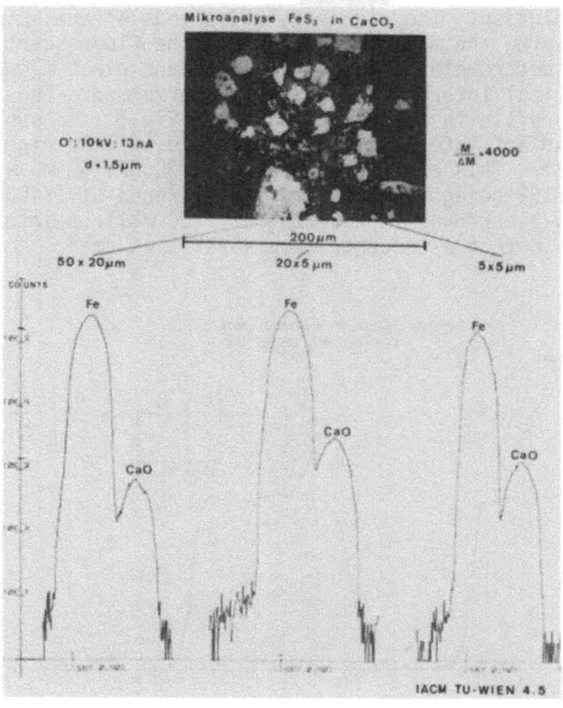
Mikroanalyse FeS_2 in $CaCO_3$
O^-: 10kV; 13nA
d = 1.5µm
$\frac{M}{\Delta M}$ = 4000
200µm
50 x 20µm
20 x 5 µm
5 x 5 µm
COUNTS
Fe
Fe
Fe
CaO
CaO
CaO
IACM TU-WIEN 4.5

Fig.3

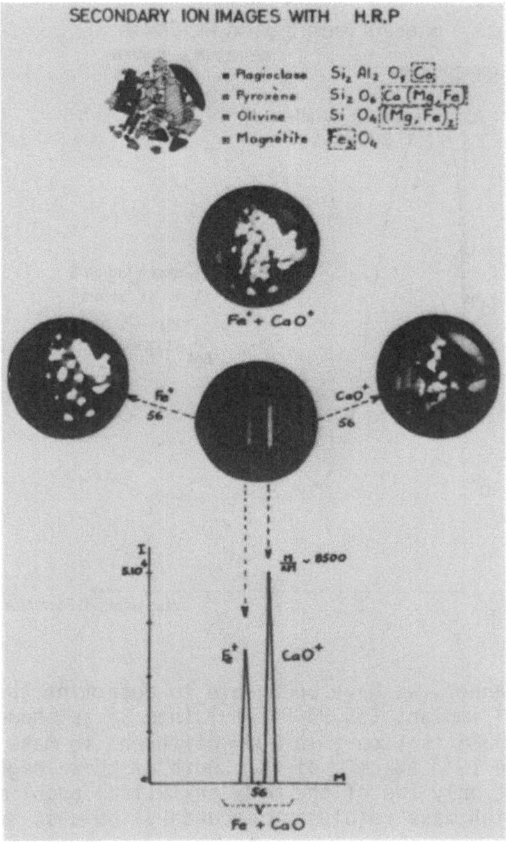
SECONDARY ION IMAGES WITH H.R.P
Plagioclase Si2 Al2 O8 Ca
Pyroxène Si2 O6 Ca (Mg, Fe)
Olivine Si O4 (Mg, Fe)2
Magnétite Fe3 O4
Fe+ + CaO+
Fe+ 56
CaO+ 56
I
5.10⁴
M/ΔM ~ 8500
Fe+
CaO+
M
56
Fe + CaO

Fig.4

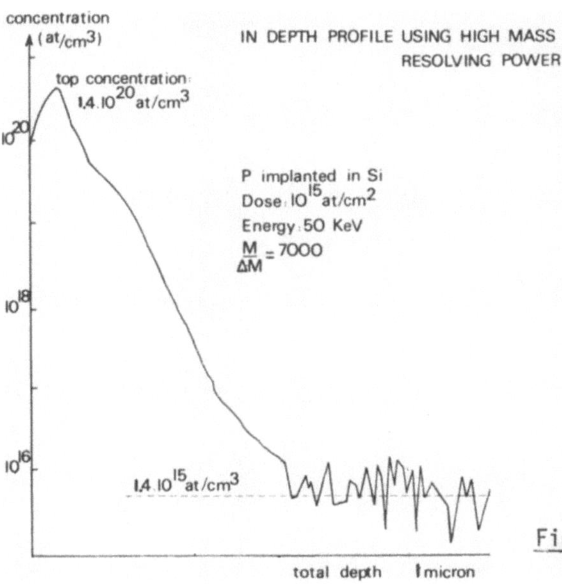
concentration (at/cm3)
IN DEPTH PROFILE USING HIGH MASS RESOLVING POWER
top concentration: 1.4.10²⁰ at/cm³
10²⁰
10¹⁸
10¹⁶
P implanted in Si
Dose: 10¹⁵ at/cm²
Energy: 50 KeV
M/ΔM = 7000
1.4.10¹⁵ at/cm³
total depth 1 micron

Fig.5

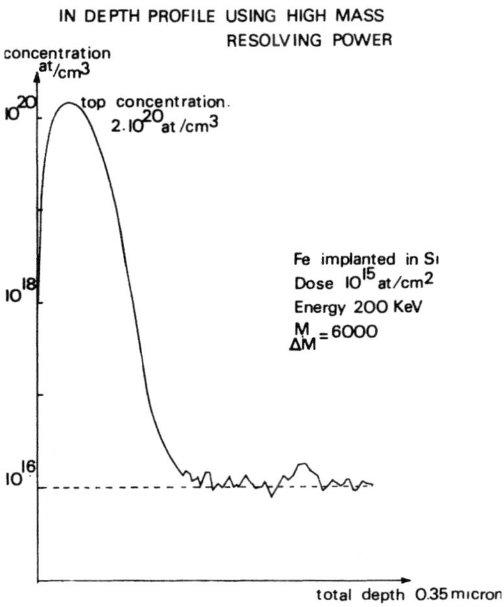

Fig.6

In a similar manner, we have been able to determine the in-depth distribution of an $^{56}Fe^{+}$ implant (55.93494 amu) into Si as shown in Fig.6. The $^{28}Si_2^{+}$ ion at 55.95386 is 1 part in 3000 different in mass and has an intensity equivalent to 10^{21} at-cm^{-3} of Fe. Both of these high resolution examples represent only two of the many analytical problems which can now be solved using high mass resolution secondary ion mass spectrometry.

The Bombardment Angle Dependence of the Sputtering and Secondary Ion Yield for Oxygen Ion Bombardment of Silicon

N. Warmoltz, H.W. Werner and A.E. Morgan
Philips Research Laboratories
Eindhoven, The Netherlands

Abstract

From theoretical considerations, we have derived a mathematical relationship which correlates the angle of bombardment with the sample sputtering yield, the incorporated concentration of the bombarding ion and secondary ion intensities. Data from other researchers was then used to provide points of reference for the derived relationships.

If silicon is bombarded with oxygen and no additional oxygen flooding takes place, the following implications on instrument performance can be concluded. It appears from experiment that instruments with a large deviation of the angle of incidence from the target normal give a relatively large sputter yield S together with small values of the ion yield S^+. Here the steady state is characterized by a relatively large sputter yield of 1,8 and a correspondingly large value of the sputter rate $\dot{z}$. On the other hand, the oxygen coverage is small, $\theta = 0,6$. The latter as has been reported before (MORGAN and WERNER [1], DELINE, et al., [2], implies a low degree of ionization β^+. Consequently, the secondary ion current will be relatively low. Instruments which are operated with normal incidence were found to have a small sputtering yield, a relatively high ion yield and secondary ion intensity. This can be understood from the deviations which show a low value for the sputtering rate and a large value for the surface concentration of oxygen. Therefore, a decrease in the rate of atom ejection as determined by the sputtering yield will lead to a significantly larger ion yield and secondary ion intensity due to the power function dependence of secondary ion yields on the implanted oxygen concentration. With oxygen bombardment oxygen flooding, the trend is the same, but the difference between the two modes is not as pronounced.

References

1. A.E. Morgan and H.W. Werner, Anal. Chem. 48 699 (1976).
2. V.R. Deline, W. Katz and C.A. Evans, Jr., Appl. Phys. Lett. 33 832 (1978).

Secondary Ion Mass Spectrometry of Small Molecules Held at Cryogenic Temperatures

H.T. Jonkman and J. Michl
Department of Chemistry
University of Utah
Salt Lake City, Utah 84112

In recent years more emphasis has been placed on the analytical potential of Secondary Ion Mass Spectrometry (SIMS) for the study of organic materials [1,2,3]. We wish to report on static SIMS experiments on a few simple molecules condensed on a copper substrate held at low temperatures. For the experiments we used an Extranuclear Inc., quadrupole massfilter equipped with a Bessel box type energy filter. The bandwidth of the ions accepted in the massfilter was set at 1 eV. Our ion gun is differential pumped and during the experiments a pressure in the 10^{-9} torr range could easily be maintained in the main chamber. Our primary ion beam current density was between 1 and 10 nA/cm^2 and the beam energy was varied between 500 and 5000 eV. As primary

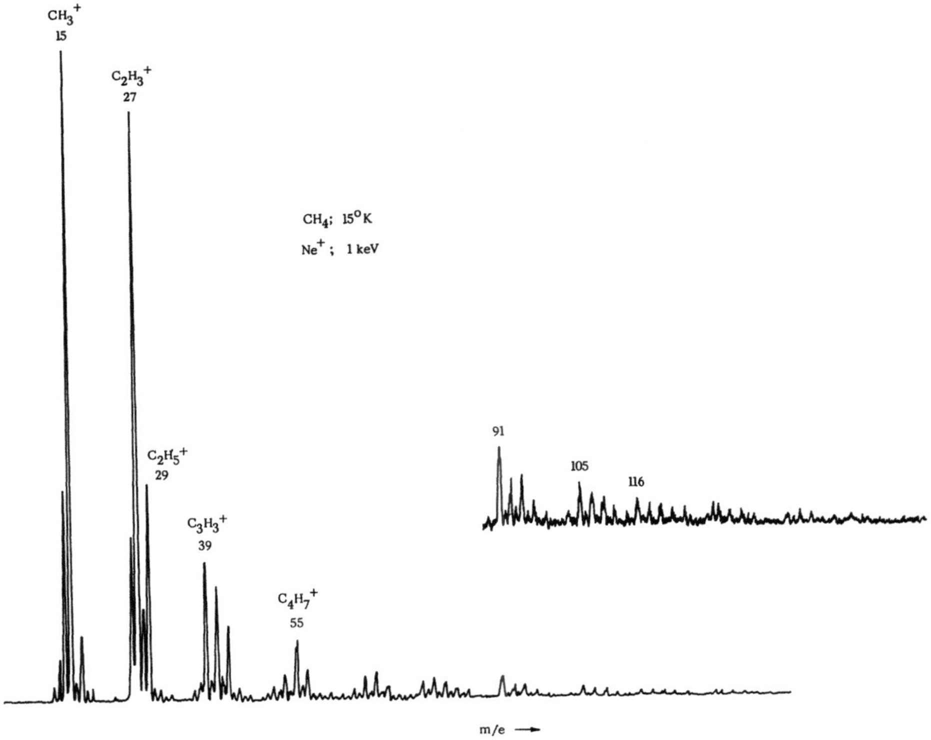

Fig.1 SIMS spectrum of neat methane at 15°K using 1 keV Ne^+ ions

ion we used He^+, Ne^+, Ar^+, Kr^+, and Xe^+. The nature of our samples made it necessary to use an electron floodgun to compensate for charge buildup on the sample. The energy of the flooding electrons was kept below 7 eV. In all the experiments the temperature of the sample was held at 15°K. In Fig.1 the positive SIMS spectrum of neat methane using a 1 keV Ne^+ beam shows fragments up to C_{11}.

The spectrum looks very similar to the ones of propane and pentane reported before [2]. The type of fragments observed are independent of the type of primary ion used and its kinetic energy, although higher beam energies and/or heavier type ions give rise to a higher abundance of the larger fragments. Fig.2 shows a SIMS spectrum of methane diluted in an argon matrix [1:500].

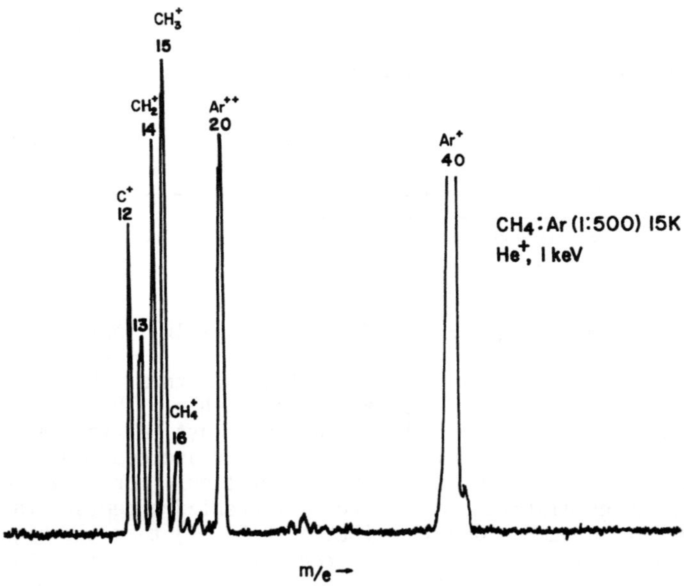

Fig.2 SIMS spectrum of methane diluted in an argon matrix [1:500] at 15°K using 1 keV He^+ ions

In this spectrum there is no evidence of ion-molecule reaction or cluster formation and only the C_1 group is visible. It is important to observe that the fragment distribution in this case is different from that of the C_1 group in neat methane. Fig.3 shows the SIMS spectrum of N_2 with a 4 keV Ar^+ beam. Here we observe N^+, N_2^+, N_3^+, and a large number of clusters with an odd and even number of nitrogen atoms up to the limit of the mass range of our spectrometer. These clusters can be considered as consisting of N^+ or N_3^+ ions for the odd ones and N_2^+ ions for the even ones, solvated in clusters of neutral N_2 molecules. The existence of these clusters and their stability of at least several milliseconds, which is their estimated flight time through the mass spectrometer, suggests that they must be extremely cold since they can only be held together by weak induced dipole and Van Der Waals forces. This suggests a rapid cooling of the clusters upon expansion in the vacuum.

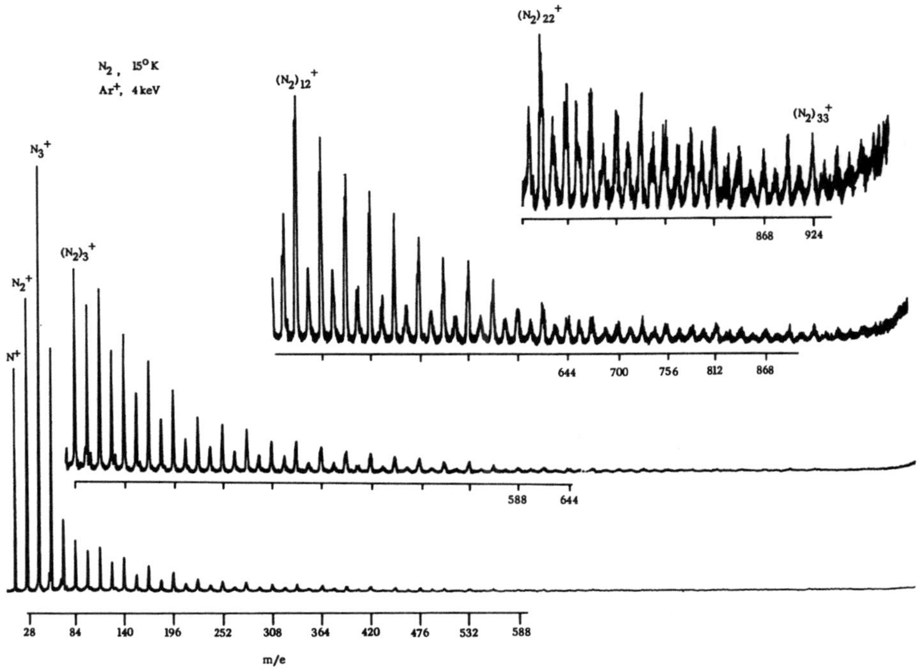

Fig.3 SIMS spectrum of neat N_2 at 15°K using 4 keV Ar^+ ions

Figure 4 shows the SIMS spectra of N_2 obtained with a 4 keV He^+ beam. Here only N^+, N_2^+, N_3^+, and N_4 fragments are visible. This can be attributed to the fact that because of their low mass, most He^+ ions are scattered and are less able to penetrate deep into the solid and consequently give less rise to ion molecule reactions and cluster formation. For N_2 we recorded the secondary ion energy distribution for the different fragments. The N^+ and N_2^+ fragments have a much wider distribution [±20 eV] than the higher clusters [±7 eV]. It was observed that heavier fragments came off with smaller average kinetic energies than lighter ones. It is impossible to measure absolute energies because of the interaction between the emission current setting of the floodgun (slightly under or over charge compensation) and the center bandpass setting of the prefilter. This contradicts observations in [3]. The fact that the lower fragments came off with a wide energy distribution causes their underrepresentation in the recorded spectra because of the relatively small bandwidth admitted to the massfilter. In addition to the molecules mentioned before, we investigated a series of two and three atomic systems. These results will be reported elsewhere.

In general, we believe that the first step upon ion impact is ionization and dissociation by charge transfer and kinetic processes followed by ion molecule reactions and cluster formation in the solid. The clusters then expand in the vacuum accompanied by a rapid cooling, which gives these clusters their remarkable stability.

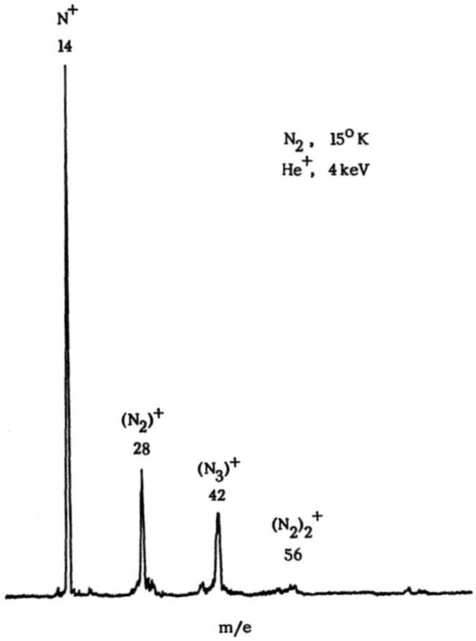

Fig.4 SIMS spectrum of neat N_2 at 15°K using 4 keV He^+ ions

Acknowledgement

This work was supported by National Science Foundation grant CHE 78-27094.

References

1. A. Benninghoven, CRC Crit. Rev. Solid State Sc., 291 (1976); Surface Sci., 35, 427 (1973); H. Grade, et al., J. Am. Chem. Soc., 99 7725 (1977).
2. H.T. Jonkman, et al., Anal. Chem., 50 2078 (1978); H.T.Jonkman and J. Michl, Chem. Commun., 751 (1978).
3. G.M. LaCosta, et al., J. Am. Chem. Soc., 101 2951 (1979).

Index of Authors

MIX
Papier aus verantwortungsvollen Quellen
Paper from responsible sources
FSC® C105338

If you have any concerns about our products,
you can contact us on
ProductSafety@springernature.com

In case Publisher is established outside the EU,
the EU authorized representative is:
Springer Nature Customer Service Center GmbH
Europaplatz 3, 69115 Heidelberg, Germany

Printed by Libri Plureos GmbH
in Hamburg, Germany